FLOW-INDUCED VIBRATIONS

An Engineering Guide

Eduard Naudascher

University of Karlsruhe, Germany

Donald Rockwell

Lehigh University, Bethlehem, Pennsylvania, USA

DOVER PUBLICATIONS, INC.

Mineola, New York

Bibliographical Note

This Dover edition, first published in 2005, is an unabridged and slightly corrected republication of the first edition of the work, originally published by A.A. Balkema Publishers, Rotterdam, in 1994 as Volume 7 in the International Association for Hydraulic Research (IAHR) series of Hydraulic Structures Design Manuals. An errata list has been added to the present edition on page 414.

Library of Congress Cataloging-in-Publication Data

Naudascher, Eduard.
    Flow-induced vibrations : an engineering guide / Eduard Naudascher, Donald Rockwell.
        p. cm.
    Reprint. Originally published: Rotterdam : A.A. Balkema, 1994.
    ISBN 0-486-44282-9 (pbk.)
    1. Vibration. 2. Fluid dynamics. 3. Structural dynamics. I. Rockwell, Donald. II. Title.

TA355.N38 2005
511'.43—dc22

                                                                2005045426

Manufactured in the United States by Courier Corporation
44282903
www.doverpublications.com

Contents

V

Preface

The work presented herein is based on a research project sponsored by the VOLKSWAGEN FOUNDATION of Hannover, Germany, and carried out, cooperatively, at the University of Karlsruhe, Germany, and Lehigh University, USA. The authors gratefully acknowledge the generous support of the Volkswagen Foundation for this joint program and their commitment to ensuring its success. In addition, projects sponsored by other agencies have had an impact on this undertaking. These agencies include the Deutsche Forschungsgemeinschaft, the National Science Foundation, the National Aeronautics and Space Administration, and the Office of Naval Research. Parts of the presented material were prepared as notes for Intensive Courses held at various locations including Quito, Ecuador; Sao Paulo, Brazil; Mendoza, Argentina; Zürich, Switzerland; Denver, Colo., Worcester, Mass., and Bethlehem, Pa., USA; Caracas, Venezuela; Cranfield, UK; Tokyo, Japan; Delft, Netherlands; Moscow, Russia; Lisbon, Portugal; Karlsruhe, Germany, The Sonderforschungsbereich 210, an interinstitutional research center at the University of Karlsruhe, supported the review process by bringing out preprints of most of the book manuscript in its report series in 1990.

The authors are indebted to many of their colleagues and collaborators who have contributed ideas and criticism, and who have helped in the preparation of the manuscript and its improvement before publication. Particular acknowledgement is due Professor J.S. McNown of the IAHR editorial committee for reviewing thoroughly the entire text and revising the English and Professor M.P. Paidoussis for his exhaustive responses and abundant advice. The drawings were prepared by Mrs I. Weber with some assistance by J. Helbing, I. Kastner, and E. Staschewski. Dr R. Ermshaus prepared the photographs and Mrs Yinan Wang, I. Kastner, and Mr S. Kanne helped with reading the proofs and related efforts. We extend our coridal thanks to all of them as well as to Mrs C. Echte, G. Krause, R. Böser, and H. Meyer who typed the text. Their devoted collaboration made it a joy to undertake this endeavor.

Eduard Naudascher
University of Karlsruhe, 1994

Donald Rockwell
Lehigh University, 1994

Reviewers

Peter W. Bearman
Department of Aeronautics
Imperial College of Science,
Technology & Medicine
London SW7 2BY, UK

Ralph Parker
Department of Mechanical Engineer-
ing
University College of Swansea
Swansea SA2 8PP, UK

Paul A. Kolkman
Rivers, Navigation and Structures
Department
Delft Hydraulics
8300 AD Emmeloord, Netherlands

Geoffrey V. Parkinson
Department of Mechanical Engineer-
ing
University of British Columbia
Vancouver, B.C., V6T 1W5, Canada

John S. McNown
Department of Hydraulic Engineering
Royal Institute of Technology
10044 Stockholm, Sweden

Frederic Raichlen
Department of Civil Engineering
California Institute of Technology
Pasadena, CA 91125, USA

Michael P. Paidoussis
Department of Mechanical Engineer-
ing
McGill University
Montreal, PQ, H3A 2K6, Canada

Thomas Staubli
Institut für Energietechnik
Eidgenössische Technische Hochs-
chule
CH-8092 Zürich, Switzerland

PARTS OF THE BOOK WERE REVIEWED BY

Robert D. Blevins
Rohr Industries, Inc.
Chula Vista, CA 92012, USA

Henry T. Falvey
Consulting Engineer
Conifer, CO 80433, USA

List of symbols

a	length dimensions
$a_1 = (dC_y/d\alpha)_{\alpha=\bar{\alpha}}$ (Equation 7.23)	
A	area
$\mathcal{A}$	amplification factor
A'	added mass ($A'\ddot{x}$ = fluid force acting equivalent to an inertia force, Equation 3.1)
$\mathcal{A}_R$	amplification factor at resonance
A'_{yx}	contribution of body motion in x direction to added mass in y direction
$A'_{y_b x_a}$	contribution of motion of tube a in x direction to fluid force on tube b in y direction
AEVS	alternate-edge vortex shedding (Figure 9.11)
b	length dimension
B	coefficient of mechanical damping (Equation 2.8); transverse dimension
B'	added damping ($B'\dot{x}$ = fluid force acting equivalent to a damping force, Equation 3.1)
$B_{cr} = 2m\omega_n$	critical damping coefficient (Equation 2.10)
B'_{yx}	contribution of body motion in x direction to added damping in y direction
B_θ	torsional damping coefficient (Equation 2.18)
c	wave celerity
c_a	acoustic wave celerity
C	spring constant (Equation 2.4); coefficient
C'	added stiffness ($C'x$ = fluid force acting equivalent to a restoring force, Equation 3.1)
C_c	contraction coefficient
C_d	coefficient of fluid force acting in phase with body velocity (Equations 3.18, 3.38); discharge coefficient (Equation 7.62)
C_D	coefficient of mean drag
C_L	coefficient of mean lift

XIII

$C_{L_{rms}}$	normalized root-mean-square value of lift fluctuations
C_m	coefficient of fluid force acting in phase with body deflection (Equations 3.18, 3.38)
C_M	moment coefficient
$C_{M_{rms}}$	normalized root-mean-square value of moment fluctuations
C_N	constant characterizing mode N
C'_{yx}	contribution of body motion in x direction to added stiffness in y direction
C_y	coefficient of fluid force acting transverse to approach flow (Equation 3.16)
$\mathring{C}_y = dC_y/d\tau;\ \mathring{\mathring{C}}_y = d^2C_y/d\tau^2$ (Equation 3.32)	
C_θ	torsional spring constant (Equation 2.18)
d	length dimension; characteristic dimension of body
D	width of body; diameter
e	length dimension; chord length
E	bulk modulus of elasticity of fluid
E_s	modulus of elasticity of structural material
EI	bending stiffness
EIE	extraneously induced excitation (Sections 1.2 and 5)
f^*	dominant frequency
f	frequency in cycles per second
f_o	frequency of vortex formation; dominant frequency of IIE
f_d	frequency of disturbance
f_n	natural frequency of body oscillator
f_N	natural frequency of mode N of body oscillator
f_R	resonant frequency of fluid oscillator
f_s	forcing frequency (Equation 2.12); frequency of controlled vibration (Figures 3.17, 3.21)
$F = \bar{F} + F'$	fluid force on body composed of mean and fluctuating components
F_o	force amplitude (Equation 2.12)
F_D	drag
F_L	lift
F_y	fluid force on body acting transverse to approach flow
$Fr = V/\sqrt{gh}$	Froude number
g	acceleration of gravity
G	gap width
h	piezometric head (Figure 3.9); depth of flow
$\tilde{h}$	amplitude of surface wave
h_a	added-mass coefficient (Equations 3.34, 3.38)
h_s	depth of submergence (Figure 7.27a)
H	head; height
$i = \sqrt{-1}$	

I	area moment of inertia
I_r	reduced moment of inertia (Equation 7.33)
I_θ	mass moment of inertia (Equation 2.18)
IIE	instability-induced excitation (Sections 1.2 and 6)
ILEV	impinging leading-edge vortices (Figures 9.1, 9.11)
J_N	joint acceptance (Equation 5.16)
$k = c_p/c_v$	where c_p and c_v are the specific heats of a gas at constant pressure and volume, respectively
k_a	excitation coefficient (Equations 3.34, 3.38)
K_r	reflection coefficient
K_t	transmission coefficient
Ka	cavitation number
$KC = V_o T/d$	Keulegan/Carpenter number (Figure 3.15)
l, L	length
L_c	correlation length
L_v	length scale of turbulence (Equation 5.2)
LEVS	leading-edge vortex shedding (Figure 9.1)
m	mass of body (per unit length in case of 2D body); lateral mode (Figure 7.42)
$\dot{m}$	mass rate of flow
m_r	reduced mass [$m_r = m/(\rho d^2)$ for 2D cylindrical body]
M	moment; momentum flux
M_o	moment amplitude
M_x	axial momentum flux
M_θ	angular momentum flux
$Ma = V/c_a$	Mach number
MIE	movement-induced excitation (Sections 1.2 and 7)
n, N	integer corresponding to mode number
p	intensity of pressure
$\tilde{p}$	amplitude of fluctuating pressure
$\tilde{p}_o$	response pressure amplitude (Figure 8.1)
$\tilde{p}_{oi}$	input pressure amplitude (Figure 8.1)
p_v	vapor pressure
P	power (Equation 6.13)
q	load per unit length (Equation 2.57)
Q	rate of flow or discharge; quality factor $Q = \frac{1}{2}\zeta$ (Equation 2.20)
$\bar{Q}$	mean rate of flow
r	radius
r_g	radius of gyration of a body about elastic center (Equation 7.7)
R	radius; correlation coefficient (Equations 5.2, 5.7)
R_*	gas constant (Equation 4.5)

Re	Reynolds number
s	gate opening; center-to-center spacing
s_o	height of conduit (Figure 7.27d)
S	static imbalance (Equation 7.7); measure of taper (Figure 7.40d)
S_{12}	co-spectral density (Equation 5.6)
S_F	power spectral density of exciting force (Figure 2.8; Equations 5.12, 5.19)
S_g	response parameter (Equation 6.5)
S_p, S_v	power spectral density of pressure and velocity (Equation 5.4)
S_x	power spectral density of response in x direction (Figure 2.8; Equations 5.12, 5.19)
$Sc = 2\zeta m_r$	Scruton number or mass-damping ratio (Equation 3.36)
Sc_θ	Scruton number for torsional-vibration system (Equation 7.25)
$Sh = f_o d/V$	Strouhal number
Sh_n	Strouhal number of nth mode of leading-edge vortex formation (Equation 6.8)
t	time; thickness
$T = 1/f$	period of vibration (Equation 2.43)
T_*	temperature referred to absolute zero
T_{js}	delay time due to jet switching (Figure 7.31d; Equation 7.43)
$Tu = v'_{rms}/\bar{v}$	turbulence level
TEVS	trailing-edge vortex shedding (Figure 9.1)
U	nondimensional velocity $V/(\omega_n d)$; rotor speed
U_e, U_i	nondimensional form of external and internal velocity (Equation 7.77)
U_r	reduced rotor speed (Equation 8.11)
$v = \bar{v} + v'$	velocity at given point composed of mean and fluctuating part
V	mean velocity of flow
V_o	velocity amplitude
V_c	convection velocity
V_D	divergence velocity (Equation 3.50)
V_j	jet velocity
V_r	reduced velocity, e.g., $V_r = V/(f_n d)$
$(V_r)_{cr}$	critical reduced velocity (V_r for onset of vibration)
$\forall$	volume
w, W	width
W_d	work per cycle done by damping force (Equation 2.34)
W_e	work per cycle done by exciting force (Equation 2.33)
x	streamwise displacement (Equation 2.1); position along x direction
x_o	vibration amplitude in x direction

x_a	displacement of tube a in x direction
$\dot{x} = dx/dt$	velocity of body in x direction
$\ddot{x} = d^2x/dt^2$	acceleration of body in x direction
X	distance between bodies in x direction
y	transverse displacement
y_a	displacement of tube a in y direction
$\tilde{y}$	mode shape (Equations 2.60, 2.61)
y_o	vibration amplitude in y direction
y_{os}	amplitude of controlled structural vibration (Equation 3.17)
Y	transverse deflection (in Section 2.7); distance between bodies in y direction
z	displacement in z direction
Z_m	mechanical impedance (Equation 2.25)
α	angle; angle of flow incidence; added-mass coefficient (Equation 3.5; Figure 3.19)
$\bar{\alpha}$	mean angle of flow incidence
α_i	amplification factor (Figure 6.3c)
β	angle; coefficient of fluid damping (Equation 3.5; Figure 3.19)
$\gamma = \rho g$	specific weight of fluid
$\delta = 2\pi\zeta/\sqrt{1 - \zeta^2}$	logarithmic decrement of mechanical damping (Equation 2.11); thickness of boundary layer
δ_m	momentum thickness
ε	number < 1
$\zeta = B/(2m\omega_n)$	damping ratio (Equation 2.10)
$\zeta_\theta = B_\theta/(2\,l_\theta\omega_n)$	torsional damping ratio
$\eta = y/d$	nondimensional deflection in y direction
$\overset{\circ}{\eta} = d\eta/d\tau;\ \overset{\circ\circ}{\eta} = d^2\eta/d\tau^2$ (Equation 3.24)	
$\eta_o = y_o/d$	nondimensional amplitude in y direction
θ	angular or torsional displacement; angle of yaw
θ_o	amplitude of torsional vibration
$\dot{\theta} = d\theta/dt;\ \ddot{\theta} = d^2\theta/dt^2$	
κ	energy parameter (Figure 7.34)
λ	wavelength
λ_R	wavelength of resonant wave
μ	mass per unit length; dynamic viscosity of fluid
ν	kinematic viscosity of fluid, $\nu = \mu/\rho$
ν_P	Poisson ratio
ρ	fluid density
ρ_g	density of gas
ρ_s	density of structural material
ρ_w	density of water

σ	void fraction (Figure 3.19); surface tension
τ	nondimensional time $\omega_n t$; delay time (Equation 5.2)
φ	phase angle (Equations 2.5, 2.14)
Φ	phase angle (Equation 3.22b); specific speed of turbine (Equation 6.14)
χ_{fl}	fluid-dynamic admittance (Equation 5.14)
χ_m	mechanical admittance (Equations 2.24, 5.17)
ω	circular frequency, $\omega = 2\pi f$
$\omega_d = \omega_n\sqrt{1 - \zeta^2}$	circular frequency of damped oscillator (Equation 2.9)
ω_n	circular natural frequency, $\omega_n = 2\pi f_n$ (Equation 2.6)
ω_n'	nondimensional form of ω_n (Equation 7.71)
ω_R	rotational speed
ω_s	circular forcing frequency (Equation 2.12)
ω_T	angular velocity of flow (Figure 6.42)
ω_y	circular natural frequency of transverse vibration
$\omega_\theta, \omega_{\theta n}$	circular natural frequency of torsional vibration
Ω	swirl parameter (Equation 6.12)

CHAPTER 1

Introduction

1.1 GOALS AND SCOPE OF MONOGRAPH

To this day, flow-induced structural vibrations and fluid oscillations or noise are being treated by different disciplines using different kinds of descriptions that depend generally on the type of structure or system involved. Moreover, methods of vibration and noise control are usually being obtained in an ad-hoc fashion as the problems arise. We are thus faced with a bewildering diversity of information and are tempted to reinvestigate a problem instead of applying results to resolve it.

The primary goal of this monograph is a synthesis of research results and practical experience from a variety of fields in the form of engineering guidelines. Guidance is intended, specifically, in assessing the possible sources of excitation in a flow system, in identifying the actual danger spots, and in finding appropriate remedial measures or cures. To this end, the material follows a unified approach, broad enough to permit treatment of the major excitation mechanisms, yet simple enough to be of practical use.

Flow-induced vibrations of structural or fluid masses can occur in so many different forms that several volumes would have been required had they been presented in terms of the variety of possible structures or systems affected. Since a given source of excitation can take a multitude of forms, moreover, this approach would still not have guaranteed reliable detection under all circumstances. Even well-understood vibration problems often come in disguise. Hardly any two gates, valves, or seals are alike in geometry, dynamic characteristics, and integration into an overall system; hence, they may be susceptible to quite different excitation mechanisms that call for different curative measures. On the other hand, completely different types of structures can be exposed to the same type of excitation; information and experience gathered from one of these structures, therefore, can well apply to the other, provided the basic ingredients of this excitation are recognized.

For these reasons, flow-induced vibrations are presented in terms of their basic elements: body oscillators, fluid oscillators, and sources of excitation. By stress-

1

ing these basic elements, we wish to provide a basis for the transfer of knowledge from one system to another as well as from one engineering field to another. In this way, well-known theories on cylinders in cross-flow or well-executed solutions from the field of wind engineering, to name just two examples, may be useful in other systems or fields on which information is scarce.

In the area of flow-induced vibration, research is being performed so intensively and practical experiences are being gathered so abundantly that it is impossible to present today all material in a final form. All colleagues working in this field are therefore invited to report both criticism and findings from research and practical experience to the IAHR Editorial Committee (c/o IAHR Secretariat, Delft Hydraulics, Delft, the Netherlands) in order that corresponding corrections and supplementary material may be incorporated in the revised edition that is planned to appear in five to ten years.

Publication of the IAHR Hydraulic Structures Design Manual was undertaken with the intent to consolidate hydraulic-design information that is normally scattered among a vast number of journals and books. In spite of such consolidation, the manual will most likely be regarded as too extensive. The inevitable, and unfortunate, course of events may be that only a small part of the information will be extracted, simplified, and reduced to straightforward 'design criteria'. The dangers inherent in such an outcome are obvious.

The material in this monograph does not codify existing knowledge or present binding design guidelines. The authors' goal was simply to give the interested engineer an overview on a subject which is still little understood.

1.2 SOURCES AND ASSESSMENT OF FLOW-INDUCED VIBRATIONS

Flow-induced vibration phenomena have been treated by a variety of engineering disciplines, each having its particular terminology. In an attempt to provide a unified overview, we propose the following definition of basic elements of flow-induced vibrations:
  a) Body oscillators;
  b) Fluid oscillators; and
  c) Sources of excitation.
Oscillators are defined herein as systems of structural or fluid mass that are acted upon by restoring forces if deflected from their equilibrium positions and undergo vibrations in conjunction with appropriate types of excitation. An engineering system will usually possess several potential oscillators and several sources of excitation. The first and most important task in the assessment of possible flow-induced vibrations is therefore to identify them.

A *body oscillator* consists of either a rigid structure or structural part that is elastically supported so that it can perform linear or angular movements (e.g., a rod or a sluice gate), or a structure or structural part that is elastic in itself so that it can perform flexural movements (e.g., a thin-walled pipe or the skinplate of a gate). No matter how simple the system may seem, it may contain a number of body oscillators, each of which can lead to vibration problems either by itself or in combination with other body or fluid oscillators. A brief review of the dynamics of basic body oscillators is given in Chapter 2, and the types of fluid loading and response of such oscillators are summarized in Chapter 3.

A *fluid oscillator* consists of a passive mass of fluid that can undergo oscillations usually governed either by fluid compressibility or by gravity. In both cases, the oscillating fluid mass can be discrete (e.g., the fluid in a pipe oscillating due to changes in fluid volume or changes in free-surface elevation in an adjacent chamber, Figures 4.3a, b), or it can be distributed (e.g., the fluid in a pipe or an open channel oscillating in the form of an acoustic or a gravity wave, respectively, Figures 4.6a, b). Fluid-flow systems may contain a number of fluid oscillators. They may give rise to undesirable fluid pulsations when excited (e.g., surging or wave oscillations); and they may amplify the vibration of a body oscillator if one of their natural frequencies coincides with the natural body-oscillator frequency. An overview of common fluid oscillators is presented in Chapter 4.

Sources of excitation for either body or fluid oscillators are numerous and may be difficult to detect. It is therefore useful to treat them within a basic framework. The following material distinguishes three types:
  - Extraneously induced excitation (EIE);
  - Instability-induced excitation (IIE); and
  - Movement-induced excitation (MIE).

Extraneously induced excitation (EIE) is caused by fluctuations in flow velocities or pressures that are independent of any flow instability originating from the structure considered and independent of structural movements except for added-mass and fluid-damping effects. Examples are the bluff body in Figure 1.1a, being 'buffeted' by turbulence of the approach flow, and the pipe filled with compressible fluid in Figure 1.1b, being excited by a loudspeaker. The exciting force is mostly random in this category of excitation, but it may also be periodic. A case in point is a structure excited by vortices shed periodically from an *upstream* cylindrical structure. In either case, the vibration is sustained by an extraneous energy source. More information on sources of EIE and their characteristics and control in cases of body and fluid oscillators is found in Chapters 5 and 8, respectively.

Instability-induced excitation (IIE) is brought about by a flow instability. As a rule, this instability is intrinsic to the flow system. In other words, the flow instability is inherent to the flow created by the structure considered. Examples of this situation are the alternating vortex shedding from a cylindrical structure (Figure 1.1c) and the oscillations of an impinging free shear layer near the mouth of an organ pipe (Figure 1.1d). The exciting force is produced through a flow process (or flow instability) that takes the form of local flow oscillations even in cases where body or fluid oscillators are absent. The excitation mechanism can

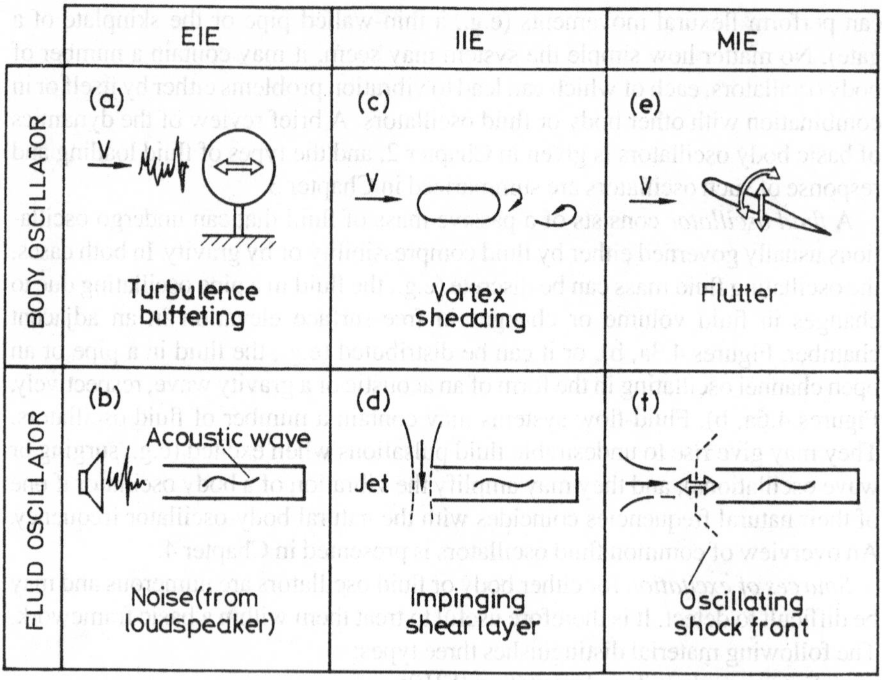

	EIE	IIE	MIE
BODY OSCILLATOR	(a) Turbulence buffeting	(c) Vortex shedding	(e) Flutter
FLUID OSCILLATOR	(b) Noise (from loudspeaker)	(d) Impinging shear layer	(f) Oscillating shock front

Figure 1.1. Examples of body and fluid oscillators excited by (a, b) extraneously induced excitation (EIE), (c, d) instability-induced excitation (IIE), and (e, f) movement-induced excitation (MIE).

therefore be described in terms of a self-excited '*flow oscillator*'. (Note that the flow rather than the body or fluid oscillator is self-excited in this instance in contrast to cases of MIE, cf. Naudascher & Rockwell, 1980a).

An important role regarding instability-induced vibrations of body or fluid oscillators is played by the type and strength of the control exerted on the flow instability. As shown in Sections 6.1 and 6.2, the nature of such control can be

- fluid-dynamic;
- fluid-elastic; or
- fluid-resonant.

In the fluid-dynamic case, the exciting force is a function of the flow conditions only. In the fluid-elastic and the fluid-resonant cases, this force depends on the dynamics of both the flow and the resonator in the system: a resonating body oscillator in the former case, and a resonating fluid oscillator in the latter. The main feature of fluid-elastic and fluid-resonant control is an amplification of the exciting force and a 'locking-in' of its frequency to that of the resonator within a certain range of flow velocities.

The main characteristics of five basic types of flow instability and methods for attenuating structural vibrations produced by these IIE are reviewed in Chapter 6. Examples of fluid oscillators affected by IIE are presented in Chapter 8.

Movement-induced excitation (MIE) is due to fluctuating forces that arise from movements of the vibrating body or fluid oscillator. Vibrations of the latter are thus self-excited. Examples are shown in Figures 1.1e and 1.1f. If the air- or hydrofoil depicted in Figure 1.1e is given an appropriate disturbance in both the transverse and torsional mode, the flow will induce a pressure field that tends to increase that disturbance. This situation can be described in terms of a dynamic instability of the body oscillator which gives rise to energy transfer from the main flow to the oscillator. Similar situations are possible with regard to fluid oscillators. As an example, Figure 1.1f shows an open pipe in supersonic flow in which a standing wave is sustained by MIE involving oscillatory movements of a shock front. In Chapter 7, different types of MIE are presented in terms of four basic groups. The main features of these MIE mechanisms are described so that they can be identified in any system. In addition, this chapter contains information on the onset of structural vibrations and methods of controlling them. Examples of MIE with fluid oscillators are given in Chapter 8.

If a system is susceptible to MIE, it is generally sufficient to ensure that the system operates below the thresholds of MIE. The latter are determined by the stability criteria presented in Chapter 7.

Frequently, the excitation of flow-induced vibration in a complex system is *mixed* in the sense that (a) both body *and* fluid oscillators are involved at the same time or (b) EIE, IIE, and MIE are present simultaneously. For example, the cylindrical structure of Figure 1.1c can be placed in a duct or channel in such a way that a standing wave is produced by the vortex shedding, in addition to structural vibration (Figure 8.14); or, it can be excited by turbulence buffeting, in addition to vortex excitation. Even in these mixed cases, however, it is advantageous to first identify all body oscillators, fluid oscillators, and sources of pure EIE, IIE, and MIE. Some examples of mixed excitation are presented in Chapter 9.

Whereas Chapters 5 to 8 focus on the basic mechanisms of flow-induced excitation, and their detection and evaluation, the focus of attention in Chapter 9 is the structure itself and the flow-induced vibrations it can undergo depending on its geometric and dynamic characteristics. Prismatic bodies and grids of prisms, on the one hand, and gate and gate components, on the other, are used as examples to illustrate the wide variety of mechanisms that can excite structural vibrations either individually or in combinations.

In rare cases, flow-induced vibrations are due to *parametric excitation*. These cases involve the variation with time of one or more parameters of the vibratory system such as mass, damping, and rigidity. As pointed out in Section 2.5, this variation may be of either the EIE or MIE variety. To safeguard against overlooking a potential source of excitation, one should search for these sources of vibration separately from the classification scheme presented in Figure 1.1.

In summary, then, the *assessment* of possible flow-induced vibrations in a system involves, first, a thorough search for

a) All body oscillators;
b) All fluid oscillators;
c) All sources of extraneously induced excitation;
d) All sources of instability-induced excitation;
e) All sources of movement-induced excitation; and
f) All sources of parametric excitation,

and, second, an assessment of all possible combinations of structural and fluid oscillations arising from (a) and (b) in conjunction with (c) through (f). What makes combinations of body and fluid oscillators dangerous is the coincidence of their natural frequencies. Estimates of these frequencies, as well as the dominant frequencies of possible EIE and IIE, should therefore be an integral part of the assessment.

A useful aid in preliminary investigations of 'danger spots' and dangerous operating conditions of a system is a global or lumped-parameter analysis as suggested, e.g., by Naudascher & Rockwell (1980a). If the magnitudes of loads on, and strains in, certain structures or structural parts are of interest, one has to resort to the methods of analysis described in Sections 3.3, 5.2 and 7.2 or to specific model tests. Nevertheless, prior assessment and global analysis will be important even in these cases in order that no essential element of the system and none of the possible types of excitation and their dangerous combinations is disregarded in the analytical or experimental investigation. In the case of model tests, such assessment will also help to exclude spurious excitations such as those caused by peculiar resonance conditions in the laboratory set-up that have no counterpart in the prototype.

CHAPTER 2

Body oscillators

2.1 OVERVIEW AND DEFINITIONS

Because the writers' main goal is to identify the variety of mechanisms by which flow-induced vibrations are excited, the structural dynamics are presented in the simplest way possible throughout this monograph. In most cases, this means representing the vibrating structure or structural part as a discrete mass, free to oscillate with one degree of freedom, linearly damped, and supported by a linear spring (Figures 2.1 and 2.2). The following sections contain a brief review from the field of mechanical vibrations concerning these simple body oscillators that is sufficient for the understanding of the monograph. A method of generalization is presented, finally, by which simple-oscillator concepts become applicable to more complex systems with continuous or distributed masses such as beams, plates, and shells.

Any vibration is describable in terms of sinusoidal functions. The simplest vibration is a harmonic motion of the form

$$x = x_o \cos \omega t, \qquad \omega = 2\pi f \tag{2.1}$$

where x = body deflection from its time-mean position, x_o = amplitude, t = time, ω = circular frequency, and f = frequency in cycles per second or Hertz. One of the most useful ways of describing simple harmonic motion is obtained by regarding it as the projection on the horizontal axis of a vector of length x_o rotating counterclockwise with uniform angular velocity ω (Figure 2.1). This rotating-vector description is commonly represented by the complex exponential function

$$x(t) = x_o e^{i\omega t} = x_o (\cos \omega t + i \sin \omega t) \tag{2.2}$$

for which Ox is the 'real' and Oy is the 'imaginary' axis ($i \equiv \sqrt{-1}$). Thus, the real part of this expression may be considered the horizontal projection and the imaginary part the vertical projection; and again, it is the former which represents the harmonic motion or vibration. In Figure 2.1b, $T = 1/f$ denotes the period of vibration.

7

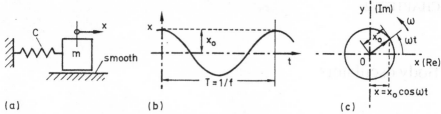

Figure 2.1. Definition sketch. (a) Simple undamped body oscillator. (b) Histogram of harmonic motion. (c) Vector respresentation of harmonic motion.

2.2 FREE VIBRATION

The simplest body oscillator consists of a discrete mass m free to vibrate in one direction (Figure 2.1a). In the absence of damping and exciting forces, an initial displacement x_o will produce free vibrations as shown in Figure 2.1b. They can be analyzed with the aid of the equation of motion or the energy method. In the equation of motion, the product of mass m and body acceleration $d^2x/dt^2 \equiv \ddot{x}$ (the dots denote derivatives with respect to time) is set equal to the sum of all forces

$$m\ddot{x} = \Sigma F_x \tag{2.3}$$

which in case of a linear spring in the undamped system of Figure 2.1a is simply the restoring force, $\Sigma F_x = -Cx$. The larger the spring constant C, the greater is the stiffness of the oscillator. The solution of the equation of motion

$$m\ddot{x} + Cx = 0 \tag{2.4}$$

is

$$x = x_o \cos(\omega_n t - \varphi) \tag{2.5}$$

with

$$\omega_n = 2\pi f_n = \sqrt{C/m} \tag{2.6}$$

where ω_n is the natural circular frequency and f_n is the natural frequency of the discrete-mass system. For the particular initial condition cited above, the phase angle φ is zero.

For systems of greater complexity, determination of the natural frequency from the equation of motion becomes so complicated that it is advisable to use an energy approach. Since the spring force Cx is the only external force doing work on the system, the sum of elastic and kinetic energy is constant. Consequently, the kinetic energy $m\dot{x}^2/2$ in the middle of the stroke ($x = 0$, $\dot{x} = \omega x_o$) must be equal to the elastic energy in an extreme position ($x = x_o$, $\dot{x} = 0$), i.e.,

$$\int_o^{x_o} Cx\,dx = \frac{1}{2}Cx_o^2 = \frac{1}{2}m\omega^2 x_o^2 \tag{2.7}$$

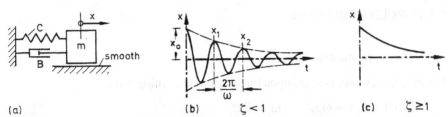

(a)   (b) $\zeta < 1$ (c) $\zeta \geq 1$

Figure 2.2. (a) Simple body oscillator with linear damping. (b, c) Histograms of responses for an underdamped ($\zeta < 1$) and an overdamped ($\zeta \geq 1$) case.

Thus, $\omega^2 = \omega_n^2 = C/m$, independent of the amplitude x_o. A method for evaluating ω_n in cases of nonlinear systems is described in conjunction with Figure 2.10.

With linear damping (i.e., resistance proportional to velocity) included, Equation 2.3 takes the form

$$m\ddot{x} + B\dot{x} + Cx = 0 \tag{2.8}$$

in which $-Bdx/dt \equiv -B\dot{x}$ is the damping force (Figure 2.2a). The solution of this equation is

$$x = e^{-\zeta\omega_n t} x_o \cos(\omega_d t - \varphi) \quad \text{with } \omega_d = \omega_n\sqrt{1 - \zeta^2} \tag{2.9}$$

as long as the damping factor or damping ratio ζ is smaller than one ($\zeta < 1$). The latter is defined as

$$\zeta = \frac{B}{B_{cr}} = \frac{B}{2m\omega_n} = \frac{B}{2\sqrt{mC}} \tag{2.10}$$

and ω_d is the frequency of the damped free vibration. (Note that for a given damper, $B = $ const., and ζ decreases with increasing mass m and stiffness C.) In Figure 2.2b, this exponentially decaying response is represented for the initial condition $t = 0, x = x_o$; thus, $\varphi = 0$. For $\zeta \geq 1$, the displaced body simply returns to its equilibrium position in an exponential fashion (Figure 2.2c). The damping for the limiting case ($\zeta = 1$) is called critical damping.

In the underdamped case, $0 < \zeta < 1$, the ratio of any two consecutive amplitudes is obtained from Equation 2.9 as $x_n/x_{n+1} = \exp(2\pi\zeta/\sqrt{1 - \zeta^2})$. The logarithm of this ratio is called the logarithmic decrement

$$\delta = \ln\frac{x_n}{x_{n+1}} = \frac{2\pi\zeta}{\sqrt{1 - \zeta^2}} \tag{2.11}$$

For small damping, $\delta \simeq 2\pi\zeta$.

With the aid of Equations 2.6 and 2.10, the equation of motion (Equation 2.8) may also be written as

$$\ddot{x} + 2\omega_n\zeta\dot{x} + \omega_n^2 x = 0$$

2.3 FORCED VIBRATION

2.3.1 *Harmonic exciting force*

If a simple oscillator is acted upon by a harmonic exciting force

$$F(t) = F_o \cos \omega_s t, \qquad \omega_s = 2\pi f_s \tag{2.12}$$

(Figure 2.3), the equation of motion (Equation 2.3) takes the form

$$m\ddot{x} + B\dot{x} + Cx = F(t)$$

or

$$\ddot{x} + 2\omega_n \zeta \dot{x} + \omega_n^2 x = \frac{F(t)}{m} \tag{2.13}$$

The solution of this equation is composed of the solution for the homogenous equation presented in Equation 2.9, and the particular solution given by

$$x = x_o \cos (\omega_s t - \varphi) \tag{2.14}$$

Since the former or transient solution dies out with time on account of damping (Figure 2.3b), only the latter or steady-state solution is of general interest. Its frequency is equal to the forcing frequency $f_s = \omega_s / 2\pi$; its amplitude x_o is obtained as

$$x_o = \frac{F_o/C}{\sqrt{[1 - (\omega_s/\omega_n)^2]^2 + (2\zeta\omega_s/\omega_n)^2}} \tag{2.15}$$

(Figures 2.4a, b); and the phase angle φ by which the response x lags the exciting force F follows from

$$tg\,\varphi = \frac{2\zeta\omega_s/\omega_n}{1 - (\omega_s/\omega_n)^2} \tag{2.16}$$

(Figure 2.4c); $\omega_n = \sqrt{C/m}$ is the natural frequency of the undamped system. The ratio of x_o and the static deflection F_o/C of the body due to F_o, depicted in Figures 2.4a and b, is called the magnification factor. Note that for lightly damped cases,

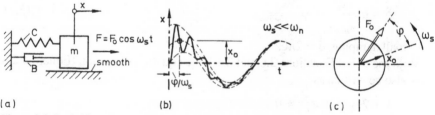

(a) (b) (c)

Figure 2.3. Definition sketch of forced vibration of a simple body oscillator.

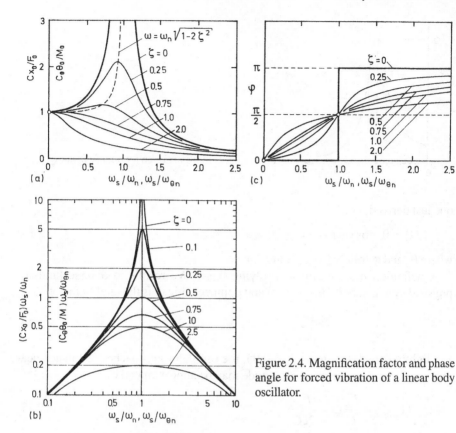

Figure 2.4. Magnification factor and phase angle for forced vibration of a linear body oscillator.

corresponding to values of ζ of the order of 0.05 or less, this ratio can be much larger than unity. The maxima of the response occur at

$$\omega = \omega_n\sqrt{1 - 2\zeta^2} \qquad (2.17)$$

The symmetric representation of the magnification factor in Figure 2.4b has advantages, for example, if the damping is to be determined from a few measured points on that diagram.

For a body with a torsional degree of freedom, Equation 2.13 takes the form

$$I_\theta\ddot{\theta} + B_\theta\dot{\theta} + C_\theta\theta = M(t)$$

or

$$\ddot{\theta} + 2\omega_n\zeta_\theta\dot{\theta} + \omega_{\theta n}^2\theta = M(t)/I_\theta \qquad (2.18)$$

where I_θ = mass moment of inertia of the body, $M(t)$ = exciting moment or torque, $-B_\theta\dot{\theta}$ = damping moment, $-C_\theta\theta$ = restoring moment, $\zeta_\theta = B_\theta/(2I_\theta\omega_n)$ = damping ratio, and $\omega_{\theta n} = \sqrt{C_\theta/I_\theta}$ = undamped circular natural frequency. The response to a harmonic exciting moment $M(t) = M_o \cos \omega_s t$ is equivalent to the

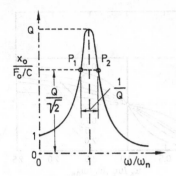

Figure 2.5. Definition sketch for quality factor Q.

one just derived, i.e.,

$$\theta = \theta_o \cos (\omega_s t - \varphi) \tag{2.19}$$

where θ_o and φ follow from Figure 2.4.

A definition most useful in applying the above results to diverse kinds of physical systems, both mechanical and nonmechanical, is the quality factor

$$Q = \frac{1}{2\zeta} \tag{2.20}$$

For lightly damped systems, $\zeta < 0.05$, the oscillator approaches resonance near $\omega/\omega_n = 1$ and the maximum amplitude may be approximated by

$$(x_o)_{max} \simeq \frac{F_o}{C} Q = \frac{F_o}{2\zeta C} \tag{2.21}$$

The points P_1 and P_2 in the amplification diagram, where the relative amplitude $x_o/(F_o/C)$ falls to $Q/\sqrt{2}$, are called half-power points, because the power absorbed by the damper in a system responding harmonically is proportional to the square of the amplitude (Figure 2.5). The increment $\Delta\omega$ of the circular frequency ω associated with the half-power points, sometimes called resonance width, takes the form

$$\Delta\omega = \omega_1 - \omega_2 \simeq 2\zeta\omega_n \tag{2.22}$$

for light damping. Consequently,

$$Q = \frac{1}{2\zeta} \simeq \frac{\omega_n}{\omega_1 - \omega_2} \tag{2.23}$$

This relationship can be used as a convenient way of determining the damping ratio ζ.

2.3.2 *Mechanical admittance and impedance*

A useful representation of forced vibration is obtained with the aid of input-output functions such as: (a) mechanical admittance

$$\chi_m(\omega) = \frac{\text{Body displacement}}{\text{Exciting force}} = \frac{x(t)}{F(t)} \tag{2.24}$$

or (b) mechanical impedance

$$Z_m(\omega) = \frac{\text{Exciting force}}{\text{Body displacement}} = \frac{F(t)}{x(t)} \tag{2.25}$$

Response and excitation in Equations 2.24 and 2.25 are commonly presented in complex form,

$$F(t) = F_o e^{i\omega t} = F_o (\cos \omega t + i \sin \omega t) \tag{2.26}$$

with the tacit understanding that the excitation is given by the real part of $F(t)$, just as the response is given by the real part of $x(t)$ in Equation 2.2. Introducing these expressions in Equation 2.13 and deriving the particular solution, one obtains

$$\chi_m(\omega) = \frac{x(t)}{F(t)} = \frac{1/C}{1 - (\omega/\omega_n)^2 + i\,2\zeta\omega/\omega_n} \tag{2.27}$$

which gives information concerning both phase and amplitude. From complex algebra, the absolute value of $\chi_m(\omega)$ is equal to the ratio of response amplitude x_o to excitation amplitude F_o, i.e.,

$$\left| \chi_m(\omega) \right| = \frac{x_o}{F_o} = \frac{1/C}{\sqrt{[1 - (\omega/\omega_n)^2]^2 + (2\zeta\omega/\omega_n)^2}} \tag{2.28}$$

This, evidently, is the magnification factor of Figure 2.4a divided by the spring constant C (Equation 2.15).

Since $e^{i\omega t} \equiv \cos \omega t + i \sin \omega t$ can be regarded as a unit vector rotating in the complex plane with angular velocity ω (Figure 2.1c), one can deduce from Equations 2.27 and 2.28

$$\chi_m(\omega) = \left| \chi_m(\omega) \right| e^{-i\varphi} \tag{2.29}$$

with φ as given by Equation 2.16. If one substitutes this and Equation 2.26 into Equation 2.27, one obtains

$$x(t) = \left| \chi_m(\omega) \right| F_o e^{i(\omega t - \varphi)} \tag{2.30}$$

This illustrates clearly that the response lags the exciting force by the phase angle φ (Figure 2.3c).

2.3.3 *Vector and force-displacement diagrams*

The complex description introduced by Equations 2.2 and 2.26 permits an interesting geometric interpretation of the equation of motion (Equation 2.13) in the complex plane. Through differentiation, one arrives at

$$x(t) = x_o e^{i\omega t}, \quad \dot{x}(t) = i\omega x(t), \quad \ddot{x}(t) = -\omega^2 x(t) \tag{2.31}$$

Since a factor i represents an advance in phase angle of $\pi/2$, the body velocity $\dot{x}$ and acceleration $\ddot{x}$ lead the body displacement x by phase angles of $\pi/2$ and π, respectively. Therefore, the restoring force $-Cx$, the damping force $-B\dot{x}$, and the intertia force $-m\ddot{x}$ can be represented by vectors as shown in Figure 2.6a, all rotating with the same angular velocity ω, equal to the driving circular frequency ω_s. The projections of these vectors onto one of the axes give the instantaneous values of the respective forces. If plotted as functions of the instantaneous values of the displacement x, they yield the force-displacement diagram shown in Figure 2.6b. (Note that 0, $\pi/2$, π ...in this figure refer to the instantaneous positions 0, $\pi/2$, π ... of the vector $\vec{x}$ in Figure 2.6a.) An advantage of such diagrams is their applicability to nonlinear oscillators or non-harmonic excitations (Sections 2.3.5 and 2.3.6).

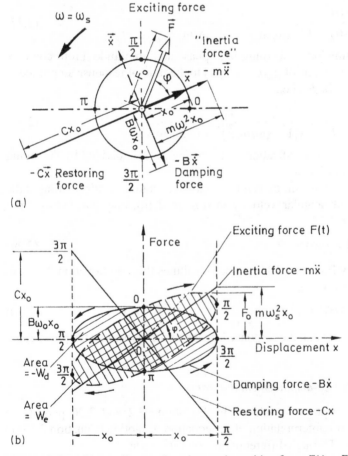

Figure 2.6. (a) Vector diagram for a harmonic exciting force $F(t) = F_o \cos \omega_s t$ acting on a simple body oscillator (Equation 2.13). (b) Corresponding force-displacement diagram.

The following features of the force displacement diagram have special significance in this monograph:

a) The linear restoring force $-Cx$ corresponds to a straight line with negative slope. A nonlinear restoring force would correspond to a curved line, and negative stiffness ($C < 0$) would produce a positive slope.

b) All non-conservative forces (like the exciting and damping forces) form a loop; the areas included in these loops correspond to the work done on the body during one cycle.

c) If a point on a force-displacement loop moves in the direction of ωt as time proceeds (i.e., counterclockwise), the work done on the body is negative and energy is dissipated.

The last of these statements is a result of Equation 2.32.

2.3.4 *Energy consideration*

The work done by a force $\vec{F}$ during a vibration cycle is obtained from the dot product with the displacement vector $d\vec{x}$, integrated over one cycle:

$$W_{cyc} = \oint \vec{F} \cdot d\vec{x} = \int_0^T \vec{F} \cdot \dot{\vec{x}}\, dt, \qquad T = \frac{2\pi}{\omega} \tag{2.32}$$

where T is the vibration period. Applied to the exciting force (Equation 2.12 or 2.26), this equation yields

$$W_e = \pi F_o x_o \sin \varphi \tag{2.33}$$

The result shows clearly that energy transfer to the body depends on the phase angle φ between exciting force and body displacement (Equation 2.16) and that this transfer is a maximum for $\varphi = +\pi/2$ and its multiples.

The work per cycle done by the damping force $-B\dot{x}$ is

$$W_d = \int_0^T (-B\dot{x})\, \dot{x}\, dt = -\pi B \omega x_o^2 \tag{2.34}$$

It is negative, which means that energy is being dissipated. From the plot of W_e and W_d for a given value of excitation frequency $\omega = \omega_s$ in the energy diagram one may conclude that steady-state force vibrations are obtained only at $x = x_o$ where $W_e + W_d = 0$. For any amplitude different from x_o, the imbalance of W_e and $-W_d$ is such that the amplitude will adjust with time to $x \rightarrow x_o$. Diagrams of this type facilitate illustration of nonlinear systems.

Even without a damper (B in Figure 2.3b), energy may be dissipated by what is

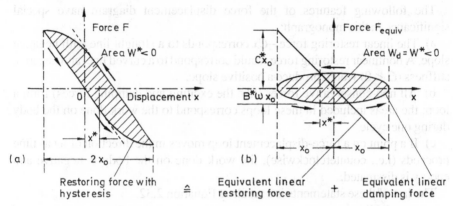

Figure 2.7. Force-displacement diagrams for (a) a spring with internal friction and (b) an equivalent linear system.

called internal friction or structural damping. Any spring made of material which is not perfectly elastic shows some hysteretic deviations from Hooke's linear stress-strain relationship. A typical force-displacement (or stress-strain) diagram for such a case is depicted in Figure 2.7a. During both loading and unloading of the spring, the force-displacement trace deviates from a straight line and forms a hysteresis loop as shown. The shape of this loop is nearly independent of amplitude x_o and strain rate (Kimball, 1929). Consequently, the energy dissipated during one cycle, which is given by the area W^* included in this loop, may be assumed to be proportional to x_o^2. W^* can be equated to the work per cycle done by an equivalent linear damping force $-B^*\dot{x}$,

$$W^* = W_d = -\pi B^* \omega x_o^2 \qquad (2.35)$$

The system with internal friction or hysteresis can thus be represented by an equivalent linear system with a damping constant $B^* = |W^*|/(\pi\omega x_o^2)$ and a spring constant $C^* = \omega_n^2 m$, where ω_n is the natural frequency of the undamped system (Figure 2.7b).

2.3.5 *Nonharmonic exciting forces*

Frequently, the exciting force is periodic but not harmonic, or it is nonperiodic or random. In the former case, it can be represented by a convergent series of harmonic functions whose frequencies are integer multiples of a fundamental frequency ω_o. Such a series of harmonic functions is known as a Fourier series and can be written in the form

$$F(t) = \sum_{N=1}^{\infty} F_N e^{iN\omega_o t}, \qquad N = 1, 2, 3, \dots \qquad (2.36)$$

with the tacit understanding that the excitation is given by the real part of $F(t)$ (Meirovitch, 1975, p. 61). If the system of the body oscillator is linear, its responses to the first, second, ... harmonics ($N = 1, 2, ...$) of Equation 2.36 can be deduced separately and then added. Thus, in analogy to Equation 2.30,

$$x(t) = \sum_{N=1}^{\infty} |\chi_m|_N F_N e^{i(N\omega_o t - \varphi_N)} \tag{2.37}$$

where $|\chi_m|_N$ and φ_N are obtained from Equations 2.28 and 2.15, respectively, after substitution of $\omega = N\omega_o$, i.e.

$$|\chi_m|_N = |\chi_m(\omega)|_{\omega = N\omega_o} \tag{2.38}$$

Clearly, the response $x(t)$ is periodic just as is $F(t)$. If the value of one of the harmonics $N\omega_o$ of the excitation is close to the natural frequency ω_n of the system and if the system is lightly damped, this harmonic will provide a relatively larger contribution to the response because it is associated with the peak value of the mechanical admittance $|\chi_m|$ (Figure 2.4a).

If the period $T_o = 2\pi/\omega_o$ of the excitation function in Equation 2.36 approaches infinity, this function becomes nonperiodic and must be represented by a Fourier integral instead of a Fourier series. In practice, such nonperiodic or random excitation is represented by means of a spectrum as shown in Figure 2.8a. If one treats the body oscillator as a linear, lightly-damped system, then the relationship between the spectra of the excitation and the response is given by

$$S_x(f) = |\chi_m(f)|^2 S_F(f), \qquad f \equiv \frac{\omega}{2\pi} \tag{2.39}$$

Here, $S_F(f) = \overline{F_f^2}/df$ and $S_x(f) = \overline{x_f^2}/df$ are the power spectral densities of the excit-

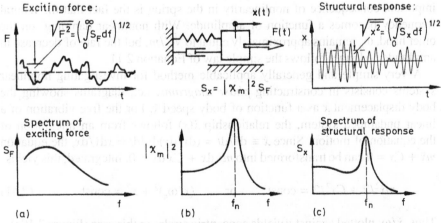

Figure 2.8. Evaluation of the response spectrum $S_x(f)$ from the excitation spectrum $S_F(f)$ with the aid of the mechanical admittance function $|\chi_m|^2$ in accordance with Equation 2.39.

ing force and the response; F_f and x_f are the amplitudes of the exciting force and the body deflection associated with frequency f, which are obtained after filtering out all frequencies outside a narrow band df centered at f; and the bars denote averaging (more details are contained in the IAHR Monograph on Hydrodynamic Forces, Naudascher, 1991, Figure 2.5). The equation implies that one has to simply multiply each ordinate value of the excitation spectrum with the corresponding value of $|\chi_m|^2$ to obtain a point on the response spectrum (Figure 2.8). The mechanical admittance function (Figure 2.4a) is hence ideally suited to determine linear-system responses to nonharmonic excitation as well.

The root-mean-square of the response or mean response amplitude is obtained by summing up the contributions to the spectrum from all frequencies as follows

$$\sqrt{\overline{x^2}} = \left[\int_o^\infty S_x(f) df \right]^{1/2} \tag{2.40}$$

If the amplitude distribution were Gaussian, the body displacement from the loaded equilibrium position, x, would exceed multiples of the mean amplitude $x' = \sqrt{\overline{x'^2}}$ (either positive or negative) as follows:

$x > x'$ or $< -x'$ 15.9% of the time each;
$x > 2x'$ or $< -2x'$ 2.3% of the time each;
$x > 3x'$ or $< -3x'$ 0.14% of the time each.

2.3.6 *Nonlinear effects*

The body-oscillator system in Figure 2.3a is called nonlinear if one or more of the coefficients m, B, or C depend on the displacement x or on its time derivatives. In mechanical cases, the most important nonlinearities occur in the damping or in the spring. Examples of nonlinear springs are shown in Figure 2.9. The most important consequence of nonlinearity in the spring is the fact that the natural frequency becomes a function of amplitude. With nonlinear damping, on the other hand, ω_n remains approximately equal to $\sqrt{C/m}$, but the rate of decrease in amplitude no longer follows the simple law of Equation 2.11.

A very simple and generally applicable method for investigating nonlinear systems consists in constructing *phase diagrams*, i.e., diagrams showing the body displacement x as a function of body speed $\dot{x}$. For the free vibration of a linear undamped system, the relationship $\dot{x}(x)$ follows from an integration of the equation of motion. Since $\ddot{x} \equiv d\dot{x}/dt = (d\dot{x}/dx)dx/dt = \dot{x}d\dot{x}/dx$, the equation $m\ddot{x} + Cx = 0$ can be transformed into $m\dot{x}\,d\dot{x} + Cx\,dx = 0$. Integrated, this yields

$$m\dot{x}^2/2 + Cx^2/2 = \text{const} \qquad \text{or} \qquad (\dot{x}/\omega_n)^2 + x^2 = \text{const} \tag{2.41}$$

Thus, $\dot{x}/\omega_n$ plotted against x yields concentric circles in this case (Figure 2.10a).

For the nonlinear system presented in Figure 2.9a, the diagram $\dot{x}/\omega_n$ versus x is

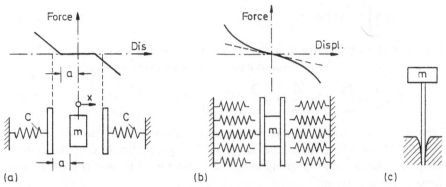

Figure 2.9. Systems with nonlinear stiffness. (a) Combination of linear springs and clearances. (b, c) Springs with more or less gradually increasing stiffness (hard springs).

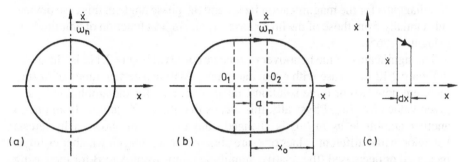

Figure 2.10. Velocity-displacement or phase diagrams for (a) linear undamped free system, (b) free system sketched in Figure 2.9a. (c) Definition sketch.

easy to construct: To the right of O_2 and to the left of O_1, the system reacts like a linear C–m-system, whereas in between, the mass moves with constant speed as it is not acted upon by any force. The result is a periodic, nonharmonic vibration completely specified by the phase diagram in Figure 2.10b. Its natural frequency can be determined graphically by carrying out the second integration in this diagram. Considering the element of curve in Figure 2.10c, one has

$$\dot{x} = dx/dt \qquad \text{or} \qquad dt = dx/\dot{x}$$

where $\dot{x}$ is the ordinate value. Thus, the time consumed by the mass progressing a distance dx is $t = \int dx/\dot{x}$. For the full period T, or one complete cycle of oscillation, hence

$$T = \oint dx/\dot{x} \qquad\qquad (2.42a)$$

or for a symmetrical spring

$$T = 4 \int_o^{x_{max}} (1/\dot{x}) \, dx \tag{4.42b}$$

This integral can be evaluated analytically or numerically. For the example shown in Figure 2.9a or 2.10b, the result is a period dependent on amplitude x_o:

$$T = \frac{1}{f_n} = \frac{2\pi}{\omega_n} = 2\pi\sqrt{\frac{C}{m}} \left(1 + \frac{2}{\pi} \frac{a}{x_o - a} \right) \tag{2.43}$$

Typical effects of nonlinearity on the forced vibration of a body are illustrated in Figure 2.11 for a body oscillator as shown in Figure 2.9b. In case the system damping is linear and the spring cubic, the equation of motion becomes, after division by the mass m,

$$\ddot{m} + 2\omega_n \zeta \dot{x} + \omega_n^2 (x + \alpha x^3) = \frac{F_o}{m} \cos \omega_s t \tag{2.44}$$

The diagrams for the magnification factor and the phase angle in this case deviate substantially from those of the linear case. Again, ω_n is a function of amplitude x_o (Magnus, 1965).

The significance of the bent-over resonance curves in Figure 2.11a is illustrated in Figure 2.12 for a case with given damping ζ. Within a certain range of ω_s/ω_n, there are now two or three solutions, rather than one, for the amplitude x_o at a given value of ω_s/ω_n. Physically, this means that there are 'jumps' from one to another amplitude as ω_s/ω_n is changed, and the system shows a hysteresis behavior in that different values of x_o are obtained depending on whether ω_s/ω_n is increased or decreased (the relative amplitude jumps from A to B for increasing ω_s/ω_n and from C to D for decreasing ω_s/ω_n). Both of these jumps as well as the hysteresis effect are typical of nonlinear systems.

Figures 2.12b and 2.12c depict the energy and phase diagram, respectively, for a value of $\omega_s/\omega_n = (\omega_s/\omega_n)^*$ for which Figure 2.12a indicates three solutions,

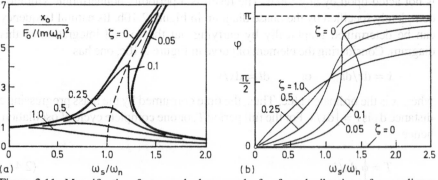

Figure 2.11. Magnification factor and phase angle for forced vibration of a nonlinear oscillator (Figure 2.9b, Equation 2.44) with hard spring ($\alpha = +0.04$) (after Magnus, 1965).

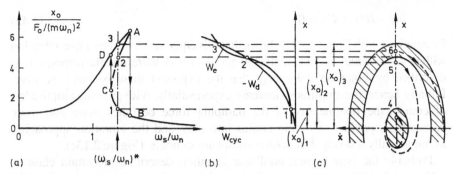

Figure 2.12. Explanation of jumps and hysteresis for a nonlinear oscillator. (a) Resonance curve; (b, c) Energy and phase diagrams for $\omega_s/\omega_n = (\omega_s/\omega_n)^*$.

namely $(x_o)_1$, $(x_o)_2$, and $(x_o)_3$. Whenever a vibration is started within the shaded areas of the phase diagram (e.g., at point 4 and 6), more energy is supplied than dissipated ($W_e > -W_d$), and the amplitude increases until the trajectories reach what is called the *limit cycles*, i.e., the $\dot{x}(x)$ curves corresponding to the solutions $x = x_o$. Clearly, then, the shaded areas correspond to regions of amplification. If x_o is below $(x_o)_2$ (say at point 5), on the other hand, dissipation will be larger than the supply of energy, and the amplitude will decay. The argument shows that only $(x_o)_1$ and $(x_o)_3$ are stable solutions, while $(x_o)_2$ marks the threshold amplitude. If x_o is larger [or smaller] than $(x_o)_2$, for example due to an initial disturbance, the amplitude will change until it approaches $(x_o)_3$ [or $(x_o)_1$]. Which of the two stable solutions is obtained for $\omega_s/\omega_n = (\omega_s/\omega_n)^*$ has thus been determined.

More detailed treatments of nonlinear oscillators, including the possibility of subharmonic responses at frequencies ω_s/N ($N = 1, 2, 3, ...$), are presented in the relevant literature (e.g., Nayfeh & Mook, 1979).

2.4 SELF-EXCITED VIBRATION

Even without an external exciting force, a body oscillator may undergo sustained vibration if there is an energy source from which the oscillator can extract energy during each cycle of free movement. Vibrations of this type are called self-excited. An example is shown in Figure 2.13. Essential for the energy transfer in this example is the negative characteristic of the dry-friction force transmitted by the belt on the body (Figure 2.13b). For small values and small changes of the relative velocity between belt and body, $v - \dot{x}$, this friction force can be approximated by $F_d = (F_d)_o - B^*(v - \dot{x}) = \text{const} + B^*\dot{x}$, and the equation of motion becomes $m\ddot{x} = \Sigma F = B^*\dot{x} - Cx + \text{const}$. Redefining the equilibrium position, one may write

$$m\ddot{x} - B^*\dot{x} + Cx = 0 \tag{2.45}$$

Comparison with Equation 2.8 shows that this equation describes a free vibration with *negative damping*. For small changes of $\dot{x}$ (or x), therefore, the response is as described by Equation 2.9 except that the exponent there becomes positive. Initially, hence, the amplitudes increase exponentially. With growing amplitudes, of course, the nonlinearity of the damping force becomes more and more pronounced, positive damping comes into play, and the vibration approaches asymptotically a steady state with constant amplitude x_o (Figure 2.13c).

Probably the best known nonlinear equation describing a certain class of self-excited oscillators (e.g., electrical circuits containing vacuum tubes) is of the Van der Pol type

$$\frac{m}{C}\ddot{x} - (\alpha\dot{x} - \beta\dot{x}^3) + x = 0 \tag{2.46}$$

The equation contains a cubic expression for the damping force with both positive and negative contributions. The phase and energy diagrams for this type of oscillator are shown in Figure 2.14. Due to nonlinearity, the limit cycle in this case deviates from an ellipse, and the work done by the damping force is such that it supplies energy to the body for $x < x_o$ and dissipates energy for $x > x_o$. A system

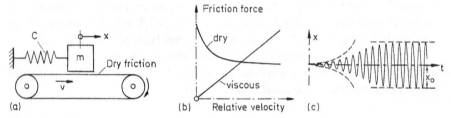

Figure 2.13. Example of self-excited vibration: Body oscillator on moving belt.

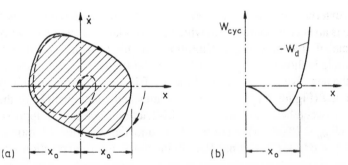

Figure 2.14. (a) Phase diagram with limit cycle and (b) energy diagram for self-excited oscillator according to Equation 2.46. (W_d = work done per cycle by damping force.)

described by Equation 2.46 will thus perform self-excited vibrations with amplitude x_o no matter whether it starts from rest or it becomes disturbed by $x > x_o$ (Nayfeh & Mook, 1979; Magnus, 1965).

2.5 PARAMETRICALLY EXCITED VIBRATION

A body oscillator is said to perform parametric oscillations if it is excited due to the fact that one or more of the parameters of the oscillatory system vary, usually periodically, with time. In mathematical terms, parametric oscillations are described by homogeneous differential equations with rapidly, and mostly periodically, varying coefficients. Hence

$$m(t)\ddot{x} + B(t)\dot{x} + C(t)x = 0 \tag{2.47}$$

or after division by $m(t)$

$$\ddot{x} + k_1(t)\dot{x} + k_2(t)x = 0 \tag{2.48}$$

The time variation of the parameters in Equation 2.48 may be either externally imposed or self-induced by the oscillator. In either case, the parametric excitation can only become effective when the oscillator moves out of the equilibrium position; it is thus movement-dependent. In contrast to the forced excitation treated in Section 2.4, where a small excitation may produce a large response only if the excitation frequency is close to a natural frequency f_n of the system, the response to a small parametric excitation can become large even if its frequency is quite different from f_n.

By introducing a new variable y, Equation 2.48 reads

$$\ddot{y} + K(t)y = 0 \qquad \text{with } x = y \exp\left[-\frac{1}{2}\int k_1(t)dt\right] \tag{2.49}$$

and

$$K(t) = k_2(t) - \frac{1}{2}\frac{d}{dt}\left[k_1(t)\right] - \frac{1}{4}k_1^2(t) \tag{2.50}$$

If the parameters k_1 and k_2 are periodic functions of time, $K(t)$ will be as well, and the solution will have the form

$$y(t) = C_1 \exp(\upsilon_1 t)y_1(t) + C_2 \exp(\upsilon_2 t)y_2(t) \tag{2.51}$$

where y_1 and y_2 are periodic functions of time, C_1 and C_2 are constants, and υ_1 and υ_2 are called the characteristic exponents of Equation 2.49a. If $K(t)$ can be expressed as

$$K(t) = K_o + \Delta K \cos \omega_k t \tag{2.52}$$

one can transfer Equation 2.49a into the standard form of Mathieu's differential equation

$$\ddot{y} + (\lambda + \kappa \cos \tau)y = 0 \qquad (2.53)$$

in which the variables are used in their dimensionless form

$$\tau = \omega_k t, \qquad \lambda = \frac{K_o}{\omega_k^2}, \qquad \kappa = \frac{\Delta K}{\omega_k^2} \qquad (2.54)$$

The stability of an oscillator governed by the Mathieu equation is determined by the sign of the exponents υ in Equation 2.51 which, in turn, depend entirely upon λ and κ, and not upon the initial conditions. The oscillator is said to be unstable if an infinitesimal disturbance initiates the build-up of vibrations. The regions of stable and unstable (i.e., bounded and unbounded) solutions can thus be represented on a $\lambda\kappa$-plane as shown in Figure 2.15. These regions are separated by boundaries for which the solutions are periodic.

The use of this stability diagram can be illustrated by considering an undamped pendulum, the suspension point of which performs periodic motion in the vertical direction as indicated in Figure 2.16a. For small oscillations, $\sin \theta \simeq \theta$, and the equation of motion becomes

$$\ddot{\theta} + \left(\frac{g}{L_r} + \frac{a_o}{L_r} \omega_k^2 \cos \omega_k t \right)\theta = 0 \qquad (2.55)$$

where $L_r = I_\theta/(ms)$ is the reduced length; I_θ, m, and s are defined in Figure 2.16a; and $\omega_n = \sqrt{g/L_r}$ is the angular natural frequency of the pendulum for a static suspension point. With the substitutions

$$\tau = \omega_k t, \qquad \lambda = \frac{g/L_r}{\omega_k^2} = \left(\frac{\omega_n}{\omega_k} \right)^2, \qquad \kappa = \frac{a_o}{L_r} \qquad (2.56)$$

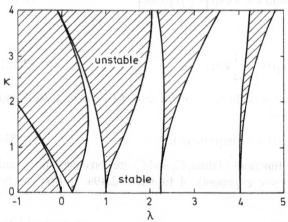

Figure 2.15. Stable and unstable regions for an oscillator governed by the Mathieu equation (Equation 2.53).

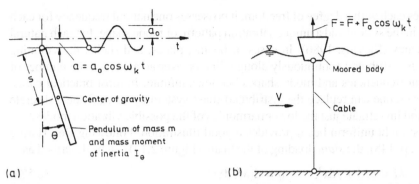

Figure 2.16. Examples for parametrically excited vibrations: (a) Pendulum with support moving periodically in the vertical direction; (b) Mooring cable with buoyant body in a tidal current with surface waves.

this equation can be transformed into the Mathieu equation, Equation 2.53, so that Figure 2.15 can be used to predict the stability of the pendulum. It shows that depending on the amplitude a_o, or rather $\kappa = a_o/L_r$, certain values of $\lambda = (\omega_n/\omega_k)^2$ cause the undamped pendulum to go into unbounded oscillations even from stable equilibrium ($\lambda > 0$). For a pendulum in unstable equilibrium, i.e., suspended below its center of gravity ($\lambda < 0$), Figure 2.15 shows the system can still be stable for certain combinations of λ and κ.

The response of a *damped* oscillator can be deduced from Equation 2.53 if a term for viscous damping is incorporated in it. For a one-degree-of-freedom system, viscosity has a stabilizing effect. Unstable regions of the type shown in Figure 2.15 occupy smaller areas of the $\kappa\lambda$ plane. Moreover, the minima of the unstable regions do not reach down to the λ axis as in Figure 2.15; in other words, a threshold value of κ is required to attain instability in this case. For a system with several degrees of freedom, on the other hand, the number of unstable regions increases, and viscous damping may have a destabilizing effect.

An example from the area of flow-induced vibration is a flexible pipe conveying pulsating flow with velocity $V = \bar{V} + V_o \cos \omega_k t$ (Paidoussis & Issid, 1974; Paidoussis, Issid & Tsui, 1980). Another example is depicted in Figure 2.16b. As was shown by Simpson (1979), a mooring cable in a tidal current can be subject to parametric excitation if the moored body gives rise to periodic forces at the upper end of the cable because of wave action.

2.6 GENERALIZATION FOR DISTRIBUTED-MASS SYSTEMS

In the preceding sections, the body oscillator was treated as a discrete-mass system with one degree of freedom. It was therefore possible to characterize this oscillator by one natural frequency. If a system contains two or more discrete masses each with a degree of freedom, it possesses one natural frequency for each mass in the system and a unique vibration pattern, or mode shape, for each natural frequency (Blevins, 1986). In cases of beams, plates, and shells, the vibrating mass is distributed continuously along a line or over an area, and the number of natural frequencies and mode shapes becomes infinite. In most practical cases, however, one can replace the distributed-mass system by an equivalent discrete one and investigate just the lower harmonics of the possible vibration modes.

A straight uniform beam provides a good illustration. Following Den Hartog (1985, p. 148), the *static* loading of the beam (Figure 2.17) can be described as

$$M = EI d^2 Y / dx^2, \qquad q = d^2 M / dx^2 \tag{2.57a}$$

or, combining the two expressions, as

$$q = \frac{d^2}{dx^2} \left(EI \frac{d^2 Y}{dx^2} \right) \tag{2.57b}$$

Here, M = bending moment, q = load per unit length, and Y = transverse deflection, all at a position x along the beam; E = elastic modulus and I = area moment of inertia of the beam cross-section (see Blevins, 1986, Table 5.1). If the cross-section of the beam is constant along its length, then EI = const., and Equation 2.57b yields

$$q = EI d^4 Y / dx^4 \tag{2.58}$$

For a beam without static load undergoing free undamped vibration at a certain mode, the *dynamic* 'loading' it experiences is an alternating inertia load. In the position of maximum downward deflection, each element of the beam is subjected to a maximum upward acceleration $d^2 Y / dt^2$. Multiplied by the mass of the element, according to the principle of action and reaction, this gives a downward inertia force on the beam. Hence, $q \hat{=} - \mu d^2 Y / dt^2$, and Equation 2.58 yields

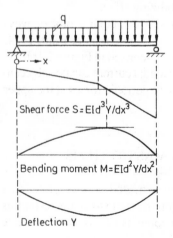

Figure 2.17. Example of beam with static loading.

$$EI \frac{\partial^4 Y}{\partial x^4} = -\mu \frac{\partial^2 Y}{\partial t^2} \tag{2.59}$$

In this differential equation for a freely vibrating slender beam of uniform cross section, μ is the mass of the beam per unit length and $Y = Y(x, t)$ is the transverse deflection at a given location x and time t. If the beam vibrates harmonically at some natural frequency $f_N = \omega_N/2\pi$, associated with a natural mode shape $\tilde{y}_N(x)$, one may express $Y(x, t)$ in the form

$$Y(x, t) = \tilde{y}_N(x) Y_N \cos(\omega_N t - \varphi_N), \qquad N = 1, 2, 3 \ldots \tag{2.60}$$

The mode shape $\tilde{y}_N$, herein, gives the distribution of beam amplitudes along x for the Nth mode in a normalized way (in general, $\tilde{y}_N = 1.0$ at a reference point); Y_N is the reference-point amplitude of the Nth mode in units of length; and φ_N is the corresponding phase angle. The magnitudes of Y_N and φ_N, of course, depend on the initial conditions used to set the beam in motion.

In the case of distributed-mass systems, therefore, one needs to know the possible mode shapes $\tilde{y}_N(x)$ in addition to f_N in order to describe a vibration. Moreover, the vibration may take place in several modes simultaneously, so that the total transverse deflection becomes, in general,

$$Y(x, t) = \sum_{N=1}^{\infty} \tilde{y}_N(x) y_N(t), \qquad y_N(t) = Y_N \cos(\omega_N t - \varphi_N) \tag{2.61}$$

The form given to Equation 2.59 by Equation 2.60, namely,

$$EI \frac{d^4 Y}{dx^4} = \mu \omega_N^2 Y \tag{2.62}$$

can now be used to determine $\tilde{y}_N(x)$. The left-hand side, according to Equation 2.58, denotes the expression for the loading q while the right-hand side is the maximum value of the inertia load. Since the latter turns out to be Y multiplied by a constant, one may state: first, any loading $q(x)$ which generates a deflection curve $Y(x)$ similar to the $q(x)$ curve can be regarded as an inertial loading during a free vibration; and, second, the amplitudes of this vibration are distributed along x in accordance with this deflection curve $Y(x)$.

According to Equation 2.62, $Y(x)$ and, hence, $y_N(x)$ (Equation 2.60) must have the property that, when differentiated four times, these functions return to their original form multiplied by $\mu(2\pi f_N)^2/EI$. From this fact and also from the boundary conditions (e.g., zero deflection and zero bending moment for simply supported ends, $\tilde{y} = 0$, $d^2\tilde{y}/dx^2 = 0$), one obtains for a beam on two simple supports (Figure 2.18a) the exact solution

$$\tilde{y}_N(x) = \sin(N\pi x/L)$$

$$f_N = \frac{N^2 \pi}{2L^2} \sqrt{\frac{EI}{\mu}}, \qquad N = 1, 2, 3 \ldots \tag{2.63}$$

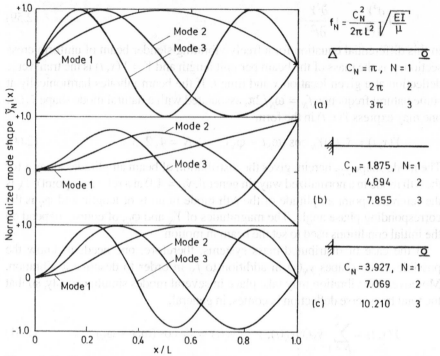

Figure 2.18. Mode shapes and natural frequencies of first three modes of straight slender beams with the following end conditions: (a) simply supported – simply supported; (b) clamped – free; (c) clamped – simply supported.

The equation of motion describing the forced vibration of a uniform slender beam is (Meirovitch, 1967)

$$EI \frac{\partial^4 Y}{\partial x^4} + \mu \frac{\partial^2 Y}{\partial t^2} = F_y = F_y^d + F_y^e \qquad (2.64)$$

where F_y denotes the sum of the damping and exciting forces per unit length acting normal to the beam axis in the direction of Y. After substitution of Equations 2.61 and 2.63, this becomes

$$\sum_{N=1}^{\infty} \left[EI \left(\frac{\pi N}{L} \right)^4 y_N(t) + \mu \ddot{y}_N(t) \right] \sin(N\pi x/L) = F_y(x, t) \qquad (2.65)$$

If one multiplies this equation by $\sin(j\pi x/L)$ and integrates over the length of the beam L, and if the beam possesses orthogonal modes with the property

$$\int_o^L \bar{y}_i \bar{y}_j dx = \begin{cases} 0 \text{ for } i \neq j \\ L/2 \text{ for } i = j \end{cases} \qquad (2.66)$$

one obtains

$$\ddot{y}_N + \omega_N^2 y_N = \frac{\int_o^L F_y(x, t)\bar{y}_N(x)dx}{\mu \int_o^L \bar{y}_N^2(x)dx}, \qquad N = 1, 2, 3 \dots \tag{2.67}$$

For the case of viscous damping, F_y^d can be approximated by $-B\partial Y/\partial t$. In terms of the equivalent damping factor for the Nth mode

$$\zeta_N = \frac{B}{2\mu\omega_N} \tag{2.68}$$

Finally, Equation 2.67 assumes the form

$$\ddot{y}_N + 2\zeta_N\omega_N \dot{y}_N + \omega_N^2 y_N = \frac{\int_o^L F_y^e(x, t)\bar{y}_N(x)dx}{\mu \int_o^L \bar{y}_N^2(x)dx}, \qquad N = 1, 2, 3 \dots \tag{2.69}$$

where the right-hand side is $1/\mu$ times the generalized exciting force of the Nth vibration mode in the Y direction.

By modal analysis, it was thus possible to reduce the complex partial differential equation describing the motion of a one-dimensional distributed-mass system to a set of simple ordinary differential equations similar to the one presented in Section 2.4 for a discrete-mass system. The derivation is strictly valid only for systems with orthogonal modes. One can show, however, that the solution of Equation 2.69 yields a close approximation to the structural response even in cases in which the structural modes are not orthogonal.

Mode shapes and natural frequencies for beams with other boundary conditions than those contained in Figure 2.18 or for other distributed-mass systems are presented in the literature (e.g., Blevins, 1986). For rings with uniform mass and stiffness, the mode shape of bending vibration is a curve consisting of sinusoids on the developed circumference of the ring, shown schematically in Figure 2.19 for modes with four, six, and eight nodes. The formula for the natural frequencies in this case is

$$f_N = \frac{N(N^2 - 1)}{2\pi\sqrt{N^2 + 1}}\sqrt{\frac{EI}{\mu R^4}} \tag{2.70}$$

where N = number of full waves along the circumference of the ring, μ = mass per unit length of the ring, R = radius, and EI = bending stiffness.

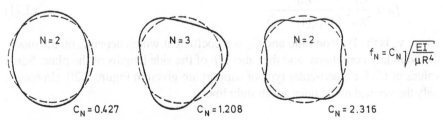

Figure 2.19. Mode shapes of a ring bending in its own plane.

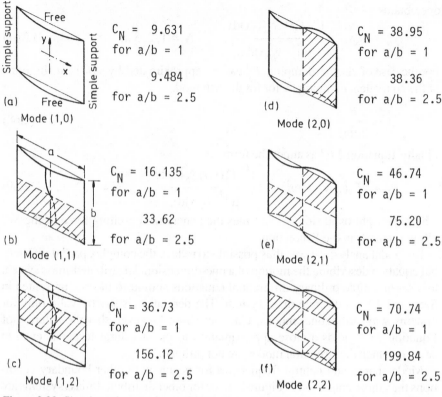

Figure 2.20. Sketches of mode shapes for a rectangular plate supported as shown in Figure 2.20a. Values of C_N (for $v_p = 0.3$) permit computation of natural frequency f_N through Equation 2.71 (after Leissa, 1973).

For two-dimensional structures such as plates and shells, two numbers are required to designate a mode shape. For plates, for example, these two numbers refer to half-waves in the x direction and y direction, respectively (Figure 2.20). The natural frequencies of rectangular plates of thickness d and density ρ_s are given by

$$f_N = \frac{C_N^2}{2\pi a^2} \sqrt{\frac{Ed^2}{12\rho_s(1 - v_p^2)}} \qquad (2.71)$$

where v_p is the Poisson ratio and C_N is a coefficient which depends on the mode, the boundary conditions, and the ratio a/b of the side lengths of the plate. Some values of C_N for a particular type of support are given in Figure 2.20. (In reality, only the vertical nodal lines are straight lines.)

CHAPTER 3

Fluid loading and response of body oscillators

3.1 DISTINCTION OF TYPES OF FLUID LOADING

A stationary body placed in a flowing fluid (Figure 3.1a) is acted upon by a force that can be subdivided into a mean and a fluctuating part, $\overline{F} + F(t)$. The former is of interest with respect to the mean hydrodynamic loading of a structure and is treated in the IAHR Monograph on Hydrodynamic Forces (Naudascher, 1991). The latter is due to fluid fluctuations around the body that are produced either by an extraneous source such as turbulence or by a flow instability such as vortex shedding. In cases of an extraneous source, $F(t)$ remains independent of body movements $x(t)$ except for added-mass and fluid-damping effects; in cases of flow instability, on the other hand, $F(t)$ is altered when the body starts moving. If a body is free to vibrate (Figure 3.1c), therefore, one must distinguish between, and treat differently, the corresponding extraneously induced and instability-induced excitations (EIE and IIE).

If a body with one degree of freedom vibrates in an otherwise stagnant fluid (Figure 3.1b), the fluid close to the body is set into motion so that again the body becomes acted upon by a fluctuating force $F(t)$. Simplified by the use of linear expressions, this force can generally be described in terms of components acting in phase with the acceleration $\ddot{x}$, speed $\dot{x}$, and displacement x of the body

$$F(t) = -A'\ddot{x} - B'\dot{x} - C'x \qquad (3.1)$$

If the equation of motion for a linear mass-spring system is written in accordance with Equation 2.13,

$$m\ddot{x} + B\dot{x} + Cx = F(t)$$

and incorporated into Equation 3.1, the result is

$$(m + A')\ddot{x} + (B + B')\dot{x} + (C + C')x = 0 \qquad (3.2)$$

This equation indicates that the system behaves like a body oscillator with added mass A', added damping B', and added stiffness C', undergoing free vibration.

An important consequence of these added quantities is that they alter the body

31

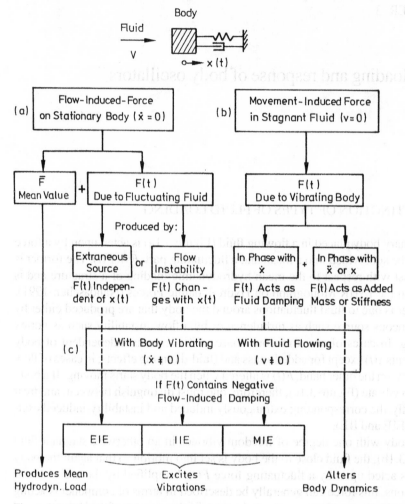

Figure 3.1. Schematic of forces exerted on a body by the surrounding fluid. (EIE, IIE, and MIE denote extraneously induced, instability-induced, and movement-induced excitations, cf., Figure 1.1).

dynamics. The natural frequency of the system, for example, becomes

$$(f_n)_{\text{Fluid}} = \frac{1}{2\pi} \sqrt{\frac{C + C'}{m + A'}} \qquad (3.3a)$$

as compared to $f_n = \sqrt{C/m}/2\pi$ in a perfect vacuum, and the total damping force is composed now of the in-vacuum damping $-B\dot{x}$ plus the fluid damping $-B'\dot{x}$.

Examples of added stiffness are presented in Section 3.4. The simplest example involves a floating body or ship. If a ship of width D and length L is displaced in the vertical direction by a distance x, a buoyancy force $-\rho DLx$ is induced acting on the body in the opposite direction. According to Equation 3.1, this force is a restoring-type of force $-C'x$ with a positive added-stiffness coefficient

$$C' = + \rho DL$$

A floating body thus forms an oscillator even without a mechanical spring ($C = 0$). Added stiffness of this kind is usually included in the spring constant of the system, C.

In a flowing fluid, however, added stiffness may also develop on account of body vibration $x = x_o \sin \omega t$. In absence of a complete theoretical model of flow-body interaction, it is impossible to separate the added stiffness C' from the added mass A' in Equation 3.1 since the body acceleration, $\ddot{x} = -x_o \omega^2 \sin \omega t = -\omega^2 x$, is in phase with $-x$. In general, therefore, the fluid forces in phase with x and $\ddot{x}$ are described by one single term, e.g.,

$$F(t) = -A'\ddot{x} - B'\dot{x} \tag{3.4}$$

The corresponding expression for the ratio of the natural frequencies in a fluid and in a vacuum is

$$\frac{(f_n)_{\text{Fluid}}}{f_n} = \sqrt{\frac{1}{1 + A'/m}} \tag{3.3b}$$

If a body vibrates in a flowing fluid (Figure 3.1c), the added, or fluid, damping can become negative as a result of mechanisms by which energy is transferred from the flow to the body. The body vibrations, consequently, become self-excited. Depending on whether a flow instability plays a significant role in the excitation process or not, the excitation in these cases is called instability-induced or movement-induced, respectively (IIE or MIE). A linearized description of fluid damping as adopted in Equations 3.1 and 3.4 is, strictly speaking, not suitable here.

3.2 ADDED MASS AND FLUID DAMPING

3.2.1 *General relationships*

For bodies with one degree of freedom vibrating in a stagnant, incompressible fluid, the dimensionless added-mass and fluid-damping coefficients

$$\alpha = A'/(\rho V), \quad \beta = B'/(\rho V \omega) \quad \text{for three-dimensional bodies} \tag{3.5a}$$

or

$$\alpha = A'/(\rho D^2), \quad \beta = B'/(\rho D^2 \omega) \quad \text{for two-dimensional bodies} \tag{3.5b}$$

are generally functions of

1. the geometry of the body and of nearby structures including the body's relative position to a free surface;
2. a Reynolds-number-like parameter, $\omega D^2/\nu$;
3. a Froude-number-like parameter, $\omega D/c$, in cases of bodies in proximity to a free surface; and
4. the relative amplitude x_o/D and the direction of vibration.

Here, ρ = density of fluid, V = characteristic fluid volume, $\omega = 2\pi f$ = circular frequency of vibration, D = significant length of structure, ν = kinematic viscosity of fluid, c = celerity of free-surface waves. (Note the difference in the dimensions of A' and B' for three- and two-dimensional bodies.) In cases of stationary bodies in oscillating fluids, the parameters 2, 3, and 4 are usually replaced by V_oD/ν, V_o/c, and $V_o/(\omega D)$, where V_o and ω are the amplitude and circular frequency of the fluid velocity, respectively.

If a two-dimensional body has a symmetric cross-section with respect to two perpendicular axes, the coefficients α and β for translation along these axes as well as for rotation about their intersection can be determined separately. In case the cross-section is not symmetric (Figure 3.2) or the body is near an unsymmetric surface, the fluid forces will generally couple translation and rotation. For this general case, the fluid forces in a stagnant fluid due to added mass must be written in matrix form (Blevins, 1986) as:

$$
\begin{bmatrix} F_x \\ F_y \\ M_\theta \end{bmatrix} = - \begin{bmatrix} A'_{xx} & A'_{yx} & A'_{\theta x} \\ A'_{xy} & A'_{yy} & A'_{\theta y} \\ A'_{x\theta} & A'_{y\theta} & A'_{\theta\theta} \end{bmatrix} \begin{bmatrix} \ddot{x} \\ \ddot{y} \\ \ddot{\theta} \end{bmatrix} \tag{3.6}
$$

According to this equation, acceleration of the body in the x direction ($\ddot{y} = \ddot{\theta} = 0$) will induce not only an added-mass force in that direction,

$$
F_x = -A'_{xx}\ddot{x} \tag{3.7}
$$

but also a force in the y direction and a moment about the origin

$$
F_y = -A'_{yx}\ddot{x} \tag{3.8}
$$

$$
M = -A'_{\theta x}\ddot{x} \tag{3.9}
$$

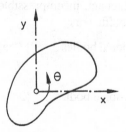

Figure 3.2. Definition sketch.

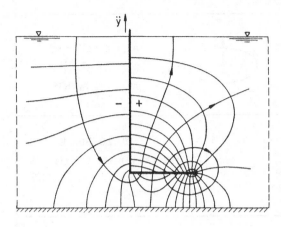

Figure 3.3. Instantaneous flow pattern for vertically accelerated L-shaped gate (after Kolkman, 1976).

A similar cross-coupling can take place for the structural mass as well (Blevins, 1986). In Equations 3.8 and 3.9, A'_{yx} is the added-mass translation cross coupling, which can be determined by potential-flow theory (e.g., Sedov, 1965). According to Ackerman et al. (1964) and McConnell & Young (1965), predictions by potential-flow theory are within about 10% of the experimentally measured values if, typically, $\omega D^2/\nu > 10\,000$ and $x_o/D < 1$.

An illustration of the coupling effect is given in Figure 3.3. For an L-shaped structure acclerated vertically upward, the potential-flow solution indicates a pressure difference across the vertical plate. Hence, this motion will not only induce a force, $F_y = -A'_{yy}\ddot{y}$, acting in the y direction, but also a force, $F_x = -A'_{xy}\ddot{y}$, perpendicular to it. If the structure has degrees of freedom in both the x and y directions, therefore, vibration excited in one may cause a coupled response in the other. The same applies to a rotational degree of freedom.

Added-mass (and structural-mass) cross coupling may also occur in cases of structures with distributed mass, e.g., a straight uniform *beam* with a cross-section according to Figure 3.2. Only if the cross-section is symmetric about the plane of transverse vibration will cross coupling be absent. The added-mass effect on the natural frequencies f_N of the beam (Figure 2.17) is then

$$\frac{(f_N)_{\text{Fluid}}}{f_N} = \sqrt{\frac{1}{1 + A'_{xx}/\mu}} \quad \text{(beams)} \tag{3.10}$$

where A'_{xx} is the added mass per unit length for the direction x of transverse vibration (in the Nth mode) and μ is the structural mass per unit length of the beam. Values of A'_{xx} for some cross-sections as obtained from potential-flow theory are presented in Figure 3.4.

For an elastic *plate*, the corresponding relationship is

$$\frac{(f_N)_{\text{Fluid}}}{f_N} = \sqrt{\frac{1}{1 + A'_N/m}} \quad \text{(plates)} \tag{3.11}$$

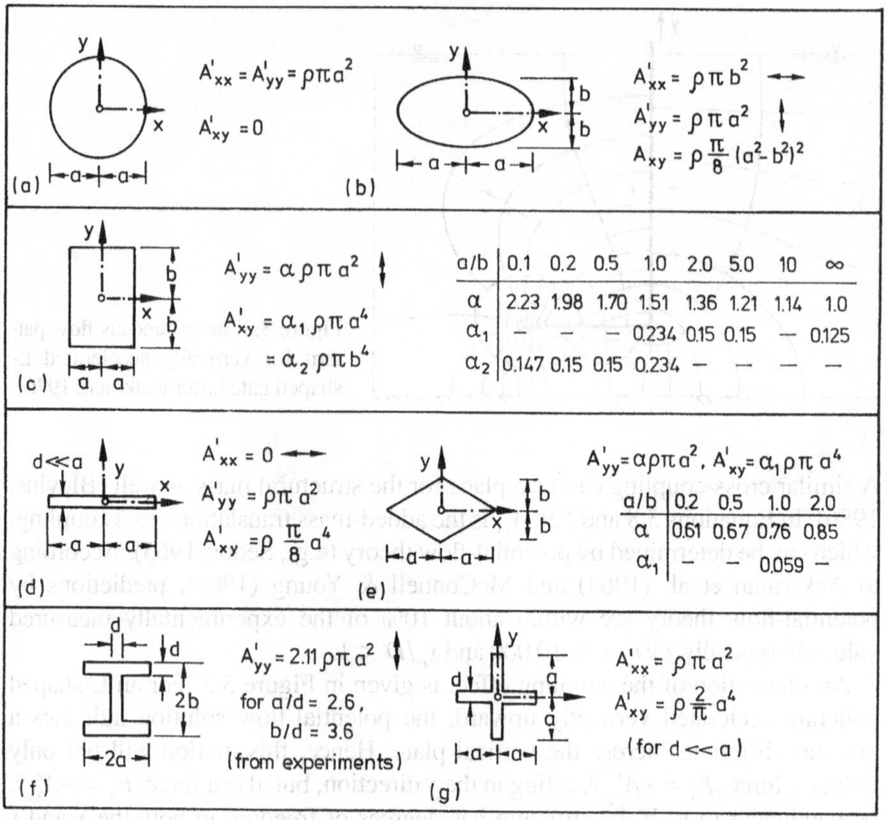

For part (a), circle:
$$A'_{xx} = A'_{yy} = \rho \pi a^2$$
$$A'_{xy} = 0$$

For part (b), ellipse:
$$A'_{xx} = \rho \pi b^2$$
$$A'_{yy} = \rho \pi a^2$$
$$A_{xy} = \rho \frac{\pi}{8}(a^2 - b^2)^2$$

For part (c), rectangle:
$$A'_{yy} = \alpha \rho \pi a^2$$
$$A'_{xy} = \alpha_1 \rho \pi a^4$$
$$= \alpha_2 \rho \pi b^4$$

a/b	0.1	0.2	0.5	1.0	2.0	5.0	10	∞
α	2.23	1.98	1.70	1.51	1.36	1.21	1.14	1.0
α_1	–	–	–	0.234	0.15	0.15	–	0.125
α_2	0.147	0.15	0.15	0.234	–	–	–	–

For part (d), $d \ll a$:
$$A'_{xx} = 0$$
$$A'_{yy} = \rho \pi a^2$$
$$A'_{xy} = \rho \frac{\pi}{8} a^4$$

For part (e), diamond:
$$A'_{yy} = \alpha \rho \pi a^2, \quad A'_{xy} = \alpha_1 \rho \pi a^4$$

a/b	0.2	0.5	1.0	2.0
α	0.61	0.67	0.76	0.85
α_1	–	–	0.059	–

For part (f), I-beam:
$$A'_{yy} = 2.11 \rho \pi a^2$$
for a/d = 2.6,
b/d = 3.6
(from experiments)

For part (g), cross:
$$A'_{xx} = \rho \pi a^2$$
$$A'_{xy} = \rho \frac{2}{\pi} a^4$$
(for d ≪ a)

Figure 3.4. Added masses A'_{xx}, A'_{yy} and added-mass moment of inertia A'_{xy} per unit length for two-dimensional bodies accelerated along x, y, and about an axis through the origin (for more information, see Blevins, 1986).

unless the added mass changes the mode shape of the plate. Here, m is the total mass of the plate and A'_N is the added mass which depends on the mode shape in addition to the parameters listed under Equation 3.5. Values of A'_N for circular plates in an infinite rigid baffle, for example, are given in Figure 3.5. Note that these values have to be doubled if the two sides of the plate are exposed to a fluid. Figure 3.6 shows A'_N values for plates obtained from strip theory on the assumption that the movement-induced flow field over slender cantilevered plates will be locally two-dimensional. A correction factor for non-slender plates is given by Lindholm et al. (1965). Data for plates with other boundary conditions have been evaluated by Greenspan (1961). A review of literature on fluid-shell interaction has been presented by Chen (1977). A simple scheme for calculating the added mass of hydraulic gates is given by Kolkman (1988).

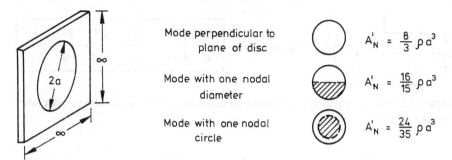

Figure 3.5. Added mass of a circular plate in an infinite rigid baffle with one side exposed to a fluid (after McLachlan, 1932). (Shaded areas are explained in Figure 2.20.)

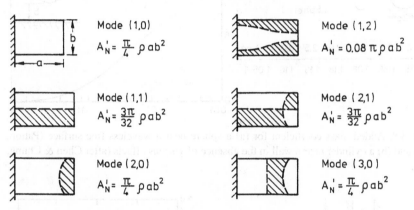

Figure 3.6. Added masses of slender cantilever plates submerged in a fluid (after Lindholm et al., 1965). (Shaded areas are explained in Figure 2.20.)

3.2.2 *Effects of geometry*

For a body vibrating with small amplitudes and high frequencies in stagnant fluid in the absence of wave radiation, fluid damping is negligible and the added mass can be deduced from potential-flow theory. Numerical methods now exist for more complex configurations (e.g., Oden et al., 1974).

The effect of the cross-sectional shape of two-dimensional bodies is shown in Figure 3.4. As the length L of a cylindrical body decreases with respect to its diameter $2a$, the added mass diminishes (Figure 3.7).

Large changes in added mass may occur if a structure is adjacent to either a free surface or a solid wall. Figure 3.8 shows these effects for a sphere near a free surface (without wave effects) and for a circular cylinder near a wall. The curve in Figure 3.8b was found to be independent of the direction of motion, in the absence of viscosity. Any additional confinement, e.g., by other cylinders in the proximity,

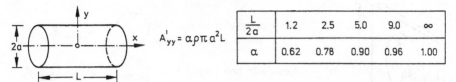

Figure 3.7. Effect of aspect ratio $L/(2a)$ on the added mass of a right circular cylinder (after Wendel, 1950).

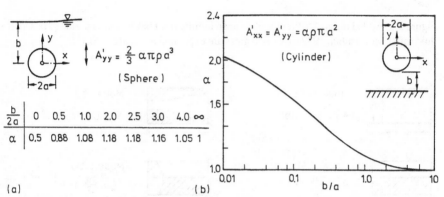

Figure 3.8. Added-mass coefficient for (a) a sphere near a waveless free surface (Patton, 1965) and (b) a cylinder near a wall in the absence of viscous effects (after Chen & Chung, 1976).

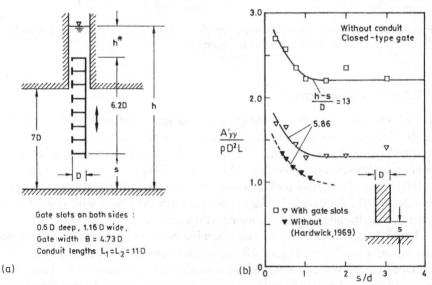

Figure 3.9. (a) Typical high-head gate (extension of skinplate removed). (b) Added mass A'_{yy} for closed-type gate without conduit top (after Nguyen, 1982): Effect of confinement by the floor.

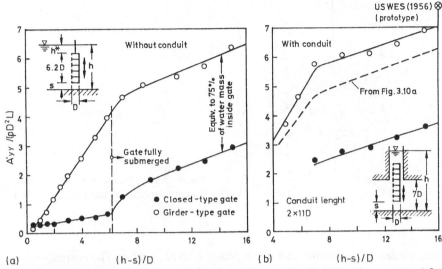

Figure 3.10. Added mass A'_{yy} for the gate shown in Figure 3.9a with $s/D = 4$ = constant (after Nguyen, 1982). (a) Effect of submergence ratio $(h - s)/D$ and gate type; (b) Same with confinement effect by the conduit.

results in a further increase in the added mass (cf. Figure 3.14). The reason for this increase is the larger build-up in pressure due to the obstruction to the movement-induced flow.

Figure 3.9 shows the added mass of a vertically vibrating high-head gate as a function of the submergence ratio $(h - s)/D$. The influence of floor proximity is limited to a range of gate openings $s/D < 2$ (Figure 3.9b). The confinement resulting from typical gate slots on the two sides (Figure 3.9a) was found to increase the added mass by 20% on the average. The effect of additional confinement by a conduit and the effects of submergence and gate geometry (closed upstream face versus open girders) are presented in Figure 3.10. With the gate fully submerged $[(h - s)/D > 6.2]$, A'_{yy} for the open-type gate is approximately equal to A'_{yy} of the closed-type gate plus 75% of the water mass contained between the girders. There is a possibility that the data in Figures 3.9 and 3.10 are not completely free of scale effects ($\omega D^2/\nu \approx 7.3 \times 10^4$; $\omega\sqrt{D/g} \approx 4.5$, $x_o/D = 0.2$), although the fair agreement with a prototype measurement in Figure 3.10b is reassuring.

3.2.3 *Effect of wave radiation*

When a completely submerged body is accelerated ($\ddot{x}$) in an inviscid, incompressible fluid, all fluid particles surrounding it respond instantaneously to its motion (Figure 3.11). Fluid particles near the body have larger accelerations than

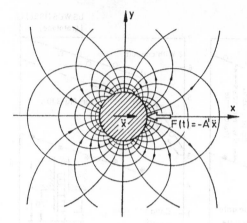

Figure 3.11. Instantaneous flow pattern for cylinder accelerating in inviscid fluid.

those farther away, but they all are in phase with each other. The corresponding force acting on the body is hence a pure added-mass force

$$F(t) = -A'\ddot{x} \qquad (3.12)$$

If the same body vibrates in a compressible fluid or near a free surface, it can induce pressure waves or gravity waves, respectively. Due to these waves, phase differences are generated between the motions of fluid elements throughout the fluid which depend, for a given geometry and vibration, on the ratio of a characteristic body speed, ωD, to the wave celerity, c. Consequently, the fluid force $F(t)$ also experiences a phase shift with respect to the body motion. If the latter is harmonic

$$
\begin{aligned}
x &= x_o \sin \omega t \\
\dot{x} &= x_o \omega \cos \omega t \\
\ddot{x} &= -x_o \omega^2 \sin \omega t
\end{aligned}
\qquad (3.13)
$$

this phase shift can be described by rewriting Equation 3.12 as

$$F(t) = -A'\ddot{x} - B'\dot{x} \qquad (3.14)$$

where A' and B' are functions of $\omega D/c$:

$$A' = A'(\omega D/c), \qquad B' = B'(\omega D/c) \qquad (3.15)$$

Evidently, fluid damping $(-B'\dot{x})$ is possible even in the absence of viscosity! It is related here to the work done in producing the energy of the radiating waves.

Examples of radiating-wave effects are given in Figure 3.12 for acoustic waves $[c = \sqrt{E/\rho}]$ and in Figure 3.13 for deep-water waves $[c \simeq g/\omega = g/(2\pi f)]$. For the vibrating piston presented in Figure 3.12a, the waves remain spherical in shape, and they are independent of the geometry of the piston face as long as the wave length $\lambda = c/f$ is much larger than the radius of the piston R $(fR/c \ll 1)$. As λ

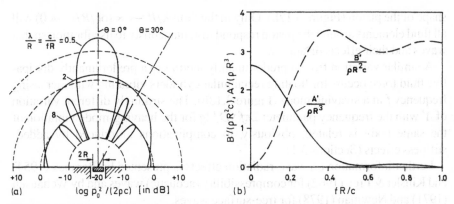

Figure 3.12. (a) Patterns of acoustic waves in terms of contours of constant pressure intensity in the fluid surrounding a flat piston of radius R vibrating at frequency f (after Kinsler & Frey, 1962); and (b) corresponding change in added-mass and fluid-damping coefficients.

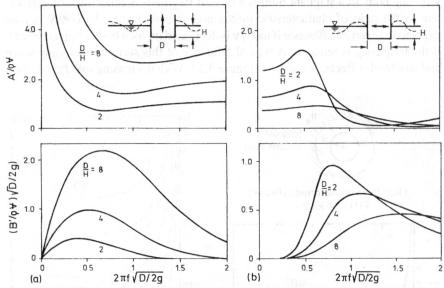

Figure 3.13. Effect of wave radiation on added-mass and fluid-damping coefficients for a rectangular cylinder vibrating at frequency f in deep water in (a) heaving mode and (b) swaying mode (after Vugts, 1968, and Wehausen, 1971). (V = volume of displaced fluid.)

approaches R, the wave patterns become quite complex; only for the special case of a pulsating sphere do they remain spherical.

For decreasing λ/R, or increasing fR/c, the fluid force $F(t)$ changes from a predominantly inertial to a predominantly dissipative force, independent of the shape of the piston (Figure 3.12b). Only in the limit $\lambda/R \to \infty$ (or $fR/c \to 0$) will all fluid elements around the piston respond instantaneously to its vibration so that waves and their effects disappear.

A similar variation from a predominantly inertial to a predominantly dissipative fluid force occurs for floating rectangular cylinders vibrating with increasing frequency f in a swaying mode (Figure 3.13b). The strikingly different variation of A' with the frequency parameter $2\pi f \sqrt{D/2g}$ for the heaving-mode vibration of the same body is related, obviously, to complications introduced by added-stiffness effects (Sections 3.1).

Further information on wave-radiation effects is presented by Beranek (1954) and Kinsler & Frey (1962) for compressibility (acoustic) waves and by Wehausen (1971) and Newman (1978) for free-surface waves.

3.2.4 *Effects of viscosity and amplitude*

The parameter describing viscous effects on the fluid force $F(t)$ exerted on a vibrating body in a stagnant fluid is a type of Reynolds number, $\omega D^2/\nu$, equivalent to the ratio of characteristic inertia and viscous forces. Viscosity is thus expected to exert an influence if $\omega D^2/\nu$ is decreased or, for a body of given size D, if the frequency is reduced. A typical influence of this sort, not including wave and amplitude effects, is shown in Figure 3.14. With decreasing $\omega R^2/\nu$, the fluid

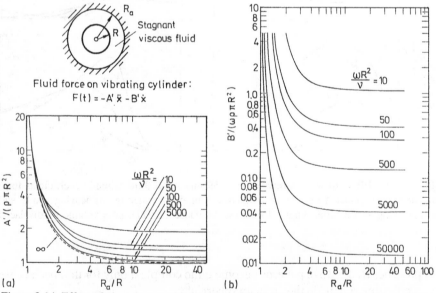

Figure 3.14. Effect of viscosity on added-mass and fluid-damping coefficient for a cylinder vibrating with small amplitudes within a rigid, concentric cylindrical shell (after Chen et al., 1976).

force $F(t)$ contains increasingly more damping (Figure 3.14b), and its added-mass component grows as well (Figure 3.14a). Evidently, $\omega R^2/\nu$ must be greater than about 5×10^3 if the inviscid result for A' from potential-flow theory ($\omega R^2/\nu \to \infty$) is to represent accurately the added mass on an unconfined cylinder in a viscous fluid. Confinement has the effect of increasing both A' and B'.

As long as vibration amplitudes x_o are so small that the flow does not separate from the vibrating body, the results in Figure 3.14 agree well with experimental results (Chen et al., 1976). For an unconfined circular cylinder of diameter d in a stagnant fluid, Sarpkaya (1979) found separation to commence at approximately $x_o/d = 0.2$ and vortex shedding at about $x_o/d = 0.7$. As the amplitude x_o increases beyond $0.7d$, one obtains patterns of newly shed and old vortices of increasing complexity, somewhat as sketched in Figure 3.15a. The corresponding effects on the added mass and the fluid damping can be inferred from Figure 3.15b. This figure was derived from Sarpkaya's study (1976) of a fixed cylinder

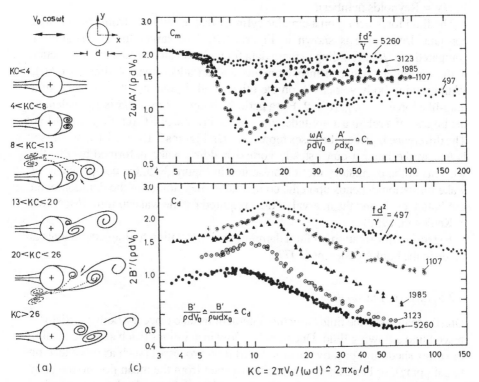

Figure 3.15. (a) Sketch of evolution of vortex patterns near a cylinder in an oscillating fluid for various ranges of Keulegan-Carpenter number $KC \equiv V_o/(fd)$ (after Sarpkaya & Isaacson, 1981). (b, c) Added-mass and fluid-damping coefficients for a smooth circular cylinder as a function of KC (after Sarpkaya, 1976), where KC can be considered equivalent to a relative cylinder amplitude.

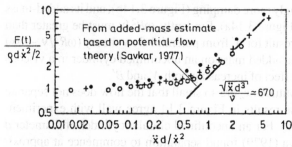

Figure 3.16. Fluid force $F(t)$ induced on a circular cylinder accelerated from rest, as first presented by Keim (1956).

in an oscillating fluid by setting the amplitude V_o of the velocity fluctuation equal to ωx_o (based on Equation 3.13b). The Keulegan-Carpenter number $KC \equiv V_o T/d = 2\pi V_o/(\omega d)$ can thus be expressed as $2\pi x_o/d$ (note that $KC \times fd^2/\nu = V_o d/\nu =$ Reynolds number).

The fluid force acting on a circular cylinder that accelerates from rest due to a constant drive force is shown in Figure 3.16. The larger the acceleration $\ddot{x}$ compared to $\dot{x}^2/d$, according to these data for a given value of $\sqrt{\ddot{x}d^3}/\nu$, the better is the agreement between measurements and an added-mass estimate of $F(t)$. With growing body velocity $\dot{x}$, on the other hand, F assumes the character of a drag-like force $F = C_D d\rho \dot{x}^2/2$. For a periodic body motion, $\ddot{x}d/\dot{x}$ is equivalent to the inverse of a relative amplitude, d/x_o, or the inverse of KC (cf. Equation 3.13). The difference between the cases represented in Figures 3.15 and 3.16, of course, is that in the latter a quasi-steady vortex-shedding wake is formed eventually, whereas in the oscillatory motion presented in Figure 3.15, the completeness of wake formation depends upon the comparative magnitudes of the duration of an oscillation cycle and the interval of time required for the wake to form (McNown & Keulegan, 1959).

Extensive information on fluid damping and its relation to drag coefficients is given in the book by Blevins (1977).

3.2.5 Effects of mean flow

Drastic changes in the fluid/structure interactions take place if a body oscillator is placed into a flowing fluid. Due to either flow instabilities (such as those leading to vortex shedding) or movement-induced flow processes (such as those leading to galloping), or both, energy can be transferred from the mean flow to the body and induce self-sustained body vibrations (Figure 3.1c). In the mathematical formulation of these problems, there is a tendency to incorporate either (a) the added mass and fluid damping determined in stagnant fluid, or (b) the still-fluid added mass alone, as constant magnitudes in the 'structural' terms of the equation of motion; the term $F(t)$ (Equation 2.13) is then referred to as purely flow-induced. Approaches of this nature appear to be adequate in cases of extraneously induced excitation (EIE due to, e.g., turbulence buffeting) in which the effects of body vibration on the exciting force are negligible. In cases of instability- or movement-induced excitation (IIE or MIE), however, such arbitrary separation of fluid-flow effects has little advantage and only confuses the physical picture. Along with several other scientists, Parkinson (1974) therefore advocates a more rigorous approach wherein the 'structural' terms are clearly separated from the 'fluid' terms. Accordingly, mass m and damping B in the equation of motion

$$m\ddot{y} + B\dot{y} + Cy = F_y(t)$$

should be interpreted as referring to the structure alone (to be determined, e.g., by exciting the structure by a push in a vacuum or, for hydraulic systems, in still air); and, if possible, the entire fluid-flow action on the body and its dependence on the body motion $y(t)$ should be described by the fluid force

$$F_y(t) = C_y Ld\, \rho V^2/2 \text{ for three-dimensional bodies}$$

$$F_y(t) = C_y d\rho V^2/2 \text{ for two-dimensional bodies} \tag{3.16}$$

Here, V = mean flow velocity of approach, Ld = characteristic area of loading, and d = characteristic dimension of cross-section of cylindrical body. For a body undergoing controlled harmonic vibration transverse to the flow,

$$y = y_{os} \sin \omega t, \quad \dot{y} = y_{os}\omega \cos \omega t, \quad \ddot{y} = -y_{os}\omega^2 \sin \omega t \tag{3.17}$$

Sarpkaya (1979) suggests decomposition of the force coefficient C_y into two components, one in phase with the body acceleration $\ddot{y}$ and the other in phase with the body velocity $\dot{y}$,

$$C_y = + C_m \sin \omega t - C_d \cos \omega t \tag{3.18}$$

Comparison with Equation 3.4 yields the following relationships with the added-mass and the fluid-damping terms A' and B'

$$C_m = \frac{2y_{os}\omega^2}{\rho dV^2} A', \qquad C_d = \frac{2y_{os}\omega}{\rho dV^2} B' \tag{3.19}$$

The great distinction between cases with and without mean flow of the ambient fluid is the fact that A' and B' (or C_m and C_d) are not constants for a system with given geometry and given parameters $\omega d^2/\nu$, $\omega d/c$, y_{os}/d. Instead, they are functions of parameters describing the mean flow and the movements of the system such as

1. Geometric conditions and roughness of the flow boundaries;
2. Approach-flow conditions, including background turbulence;
3. Flow parameters accounting for possible effects due to viscosity, gravity, surface-tension, compressibility, vapor pressure, etc.; and
4. Inertia, damping, and stiffness properties of the body oscillator or, in case of

controlled vibration, the frequency $f_s = \omega/2\pi$ and amplitude y_{os} of that
vibration.

These functional relationships (indicated further on as Equation 3.26) are commonly evaluated by measurement of fluid forces during controlled vibration of the structure. Figure 3.17 depicts typical results for the case of a circular cylinder, for which vortex shedding causes instability-induced excitation (Section 6.3). With the cylinder forced to vibrate in the transverse direction at controlled amplitude y_{os} and frequency f_s, very large changes in C_m and C_d were observed near resonance, i.e., when f_s approached the vortex shedding frequency f_o. For the conditions represented in Figure 3.17, resonance occurs at a dimensionless velocity $V_r \equiv V/(f_s d) = 1/\text{Sh} \approx 4.8$, where $\text{Sh} \equiv f_o d/V$ denotes the Strouhal number and f_o the vortex shedding frequency at $y_{os} = 0$; V_r is called the reduced velocity. As seen from Figure 3.17b, the resonance range is marked by the greatest *negative* damping ($C_d < 0$). Energy transfer from the flow to the body can thus be expected to sustain vibrations of a corresponding spring-supported body near

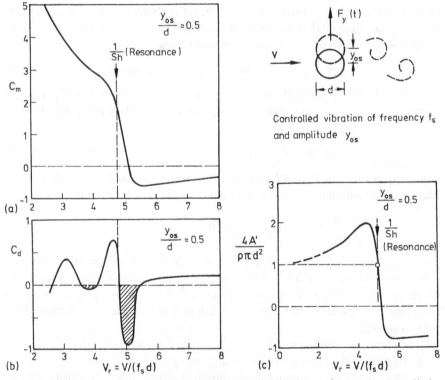

Figure 3.17. Force coefficients C_m and C_d and 'added-mass' term A' for a circular cylinder undergoing controlled vibrations in a uniform flow ($5000 < Vd/\nu < 25000$) (after Sarpkaya, 1978).

$V_r = 1/Sh$ unless its positive damping (B) is very large. The sharp drop in C_m from positive to negative values near resonance (Figure 3.17a) can be understood if it is recalled that the component $C_m \sin \omega t$ in Equation 3.18 combines dynamic fluid forces in phase with both acceleration $\ddot{y}$ and displacement y of the body (Equation 3.17). Apparently, there is a predominance of the latter, i.e., of added stiffness, for $V_r > 1/Sh$. (The negative values of C_m might also be explained by the total 'drift' mass becoming larger during the period of body deceleration than during the period of body acceleration.) Figure 3.17c was deduced from Figure 3.17a with the aid of Equation 3.19a in order to provide a basis for determining (with Equation 3.3b) the fluid-generated shift in natural frequency. By coincidence, the still-fluid value of added mass (Figure 3.4a) is obtained for both $V_r = 0$ and $V_r = 1/Sh$.

Results from an investigation involving the instability-induced vibration (IIE) of a gate during underflow are presented in Figure 3.18. They show that changes in the added-mass term A' similar to those of Figure 3.17c occur for very different flow/structure configurations. Again, the still-fluid value of added-mass is reached within the resonance range shortly before A' drops sharply and assumes negative values. The fluid damping was not measured, but one may expect C_d or B' for the vibrating gate to become negative within the range of resonance, as in Figure 3.17b.

One can infer from these results that added-mass data for a still fluid as presented in the preceding sections may well give an estimate of the correspond-

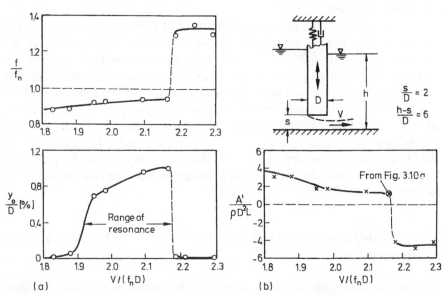

Figure 3.18. (a) Frequency and amplitude response, and (b) corresponding 'added-mass' term A' for a gate with underflow vibrating because of flow instability (after Nguyen, 1982).

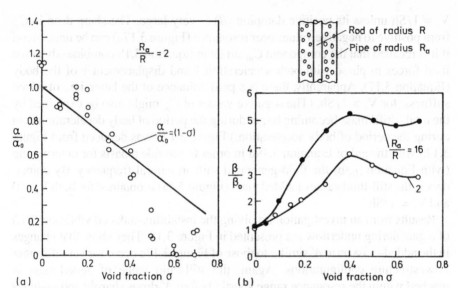

Figure 3.19. Added-mass and fluid-damping coefficients [$\alpha \equiv A'/(\rho_w \pi R^2)$ and $\beta \equiv B'/(\rho_w \pi R^2 \omega)$] for a circular rod vibrating in an air-water mixture near $\omega R^2/v_w = 2000$ (after Hara, 1982b). (Subscript w denotes water.)

ing fluid force component at a particular point within the resonance range of flow-induced vibration due to IIE; it would be wrong to use these results throughout that range.

In *two-phase flows*, an important parameter that characterizes the flow is the void fraction σ, that is, the ratio of the volume of gas to the volume of the gas-liquid mixture ($\sigma = 0$ indicates pure water flow and $\sigma = 1$ pure air flow). A number of flow regimes can occur for a given boundary configuration depending on the concentration and size of the gas bubbles and on the mass-flow rates of the two phases. It is impossible, therefore, to describe two-phase-fluid effects on added-mass and fluid damping in a simplified way. A typical trend of these effects, however, is as shown in Figure 3.19. Represented in this figure are the fluid forces acting on a vertical rod that vibrates within a concentric pipe filled with an air-water mixture. The boundary configuration is thus comparable to that presented in Figure 3.14.

3.3 DESCRIPTION OF FLOW-INDUCED LOADING AND STRUCTURAL RESPONSE

Among the three types of flow-induced loading which may excite structural vibration (Figures 1.1 and 3.1),

1. Extraneously induced excitation (EIE);

2. Instability-induced excitation (IIE);
3. Movement-induced excitation (MIE),

only EIE can be described independently of the vibration excited by it. The structural response, therefore, can be determined as being 'forced' by a given independent fluid loading (Section 2.3) only in cases of EIE. If sources for IIE exist, the flow-induced loading may be substantially altered when the structure vibrates. In cases of MIE, moreover, the loading is directly linked to the structural movements and subsides when these movements stop. Clearly, then, any loading due to IIE or MIE can be adequately described only in conjunction with the structural vibration it induces. Some of the common forms used for this description are briefly summarized below.

3.3.1 *Use of force coefficients and phase angle*

A typical example of flow-induced loading which can be of either the IIE or the MIE type is that involving a cylinder in cross-flow. If that cylinder forms a linear system with one degree of freedom, its equation of motion is

$$m\ddot{y} + 2\zeta m\omega_n \dot{y} + m\omega_n^2 y = F_y(t) \tag{3.20}$$

Here, y = cylinder displacement in the cross-flow direction, m = cylinder mass per unit length, F_y = fluid load in the y direction per unit length, ζ = damping ratio, $\omega_y = 2\pi f_n$ = natural circular frequency. If the cylinder is lightly damped, its response can be approximated by the harmonic expression

$$y = y_o \sin \omega t \tag{3.21}$$

with y_o = response amplitude and $\omega = 2\pi f$ = circular frequency of body vibration. Since only the part of the exciting force associated with the frequency of body vibration can perform work on the body, only this part needs to be considered; hence one may write

$$F_y(t) = C_y d\rho V^2/2; \qquad C_y = C_{yo} \sin(\omega t + \phi) \tag{3.22a, b}$$

or, according to Equation 3.18,

$$C_y = C_m \sin \omega t - C_d \cos \omega t \tag{3.22c}$$

The relationship between these two representations is shown in the vector diagram of Figure 3.20.

If one introduces the dimensionless variables

$$m_r = \frac{m}{\rho d^2}, \qquad \tau = \omega_n t, \qquad U = \frac{V}{\omega_n d} \tag{3.23}$$

$$\eta = \frac{y}{d}, \qquad \overset{\circ}{\eta} = \frac{d\eta}{d\tau} = \frac{1}{\omega_n d}\dot{y}, \qquad \overset{\circ\circ}{\eta} = \frac{d^2\eta}{d\tau^2} = \frac{1}{\omega_n^2 d}\ddot{y} \tag{3.24}$$

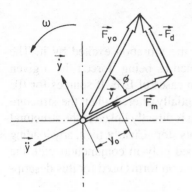

Figure 3.20. Definition sketch. (The magnitudes of $\vec{F}_{yo}$, $\vec{F}_d$, $\vec{F}_m$ are equal to $d\rho V^2/2$ times C_{yo}, C_d, C_m, respectively.)

where $^\circ$ denotes differentiation with respect to τ, one can now transform Equation 3.20 into the dimensionless form

$$\overset{\circ\circ}{\eta} + 2\zeta\overset{\circ}{\eta} + \eta = \frac{U^2}{2m_r} C_y(\tau) \tag{3.25}$$

This equation can be solved, if information is available on $C_y(\tau)$ from free- or controlled-oscillation experiments. In the latter case, this information takes the form, in conjunction with Equation 3.22b or c,

$$
\begin{array}{l}
C_{yo}, \phi \\
C_m, C_d
\end{array}
=
\begin{array}{l}
\text{Func-} \\
\text{tions} \\
\text{of:}
\end{array}
\left[
\begin{array}{l}
\text{Geometry and roughness of flow bounda-} \\
\text{ries; approach-flow conditions (e.g., Tu);} \\
\text{flow parameters (e.g., Re); } V/(f_s d), y_{os}/d.
\end{array}
\right] \tag{3.26}
$$

where f_s and y_{os} denote the frequency and the amplitude of the controlled body vibration, respectively, and Tu the turbulence level. Typical results for a circular cylinder are shown in Figures 3.17a, b and 3.21. The representation of one set of measurements in Figure 3.22 shows that C_{yo} grows continuously until a maximum is reached near what can be called a resonance point of the fluid-dynamic system ($f_s/f_o = 1$). For increasing values of $f_s d/V$, C_{yo} first decreases rapidly and then increases again on account of added-mass type forces ($\phi = 0°$). The phase angle ϕ between cylinder vibration and fluid force undergoes a rapid drop as $f_s d/V$ approaches resonance. Energy transfer from the flow to the body, of course, is possible only in the range $0° < \phi < 180°$. (Note that the reduced frequency $f_s d/V$ used in Figures 3.21 and 3.22 is the inverse of the reduced velocity V_r depicted in Figure 3.17.)

The *response* to cross-flow-induced loading of a cylinder forming an oscillator as described by Equation 3.20 or 3.25 can be calculated with the aid of the empirical information in Figures 3.17 or 3.21 as follows. If this response is quasi-harmonic, $C_y(\tau)$ takes the form

$$C_y(\tau) = C_{yo} \sin\left(\frac{\omega}{\omega_n}\tau + \phi\right) \tag{3.27}$$

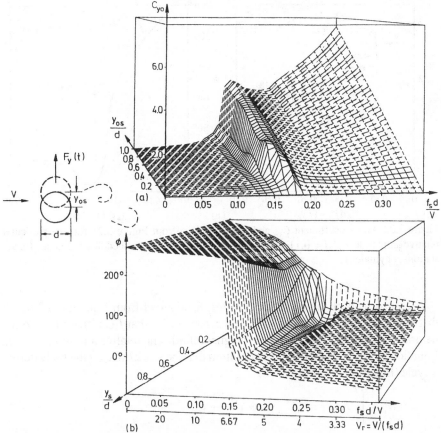

Figure 3.21. Force coefficient C_{yo} and phase angle ϕ as functions of amplitudes y_{os} and frequency f_s of controlled vibrations of a circular cylinder in cross-flow ($Vd/\nu = 6.8\times10^4$, $L/D = 12.5$, Tu $\simeq 0$) (after Staubli, 1983).

and the solution of Equation 3.25 becomes

$$\eta(\tau) = \frac{y_o}{d} \sin\left(\frac{\omega}{\omega_n}\tau\right) \tag{3.28}$$

The frequency $f = \omega/2\pi$ and amplitude y_o of this response follow, by a numerical iterative procedure, from

$$\left(\frac{\omega}{\omega_n}\right)^2 = 1 - \frac{U^2}{2m_r y_o/d}C_m, \qquad C_m = C_{yo}\cos\phi \tag{3.29}$$

$$\frac{y_o}{d} = -\frac{U^2}{4m_r\zeta}\frac{\omega_n}{\omega}C_d, \qquad C_d = -C_{yo}\sin\phi \tag{3.30}$$

plus the information on C_{yo} (η, V_r) and ϕ (η, V_r) in Figure 3.21.

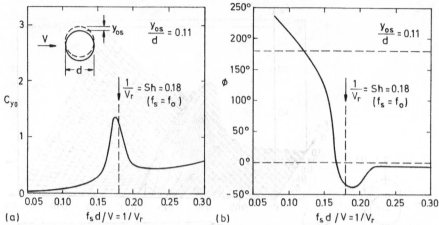

Figure 3.22. Force coefficient C_{yo} and phase angle ϕ from Figure 3.21 for one particular relative amplitude $y_{os}/d = 0.11$. (Sh $= f_o d/V$, where f_o = vortex-shedding frequency for the stationary cylinder.)

A comparison of the frequency $f = \omega/2\pi$ obtained from Equation 3.29 with that resulting from Equations 3.3b and 3.19a shows that f is rather close to the natural frequency of the cylinder in stagnant fluid. The relative amplitude y_o/d in Equation 3.30 corresponds to the condition that the work done on the body during a cycle (Equation 2.32)

$$W_e = \int_o^T F_y \dot{y} dt = \pi C_{yo} \sin \phi \; d \frac{\rho V^2}{2} y_o$$

is balanced by the work done by the damping force, $W_d = -\pi \; (2\zeta m\omega_n) \; \omega y_o^2$ (Equations 2.10 and 2.34). It turns out to be inversely proportional to a mass-damping ratio or Scruton number Sc $\equiv 2\zeta m/(\rho d^2)$ (cf. Figure 3.24b):

$$\frac{y_o}{d} = \left[\frac{C_{yo} \sin \phi}{8\pi^2} \frac{f_n}{f} \right] \frac{V_r^2}{Sc} \tag{3.31}$$

Figure 3.23 shows a comparison of measurements and computations using the above approach for an extremely low damping ratio ζ. The agreement within the range of resonance is very good. Even the hysteretic jump behavior has been predicted satisfactorily. Typically, for air-flow data, the frequency f of vibration is near f_n due to the relatively large value of m_r. With the Strouhal number defined as Sh $\equiv f_o d/V$, the abscissa is equivalent to Sh $V_r = f_o/f_n$, where f_o = vortex-shedding frequency for the stationary cylinder. Note the locking-in of the shedding frequency to the body-vibration frequency within the resonance range (Figure 3.23a).

The results reported above permit a number of insights into the physics of instability-induced excitation. Only *one* particular flow condition

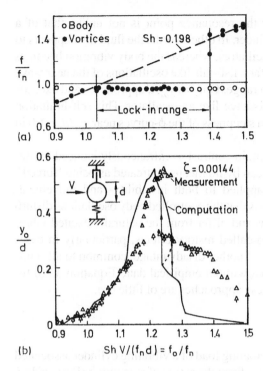

Figure 3.23. Comparison of measured (Feng, 1968) and computed (Staubli, 1983) response of a circular cylinder in air-flow ($m/\rho d^2 \simeq 195$, Tu $\simeq 0.1\%$).

(Re $= 6.8 \times 10^4$) is presented in Figure 3.21, that for which an exciting force $F_y(t)$ with fixed values of the force coefficient C_y and a frequency $f = f_o$ exists in case of a stationary cylinder. Yet, as soon as the cylinder vibrates (and only then is the presented approach applicable), the exciting force undergoes extreme changes in amplitude (proportional to C_{yo}) and frequency (equal to the body-vibration frequency, $f = f_s$)[1]. The difference between extraneously and instability-induced excitation, EIE and IIE, is striking: In case of EIE, the exciting frequency is independent of body vibration and controls the frequency of the vibration. In case of IIE, the frequency of the exciting force is controlled completely by the body vibration. Even the flow oscillations associated with the flow instability (here the vortex shedding) are controlled by the body movements within what is called the lock-in range (Figure 3.23a).

Most revealing in this connection is the variation of the 'added-mass' term A' with reduced velocity in Figure 3.17c. As resonance is approached, $V_r \rightarrow 1/\text{Sh}$, the added-mass coefficient changes from the still-fluid value 1.0 to about 2.0 which is the value obtained if the *flow* oscillates about a cylinder at rest (Figure 3.15b, KC $\rightarrow$ 0). One may thus argue, as does Sarpkaya (1978), that the net effect

---

[1] A minor part of $F_y(t)$, associated with f_o, was neglected here because it has no effect on the energy transfer to the vibrating body.

of the body/flow interaction near the resonance point is not unlike that of a periodic flow about a stationary cylinder. In other words, the fluid flow appears to become the oscillator here (flow oscillator), whereas the body vibration becomes the agent forcing it! As a matter of fact, fish-tail-like oscillations of the near wake are clearly discernible in the lock-in range. Only as y_{os}/d approaches and exceeds unity does the fluid flow cease to behave like an oscillator. This self-limitation effect is evident from the increase in steepness of the ϕ-curves near $y_{os}/d = 1.0$ in Figure 3.21a.

This discussion shows clearly that, in contrast to extraneously induced excitation, instability-induced excitation can by no means be treated as being 'forced'. The foregoing approach can be applied to both instability- and movement-induced excitation. Its disadvantage is that the controlled body oscillations disturb the complex body/flow interactions and differ from the naturally excited vibrations because of a difference in so-called history effects, particularly in cases involving hysteretic flow processes. Another disadvantage, common to all semi-empirical approaches, is the fact that without empirical data (Equation 3.26) for the practical situation of interest, these approaches are of little use.

3.3.2 *Use of oscillator models*

In the foregoing discussion, the fluctuating load of a vibrating cylinder associated with vortex shedding was seen to arise from the action of something akin to a flow oscillator. Pursuing the wake-oscillator concepts of Birkhoff (1957) and Bishop & Hassan (1964) further, Hartlen & Currie (1970) suggested a mathematical model by which this load is actually derived from an oscillator equation embodying the essential characteristics deduced from experiments: (1) an inertial and restoring term producing the correct load fluctuation for a stationary body ($\overset{\circ}{\eta} = 0$); (2) a damping term containing a negative and a positive component for the observed self-sustaining and self-limiting features, respectively; and (3) a forcing term, coupling the oscillator to the body motion. Based on Equations 3.22a, 3.23, 3.24, and 3.25, this oscillator equation takes the dimensionless form

$$\overset{\circ\circ}{C}_y + \text{(damping term)} + \Omega_o^2 C_y = \text{(forcing term)} \tag{3.32a}$$

where $\Omega_o = \omega/\omega_n = f_o/f_n$ and superscript $^\circ$ denotes differentiation with respect to $\tau = \omega_n t$. As the simplest forms of the damping and forcing terms which satisfy conditions (2) and (3), Hartlen & Currie (1970) suggested the following

$$\overset{\circ\circ}{C}_y - \beta_1 \Omega_o \overset{\circ}{C}_y + \frac{\beta_3}{\Omega_o} \overset{\circ}{C}_y^3 + \Omega_o^2 C_y = \alpha_1 \overset{\circ}{\eta} \tag{3.32b}$$

where β_1 and β_3 are the van der Pol coefficients and α_1 is the interaction parameter, to be determined empirically. In contrast to the parameters of the foregoing approach (Equation 3.26), these quantities are assumed to be independent of the body motion, i.e., independent of $V/(fd)$ and y_o/d:

$$\begin{matrix} \beta_1, \beta_3, \\ \alpha_1 \end{matrix} \begin{matrix} \text{Func-} \\ = \text{tions} \\ \text{of} \end{matrix} \left[\begin{matrix} \text{Geometry and roughness of flow bounda-} \\ \text{ries; Approach-flow conditions (e.g., Tu);} \\ \text{Flow parameters (e.g., Re).} \end{matrix} \right] \quad (3.33)$$

Another advantage is the possible application of the oscillator-model approach to the problem of flow-induced loading of a stationary body.

A comparison of oscillator-model predictions and experimental data is shown in Figure 3.24. The ratio of β_1 to β_3 in the model of Hartlen & Currie was determined from application of Equation 3.32b to the case of a stationary cylinder $(\overset{\circ}{\eta} = 0)$,

$$\sqrt{4\beta_1/3\beta_3} = (C_{yo})_{\text{stat. cyl.}}$$

and β_1 and α_1 were selected according to controlled-vibration experiments. The agreement is quite satisfactory. Even the slight frequency variation (Figure 3.23a) and the trend in the phase angle ϕ is correctly reproduced by the model. The accuracy of prediction can be improved by more sophisticated expressions for the damping and forcing terms in Equation 3.31, as several investigators, including Berger (1978a, b) and Blevins (1977), have shown. (Note the fair agreement between the trend of the curve in Figure 3.24b and the relationship $y_o/d \propto 1/\text{Sc}$ of Equation 3.31.)

The disadvantage of the oscillator-model approach lies in the fact that equations for the 'flow oscillator' cannot be readily deduced and are difficult to interpret physically. Nevertheless, they do offer promise for the assessment and treatment of complex systems composed of a variety of body, fluid, and flow oscillators (Naudascher & Rockwell, 1980a).

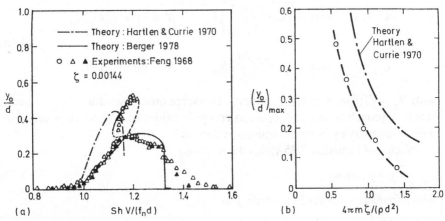

Figure 3.24. Comparison between oscillator-model predictions and experimental data on the response of a circular cylinder in cross-flow.

3.3.3 *Use of added coefficients and stability diagrams*

Several investigators, including Scruton (1963) and Bardowicks (1976), have proposed a description of the flow-induced load due to instability- and movement-induced excitation in the form of components in phase with body displacement and velocity. For a cylinder in cross flow, for example, this load per unit length is expressed as

$$F_y(t) = \rho d^2 \omega_n (\omega_n h_a y + k_a \dot{y}) \tag{3.34}$$

In terms of the dimensionless variables introduced in Equations 3.32 and 3.24, the equation of motion (Equation 3.20) may thus be written as follows

$$m_r \ddot{\eta} + (Sc + k_a)\dot{\eta} + (m_r - h_a)\eta = 0 \tag{3.35}$$

Here,

$$Sc = \frac{B}{\rho d^2 \omega_n} = 2m_r \zeta = 2\frac{m}{\rho d^2}\zeta \tag{3.36}$$

is the mass-damping parameter or Scruton number, which plays an important role with respect to the maximum possible response amplitude for a given flow/structure system (Figure 3.24b). The excitation coefficient k_a and added-mass coefficient h_a must be determined from free- or controlled-vibration experiments

$$\begin{matrix} k_a \\ h_a \end{matrix} = \begin{matrix} \text{Func-} \\ \text{tions} \\ \text{of} \end{matrix} \left[\begin{matrix} \text{Geometry and roughness of flow bounda-} \\ \text{ries; Approach-flow conditions (e.g., Tu);} \\ \text{Flow-parameters (e.g., Re); } V/(f_s d), y_{os}/d \end{matrix} \right] \tag{3.37}$$

just as in the case of C_m and C_d (Equation 3.26). In fact, there exist the following relationships between the various sets of empirical coefficients C_m, C_d; C_{yo}, ϕ; k_a, h_a; and the added coefficients A', B' (Equation 3.4):

$$-C_d = C_{yo} \sin \phi = \left(8\pi^2 \frac{y_o/d}{V_r^2}\frac{f}{f_n}\right) k_a, \qquad B' = -\rho d^2 (2\pi f_n) k_a$$

$$C_m = C_{yo} \cos \phi = \left(8\pi^2 \frac{y_o/d}{V_r^2}\right) h_a, \qquad A' = \rho d^2 \left(\frac{f_n}{f}\right)^2 h_a \tag{3.38}$$

with $V_r \equiv V/(f_n d)$ = reduced velocity. In exceptional cases for which a quasi-steady approach is justified, one can compute values of k_a from measurements with a stationary cylinder as outlined in Section 7.3.5.

Solution of Equation 3.35 yields the response

$$y = y_o \sin \omega t \tag{3.39}$$

with a frequency $f = \omega/2\pi$ according to

$$\left(\frac{f}{f_n}\right)^2 = \frac{m_r - h_a}{m_r} = 1 - \frac{h_a}{m_r} \tag{3.40}$$

and a steady-state amplitude $y_o = d\eta_o$, which follows from the condition of equilibrium between the work done by the exciting and the damping force. Inspection of the second term in Equation 3.35 yields, for this equilibrium condition,

$$|k_a|_{\eta_o} = \text{Sc} \tag{3.41}$$

The solution for y_o, in other words, requires a comparison of the given system parameter Sc with the empirically determined information on $k_a = k_a(\eta_o)$, which is mostly derived from controlled-vibration experiments (Figures 3.25 and 3.26). For air-flow problems, $h_a/m_r \simeq 0$.

The characteristic feature of this approach consists in the seemingly linear expression for $F_y(t)$ in Equation 3.34 in combination with variable terms k_a and h_a. The nonlinearity of the flow-induced excitation has thus become 'buried' in the empirical coefficients, just as with the procedures reported in Section 3.3.1. Again, both instability- and movement-induced excitations (IIE and MIE) can be described, as illustrated in Figure 3.25: For a circular cylinder, k_a exceeds zero only within the limited range of possible vortex excitation $4 < V_r < 8$ ($k_a > 0$ signifies negative damping according to Equation 3.35); note the approximate proportionality $(k_a)_{\text{max}} \propto 1/\eta_o$, which is related to the feature of self-limitation typical of IIE (Scruton, 1963; Bardowicks, 1976). For a square prism, on the contrary, k_a grows apparently without limit as V_r increases beyond the range of vortex excitation. As is shown in Section 7.3.3, this fact is related to the MIE known as galloping.

An advantage of k_a plots like those in Figures 3.25 and 3.26 is that one may transfer them into stability diagrams. The *criterion for dynamic stability* of a system described in terms of Equation 3.35 is simply

$$\text{Sc} - k_a > 0 \tag{3.42}$$

for all conditions encountered. This criterion requires that the total damping in the system should always be positive. It was Scruton (1956) who first suggested use of stability diagrams like the one shown in Figure 3.27 as design tools (a summary is given by Wootton & Scruton, 1971). A stability diagram is obtained by applying the stability criterion to the one curve $k_a = k_a (V_r)$ which corresponds to an amplitude of zero or, more realistically, to the largest relative amplitude y_o/d permissible. The diagram in Figure 3.27, for example, was constructed in this fashion from data similar to those in Figure 3.25b ($\delta \simeq 2\pi\zeta$ is the logarithmic decrement for the system damping in vacuum). Such diagrams are most useful for a quick check on the safety of a structure with respect to IIE and MIE on the basis of Sc and V_r values encountered in the prototype. Similar diagrams for T-profiles result directly from Figure 3.26 if k_a is replaced by $\text{Sc} \equiv 2\zeta m/(\rho d^2)$ and only the contours above the $V_r - \alpha$ plane are retained. Any condition falling above that contour is stable in the sense that no transverse vibration with amplitudes larger than $0.067\,d$ is then possible.

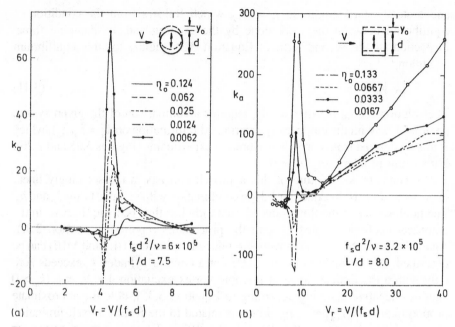

Figure 3.25. Coefficient k_a for transverse excitation as a function of reduced velocity V_r and relative amplitude $\eta_o = y_o/d$ for a circular cylinder (a) and a square prism (b) in cross-flow (Tu $\simeq$ 1%) (after Bardowicks, 1976).

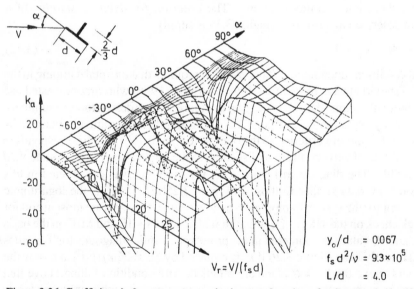

Figure 3.26. Coefficient k_a for transverse excitation as a function of reduced velocity V_r for a T-section cylinder in cross-flow (Tu $\simeq$ 1%) (after Bardowicks, 1976).

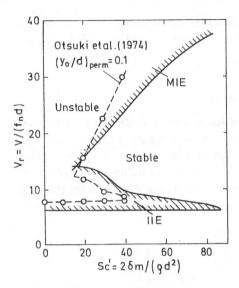

Figure 3.27. Stability diagram for transverse excitation of square prisms in cross flow as first suggested by Scruton (1963).

There are several limitations regarding the use of stability diagrams. None of the figures presented above, for example, gives any information on possible streamwise or torsional excitations. The essential parameters controlling flow-induced vibrations of any type of instability- or movement-induced excitations are Sc and V_r. There are, however, cases in which: (a) damping ζ and reduced mass m_r need to be considered separately rather than combined in Sc; or, (b) 'hard' excitation is possible due to disturbances beyond a certain threshold amplitude, even though the system is stable with respect to small-amplitude excitation from rest. The greatest limitation of stability diagrams is that they apply only to the specific conditions for which they were derived. Even small changes in such quantities as Reynolds number, uniformity, intensity, and scale of turbulence of the approach flow, surface roughness, aspect ratio or end conditions of cylindrical bodies, angle between cylinder axis and flow, etc., may drastically affect the excitation conditions, and hence k_a. Unfortunately, very few authors cite completely the conditions under which their data were obtained; and even if they do, one must still judge whether their results are applicable to the problem at hand.

One may conclude that for the engineer confronted with vibration problems in flow systems, knowledge of the fundamentals of fluid flow is as indispensable as is an understanding of the possible excitation mechanisms presented in Chapters 6 and 7.

3.3.4 *Use of quasi-steady approach*

In addition to the methods based on added coefficients or stability diagrams, one may use a quasi-steady approach to describe the flow-induced loading and structural response for certain types of movement-induced excitation. The approach is presented in Section 7.2.

3.3.5 Generalization for distributed-mass systems

Typical responses of a prismatic body oscillator with mass-damping ratios Sc are shown in Figure 3.28. They can be deduced from Figure 3.25b by drawing a horizontal line at k_a = Sc and plotting the points of intersection $(V_r, \eta_o = y_o/d)$ on a diagram of y_o/d versus V_r. Typical features of instability- and movement-induced excitation (IIE and MIE) are evident: Vibration due to vortex shedding (IIE) occurs with maximum amplitude at an almost constant reduced velocity,

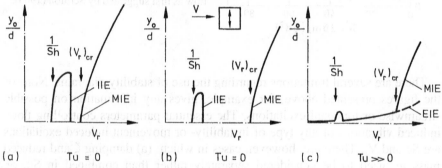

Figure 3.28. General character of transverse response of square prism to (a, b) smooth and (c) turbulent cross-flow for two mass-damping ratios Sc: (a) small Sc and (b, c) large Sc.

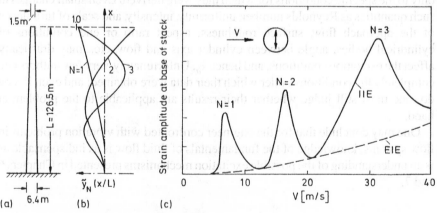

Figure 3.29. (a) Tapered stack of circular cross-section. (b) Vibration modes $N = 1, 2, 3$. (c) Strain-amplitude response obtained from aero-elastic model of tapered stack in cross-flow for low damping, $\zeta = 0.014$, for $N = 1$ (after Scruton, 1963).

$V_r = 1/\text{Sh}$, independent of Sc. Vibration due to galloping (MIE) commences at increased reduced velocities, $(V_r)_{cr}$, as Sc increases. If the approach flow contains large-scale turbulence, the prismatic body is subjected to 'buffeting', and the typical response to this extraneously induced excitation (EIE) involves a parabolically increasing amplitude with increasing V_r (Figure 3.28c).

An important distinction between discrete-mass and distributed-mass systems is the number of natural frequencies. Whereas the former has generally one value of f_n, the latter has one for each mode of vibration that can occur (Section 2.7). Hence, IIE excites ever higher modes as the flow velocity V is increased. Figure 3.29 shows an example of this multiple resonance. (Note that the tip amplitude y_o was much larger for the fundamental than for the higher modes despite the opposite trend for the strain amplitudes.)

For distributed-mass systems like the one shown in Figure 3.29a, the equation of motion (Equation 3.20) takes another form. For an orthogonal mode N of vibration, one may write, according to Equation 2.69,

$$\ddot{y}_N + 2\zeta_N \omega_N \dot{y}_N + \omega_N^2 y_N = \frac{\int_o^L F_y(x, t)\bar{y}_N(x)dx}{\int_o^L \mu(x)\bar{y}_N^2(x)dx} \tag{3.43}$$

where $y_N(x) = N$th mode shape, normalized in such a way that $y_N = 1.0$ at a reference point of the body (Figures 3.29b and 2.18); $\bar{y}_N =$ displacement of that reference point; $\zeta_N =$ damping ratio of the Nth mode; $\omega_N = 2\pi f_N =$ natural circular frequency of the Nth mode; and $\mu(x) =$ mass per unit length at the location x. The exciting force per unit length

$$F_y(x, t) = \frac{\rho}{2} d(x)\, V^2(x)\, C_y(x, t) \tag{3.44}$$

can be formulated with expressions for the local force coefficient $C_y(x, t)$ equivalent to Equations 3.22b, 3.22c, or 3.34. Equation 3.44 makes allowance, in an approximate way, for the effect of an uneven distribution of approach velocity

$$V(x) = V^* v(x) \tag{3.45}$$

where $V^* =$ flow velocity at the reference point and $v(x) =$ normalized velocity profile.

Briefly, diagrams such as the one in Figure 3.25b can be used, according to Scruton (1963), to evaluate the total energy interchange between flow and structure during vibration in the Nth mode in terms of

$$(\zeta_a)_N = \frac{\rho \int_o^L (d^2 k_a)\, \bar{y}_N^2(x)\, dx}{2 \int_o^L \mu(x)\, \bar{y}_N^2(x)\, dx} \tag{3.46}$$

where ζ_a replaces $\rho d^2 k_a/2m$ in Equation 3.34. The condition for steady-state vibration (Equation 3.41) is, then,

$$|(\zeta_a)_N|_{\eta_o} = \zeta \tag{3.47}$$

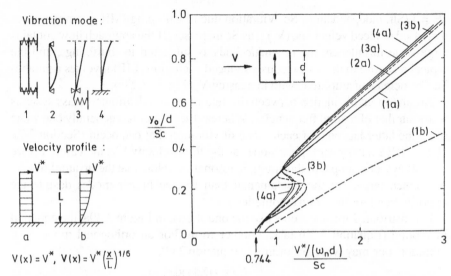

Figure 3.30. Universal galloping (MIE) response of square prisms in cross-flow for different vibration modes and profiles of approach velocity after computations of Novak (1969).

where ζ = damping ratio of the structure and $\eta_o = y_o/d$ = relative amplitude of the reference point.

An illustration of the effect of mode shape $\bar{y}_N(x)$ and velocity distribution $v(x)$ on the response of a structure due to movement-induced excitation (MIE) is given in Figure 3.30. The two MIE curves of Figures 3.28a, b reduce to one universal curve in Figure 3.30 if both y_o/d and $V_r \equiv V/(f_n d)$ are divided by Sc.

3.4 STATIC INSTABILITY (DIVERGENCE)

Dynamic instability was seen in Sections 3.3.3 and 3.3.4 to be related to the damping term of the equation of motion. A system is dynamically stable if that term is positive for all conditions of operation. If it is dynamically unstable, the system is susceptible to movement-induced excitation. *Static* instability, or divergence, in contrast, is related to the stiffness term (or restoring force) which can be written as

$$(C + C')y \qquad \text{or} \qquad (C_\theta + C'_\theta)\theta$$

for a linear system with a translational or torsional degree of freedom (Equation 3.3). The system is statically stable if the stiffness term remains positive under all conditions. Static instability leads to increased deflection; it may occur if a linear or angular displacement, y or θ, from the equilibrium position produces a force or moment which tends to increase the displacement. Hence, static instability

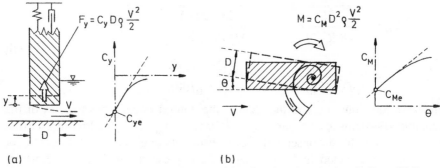

(a)   (b)

Figure 3.31. (a) Gate with transverse degree of freedom in submerged underflow; (b) Beam with torsional degree of freedom in uniform cross-flow.

requires negative stiffness in excess of the positive mechanical stiffness or, more generally, a negative value of the square of the natural frequency (Equation 3.3a).

The two systems depicted in Figure 3.31 provide illustrations of static stability. In both cases, static displacements from an equilibrium position are accompanied by a restoring force or moment consisting of a mechanical part ($-Cy$ or $-C_\theta\theta$) and a fluid-dynamic part. For small displacements, the latter may be approximated by the linear expressions

$$F_y \approx \frac{\rho}{2}V^2D[C_{ye} + (dC_y/dy)_e y]$$

$$M \approx \frac{\rho}{2}V^2D^2[C_{Me} + (dC_M/d\theta)_e \theta]$$

(3.48)

where the subscript 'e' denotes the values at the equilibrium positions $y = 0$ and $\theta = 0$, respectively. Solving now the equation for static equilibrium ($Cy = F_y$ or $C_\theta\theta = M$, i.e., an 'equation of motion' without inertia and damping terms), one obtains

$$\left[C - \frac{\rho}{2}V^2D\left(\frac{dC_y}{dy}\right)_e\right]y = \frac{\rho}{2}V^2DC_{ye}$$

$$\left[C_\theta - \frac{\rho}{2}V^2D^2\left(\frac{dC_M}{d\theta}\right)_e\right]\theta = \frac{\rho}{2}V^2D^2C_{Me}$$

(3.49)

These equations show clearly that the displacements y or θ become infinite as the total stiffness, consisting of the spring stiffness C plus the negative fluid-dynamic stiffness (second term within brackets), becomes zero. The velocities of flow V corresponding to this critical conditions are called divergence velocities:

$$V_D = \sqrt{\frac{2C}{\rho D (dC_y/dy)_e}} = \omega_n D \sqrt{\frac{2m/(\rho D^2)}{D(dC_y/dy)_e}}$$

$$= \sqrt{\frac{2C_\theta}{\rho D^2 (dC_M/d\theta)_e}} = \omega_{\theta n} D \sqrt{\frac{2I_\theta/(\rho D^4)}{(dC_M/d\theta)_e}}$$

(3.50)

In these equations, ω_n and $\omega_{\theta n}$ denote the natural circular frequencies of the undamped systems in a vacuum, $\omega_n = \sqrt{C/m}$ and $\omega_{\theta n} = \sqrt{C_\theta/I_\theta}$. In contrast to the critical velocity for the onset of dynamic instability (e.g., $V_D = 0.744\,\omega_n D$ Sc for the situation depicted in Figure 3.30), the expressions in Equation 3.50 are independent of damping. Instead, they depend strongly on the structural flexibility. The larger the stiffness or, more precisely, the natural frequency, the less likely, in general, is divergence to occur.

For a plug valve as shown in Figure 3.32a, Kolkman (1976) has shown that negative fluid-dynamic stiffness can be generated as follows. If the natural frequency and, thus, $\omega_n D/V$ is very large, valve displacements y do not change the volume rate of flow Q due to the large inertia of the (incompressible) fluid in the pipe; rather, the head $H = (Q/sb)^2/2g$ across the valve is induced to fluctuate in accordance with

$$H = \bar{H} + H(t) = \bar{H} \left(\frac{\bar{s}}{\bar{s} + y \sin \alpha} \right)^2 \approx \bar{H} \left(1 - \frac{2 \sin \alpha}{\bar{s}} y \right)$$

(3.51)

Here, b = peripheral length of the outflow gap and s = its width (Figure 3.32a). The force of the valve can therefore be expressed as

$$F_y \approx \gamma H A_v \approx \frac{\rho}{2} \frac{A_y}{(\bar{s}b)^2} \bar{Q}^2 \left(1 - \frac{2 \sin \alpha}{s} y \right)$$

(3.52)

where $\gamma = \rho g$ = specific weight of the fluid. A comparison of this expression with

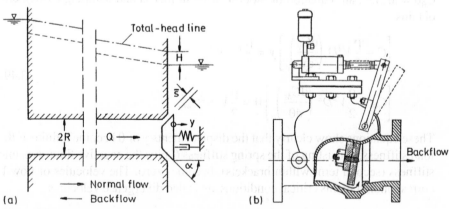

Figure 3.32. (a) Schematic of flow past a plug valve. (b) Swing check valve.

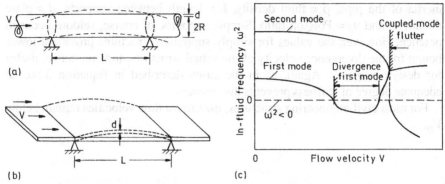

Figure 3.33. (a) Flexible pipe; (b) Flexible plate; (c) Typical effect of flow velocity on vibration frequency in cases of pipes, plates, and shells exposed to subsonic flows (after Weaver, 1974). (Coupled-mode flutter does not occur for pipes; Paidoussis, 1987.)

Equations 3.48a and 3.49a reveals that F_y contains a fluid force equivalent to positive fluid-dynamic stiffness. For the case of backflow (Figures 3.32a, b), this force produces the equivalence of negative stiffness and can lead to divergence.

Classical examples of structures for which divergence in the form of buckling can occur are pipes, plates, and shells with fluid flowing through or over them (Figures 3.33a, b). Extensive reviews have been presented by Paidoussis (1974, 1980, 1981, 1987) and Dowell (1974, 1989). As shown in a summary paper by Weaver (1974), all plates, shells, and pipes which are supported at the leading and trailing edges behave in essentially the same way (Figure 3.33c): With increasing flow velocity, the natural in-fluid frequency decreases gradually until, at some critical point, it becomes zero and the structure buckles in the first mode as if it had been axially loaded. If the flow velocity is increased further, the first-mode frequency reappears and becomes coupled with the second mode by the flowing fluid to produce a dynamic instability called coupled-mode flutter (Sections 7.2.3 and 7.5).

For long pipes or shells and plates with high aspect ratio, the behavior is similar to that of a beam, and the divergence velocities take on the simple expressions

$$V_D = \frac{\pi}{L}\sqrt{\frac{EI}{\rho\pi R^2}} \quad \text{for simply supported pipe,}$$

$$= \frac{2\pi}{L}\sqrt{\frac{EI}{\rho\pi R^2}} \quad \text{for clamped-ended pipe,} \tag{3.53}$$

$$= \frac{1.14\,\pi}{L}\sqrt{\frac{Ed^3}{12\,\rho L(1 - v_p^2)}} \quad \text{for simply supported plate}$$

(Weaver, 1974), where E = modulus of elasticity, $I = \pi R^3 d$ = area moment of inertia of the pipe, ρ = fluid density, L = length between supports, d = plate thickness, and v_p = Poisson ratio. Simple supports, of course, seldom occur in practice. However, the values for simply supported structures provide a lower bound for the divergence velocity of the actual structures and are, hence, useful for design purposes. Again, as in the cases described in Equation 3.50, an adequate degree of stiffness prevents divergence.

For most civil-engineering structures, maximum flow velocities remain below V_D.

CHAPTER 4

Fluid oscillators

4.1 OVERVIEW AND INTRODUCTION

Fluid oscillators, or resonators as they are often called, play an important role in the field of flow-induced vibration for two reasons. First, they may give rise to fluid oscillations even without a body oscillator present in the system; such oscillations may become excited either extraneously (e.g., by wave action), by flow instabilities (e.g., by vortex shedding), or by a movement-induced mechanism of self-excitation. Second, fluid oscillators may enhance the amplitude of flow-induced vibrations of body oscillators, particularly if one of their resonant frequencies coincides with the natural body-oscillator frequency, the dominant frequency of flow instability, or with both. It is essential therefore to at least identify all fluid oscillators and determine their resonant frequencies if one wishes to assess a system with regard to possible fluid oscillations or structural vibrations. This chapter aims at providing the information necessary to achieve this end.

Fluid oscillators, like body oscillators, can be classified (Figure 4.1) as

1. Discrete-mass systems; and
2. Distributed-mass systems.

In the former group, a more or less well defined, discrete mass of fluid performs oscillations in a rigid-body manner. Best known among these are: Helmholtz resonators in compressible systems; their free-surface counterparts, also called open-basin resonators (Figure 4.1c); and U-tubes. In the group of distributed-mass fluid oscillators, all elements of the fluid do not oscillate in unison, and therefore the fluid mass cannot be treated as discrete. Oscillators of this type give rise to standing waves (Figure 4.1d) in either one, two, or three dimensions.

In distinguishing the two types of oscillators, one has to consider the wavelength λ corresponding to the excitation relative to the largest dimension L of the body of fluid. The former is related to the frequency of excitation f as

$$\lambda = c/f \qquad (4.1)$$

where c is the wave celerity (celerity of pressure waves for acoustic systems and

67

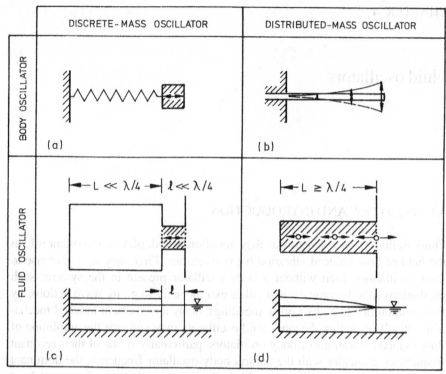

| DISCRETE-MASS OSCILLATOR | DISTRIBUTED-MASS OSCILLATOR |

Figure 4.1. Schematic representation of discrete- and distributed-mass body and fluid oscillators.

celerity of gravity waves for free-surface systems). If the wavelength λ is much larger than L, a standing wave cannot form, but a so-called pumping mode of fluid oscillation is possible as shown in Figure 4.1c. In this case one can regard the up and down motion of the free surface in the basin as the spring action of a single-degree-of-freedom oscillator and the fluid within the neck or channel as the vibrating mass. All fluid particles within the channel can be considered as moving in unison as long as the channel length l is much smaller than λ. If, on the other hand, L or l is equal to or larger than $λ/4$, fluid particles no longer move in a rigid-body fashion; instead, standing waves may form involving widely varying amplitude and phase from one fluid particle to another as shown in Figure 4.1d. Fluid systems of this kind can be excited in several modes just like distributed-mass body oscillators (Figures 4.1b and 4.1d show the fundamental mode $L = λ/4$ only).

Of course, the oscillating fluid mass does not terminate at the open ends indicated by dotted lines in Figures 4.1c, d. Fluid outside these end sections participates in the oscillations in a way quite similar to the fluid near a vibrating body (Figure 4.2). Consequently, the end effects in these cases are similar. In fact,

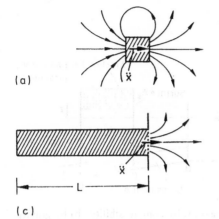

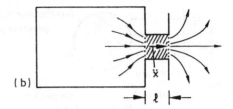

Figure 4.2. Sketches of instantaneous stream-line patterns during acceleration $\ddot{x}$ of (a) a body oscillator in still fluid and (b, c) fluid oscillators according to Figures 4.1c, d.

much of what is presented on these effects for body oscillators in Chapter 3 is applicable to fluid oscillators as well. In all cases, the end effects entail changes in natural frequency and dissipative action which can be described in terms of added mass and fluid damping, respectively. For the fluid oscillators shown in Figures 4.2b, c, one can take the added mass into account by introducing an effective length l_{eff} or L_{eff} in place of l or L.

A complication in the case of distributed-mass fluid oscillators results from the fact that ducts or channels may have more complicated end or boundary conditions than those corresponding to completely open or closed ends (Figures 4.1d and 4.2c). For example, the duct or channel may be terminated by a porous plate, by partly absorbing material, or by a sudden expansion or contraction whose reflection characteristics depend on the boundary geometry and the wavelength of the incident wave. Moreover, even in simple ducts or channels, the presence of mean flow can produce boundary conditions that differ from those for still fluids. In general, effects of this nature are accounted for by a reflection coefficient, K_r (e.g., Ippen, 1966, and Chapter 8).

Fluid oscillators may be affected by one or more of a number of fluid properties or fluid forces such as:

Compressibility or elasticity (E);

Gravity or specific weight ($\gamma = \rho g$);

Surface tension (σ).

For most engineering applications it suffices to consider the action of (a) compressibility and (b) gravity, either alone or in combination. Cases governed by compressibility occur if the fluid within a vessel or a pipe system has no free surface (e.g., Figure 4.3a). With a free surface, the fluid may perform oscillations affected by compressibility, gravity, and surface tension, but gravity effects will dominate (e.g., Figure 4.3b). An example of a fluid oscillator affected by both compressibility and gravity is shown in Figure 4.3c.

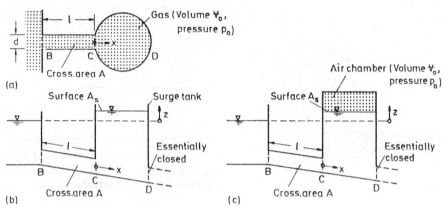

Figure 4.3. Discrete-mass fluid oscillators governed (a) by compressibility, (b) by gravity and (c) by both.

An important quantity concerning the resonant frequency and wavelength, f_R and λ_R, is the celerity c with which changes in pressure and velocity are propagated through a fluid relative to the surrounding medium. In analogy to Equation 4.1,

$$f_R = c/\lambda_R \tag{4.2}$$

the celerity c is determined by the force property which plays the predominant role in the generation of the pressure or velocity disturbance. For pressure or *acoustic* waves, the disturbances and their propagation are governed by the bulk modulus of elasticity E of the fluid. In the case of small changes in fluid density ρ and no or rigid boundaries confining the fluid, the speed of sound is

$$c = c_a = \sqrt{E/\rho} \tag{4.3}$$

(the index a stands for acoustic). For a circular duct of diameter D with thin walls of thickness d_s and elasticity modulus E_s, c_a may be approximated by

$$c_a = \sqrt{\frac{E}{\rho}} \left[1 + \frac{D\,E}{d_s\,E_s} \right]^{-\frac{1}{2}} \tag{4.4}$$

In gases, changes in density can be large compared to the density ρ of the undisturbed medium. Provided that these changes take place adiabatically, that is without loss of heat, the speed of sound is given by

$$c_a = \sqrt{k\,p_*/\rho} = \sqrt{k\,R_*\,T_*} \tag{4.5}$$

where $k = c_p/c_v$; c_p and c_v are the specific heats of the gas at constant pressure and volume, respectively; $R_* = $ gas constant; and T_* and p_* are the temperature and pressure of the undisturbed gas, both referred to the respective absolute zeros. In air, at atmospheric conditions and 20°C, $c_a \simeq 343$ m/s, whereas in water at 20°C,

$c_a \simeq 1430$ m/s. Thus, sound waves travel more than four times faster in a liquid than in a gas.

In a bubbly gas-water mixture, the speed of sound is

$$c_a = \left\{ \frac{E_w k p_*}{[(1 - \sigma) k p_* + \sigma E_w][(1 - \sigma) \rho_w + \sigma \rho_g]} \right\}^{1/2} \qquad (4.6)$$

where E_w = water modulus of elasticity, ρ_w = water density, ρ_g = gas density, and σ = volume of gas per volume of mixture. [For sound or pressure waves the frequency of which approaches the resonant frequency f_R of the gas bubbles (Equation 4.27), c_a becomes a function of frequency (cf. Blake, 1986).] According to this relationship, c_a of an air-water mixture at atmospheric pressure and only 0.1 percent air concentration is equal to c_a in air; for higher air concentrations, the sonic speed of the mixture is less than that in air!

In a fluid with clusters of solid bodies, the effective speed of sound $(c_a)_{\text{eff}}$ is less than c_a in the unobstracted fluid. From experiments with banks of heat exchanger tubes, Parker (1978) and Blevins (1986b) found good agreement with the approximate formula

$$(c_a)_{\text{eff}} = \frac{c_a}{\sqrt{1 + \beta_s}} \qquad (4.7)$$

advanced by Meyer & Neumann (1972) for cases where the acoustic wavelength is large relative to dimensions of the solid bodies in the enclosure (e.g., the tube diameter). Herein, β_s is the ratio of the volume occupied by the solid bodies to the total volume of the enclosure. [Effects of tube arrangement on $(c_a)_{\text{eff}}$ in different directions are addressed by Burton, 1980.]

For gravity or *free-surface* waves, the disturbances and their propagation are governed by the specific weight of the fluid $\gamma = \rho g$, and the celerity is a function of the depth h under the free surface and the wavelength λ (or the wave frequency f). Presuming again small amplitudes and neglecting effects of surface tension, one may use Airy's solution

$$c = \left[\frac{g\lambda}{2\pi} \tanh\left(\frac{2\pi h}{\lambda}\right) \right]^{1/2} = \frac{g}{2\pi f} \tanh\left(\frac{2\pi f h}{c}\right) \qquad (4.8a)$$

with good approximation (Sarpkaya & Isaacson, 1981). For shallow-water, this expression reduces to

$$c \simeq \sqrt{gh} \qquad \text{if } h/l < 0.05 \qquad (4.8b)$$

and for deep-water waves,

$$c \simeq \sqrt{\frac{g\lambda}{2\pi}} = \frac{g}{2\pi f} \qquad \text{if } h/\lambda > 0.5 \qquad (4.8c)$$

An estimate of the effect of wave amplitude $\hbar$ is obtained from the following expressions (Rouse, 1938):

$$c \simeq \sqrt{gh} \left[1 + \frac{3}{2} \frac{h}{h} \right]^{\frac{1}{2}} \qquad \text{if } h/\lambda < 0.05 \qquad (4.9a)$$

$$c \simeq \sqrt{\frac{g\lambda}{2\pi}} \left[1 + \left(\frac{h}{\lambda/2\pi} \right)^2 \right]^{\frac{1}{2}} \qquad \text{if } h/\lambda > 0.5 \qquad (4.9b)$$

4.2 DISCRETE-MASS FLUID OSCILLATORS

4.2.1 *Compressibility-governed systems*

A discrete-mass type of fluid oscillator governed by compressibility is the *Helmholtz resonator* shown in Figure 4.3a. It consists of a cavity of length L and volume V with a small opening (neck) to the atmosphere. The most familiar Helmholtz resonator is an empty bottle in which a tone is excited by blowing across the neck. Of special importance is the property of such resonators to absorb acoustic energy at their resonant frequency f_R. This property is used, for example, in chambered mufflers (Section 8.6).

The assumptions for the analysis of Helmholtz resonators are: (a) $f_R < c/4L$, so that no acoustic standing waves can form within the cavity; (b) fluid velocities in the cavity are much smaller than in the neck, so that it suffices to model the fluid within the neck as the mass of the oscillator; and (c) cavity and neck are formed of rigid walls. If assumption (a) is not satisfied, one must consider also the acoustic standing waves which may form within the cavity (Section 4.3).

The simplest way to determine the natural or resonant frequency f_R is by means of the equation of motion for the undamped fluid oscillator without excitation. One proceeds by giving the oscillating mass $m = \rho A l_{eff}$ a displacement x and estimating the restoring force acting on it. With the notation given in Figure 4.3a, the restoring force is $F_x = -A\Delta p$, where the pressure increase Δp for adiabatic conditions and small amplitudes is $\Delta p = k p_* A x / V$. Hence the equation of motion becomes, in combination with Equation 4.5,

$$m\ddot{x} = F_x \qquad \text{or} \qquad \ddot{x} + \frac{c_a^2 A}{l_{eff} V} x = 0 \qquad (4.10)$$

Obviously, this equation is harmonic and describes the free vibration of a simple oscillator. The solution, by analogy to Equation 2.4, is a sinusoidal variation in x with a circular frequency which is given by the square root of the coefficient of x in Equation 4.10. Hence,

$$f_R = \frac{1}{2\pi} \left[\frac{k p_* A}{\rho l_{eff} V} \right]^{\frac{1}{2}} = \frac{c_a}{2\pi} \left[\frac{A}{l_{eff} V} \right]^{\frac{1}{2}} \qquad (4.11)$$

where c_a = speed of sound (Equations 4.3 through 4.7); A = cross-sectional area of neck; V = volume of cavity; $l_{eff} = l + 2l'$; l = length of neck; l' = 'added length' accounting for the added-mass or end effects at one of the two ends (Figure 4.2b). In general, l'/l diminishes with increasing l/d as shown for the free-surface counterpart of a Helmholtz resonator in Figure 4.4. For acoustic systems, l' has been investigated by Alster (1972). Kinsler & Frey (1962) report that for low frequencies, l' is, typically, $0.3d$ for a wide-flanged termination of a circular neck of diameter d, and it is $0.42d$ for an unflanged termination. To be sure, this gives only a rough indication as all the various effects on added mass discussed in Section 3.2 apply qualitatively here as well.

Information on the resonant frequencies and modes of systems of two or more Helmholtz resonators, as well as on l', are given in the book of Blevins (1986).

4.2.2 *Free-surface systems*

The free-surface counterpart of a Helmholtz resonator is an open-basin or *surge-tank system*, as presented in Figure 4.3b. Here, the restoring force acting on the mass $m = \rho A l_{eff}$ in the x direction due to a displacement z of the liquid surface is $F_x = -\rho g A z$. Under assumptions equivalent to those cited in the preceding section, the equation of motion for free vibration becomes

$$m\ddot{x} = F_x \qquad \text{or} \qquad \rho A l_{eff}\ddot{x} = -\rho g A z$$

Substituting the mass-conservation relationship $z = xA/A_s$ into this equation, one obtains, as long as $A_s \gg A$,

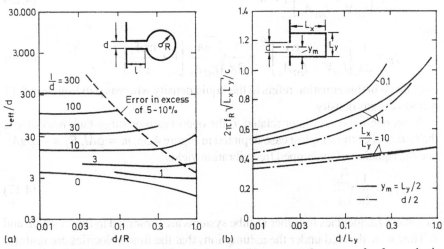

Figure 4.4. (a) Effective channel length l_{eff} and (b) resonant frequency f_R of open-basin resonators after computations of Miles (1971).

$$\ddot{x} + \frac{gA}{l_{eff}A_s}x = 0 \tag{4.12}$$

and, hence,

$$f_R = \frac{1}{2\pi}\sqrt{\frac{gA}{l_{eff}A_s}} \tag{4.13}$$

where A_s = area of the free surface and $l_{eff} = l + 2l'$ as explained after Equation 4.11.

The influence of various geometric parameters on the effective neck or channel length l_{eff} and the resonant frequency f_R of open-basin resonators are shown in Figure 4.4, obtained from inviscid-fluid computations including wave radiation. The results in Figure 4.4a are strictly valid only for $d/R \ll 1$ and $f_R \ll c/l$, but the errors corrresponding to $d/R < 1$ and $f_R l/c < 0.1$ are not likely to exceed 5 to 10%. The curve for $l/d = 0$ demonstrates clearly that an effective oscillating mass $\rho l_{eff}hd$ exists even if the length of the channel is zero. In fact, all curves in Figure 4.4b relate to resonators with zero channel length. (Information on corresponding acoustic Helmholtz resonators is provided by Alster, 1972.)

For the *air-chamber system* shown in Figure 4.3c, the restoring force is the sum of the forces due to the compression of the air volume V and to the displacement z of the water surface,

$$F_x = -\rho\frac{c_a^2 A_s}{V}Az - \rho gAz \tag{4.14}$$

Using again the mass-conservation relationship $z = xA/A_s$, one may write

$$\ddot{x} + \left[\frac{c_a^2 A}{l_{eff}V} + \frac{gA}{l_{eff}A_s}\right]x = 0 \tag{4.15}$$

and

$$f_R = \frac{1}{2\pi}\left[\frac{c_a^2 A}{l_{eff}V} + \frac{gA}{l_{eff}A_s}\right]^{1/2} = \frac{1}{2\pi}\sqrt{\frac{gA}{l_{eff}A_s}}\left[1 + \frac{c_a^2 A_s}{gV}\right]^{1/2} \tag{4.16}$$

Note that ρ in this equation refers to the liquid density whereas ρ in Equation 4.11 denotes the gas density.

A discrete-mass oscillator related to the open-basin or surge-tank resonator is the *U-tube*. For the simplest case depicted in Figure 4.5a, $m = \rho Al$, $F_x = -2\rho gAz$, and the equation of undamped free vibration yields

$$f_R = \frac{1}{2\pi}\sqrt{\frac{2g}{l}} \tag{4.17}$$

Resonator frequencies for other U-tube systems are presented in Figures 4.5b and c. They were derived under the assumptions that the fluid velocities are uniform over any cross-section at any time and that the amplitudes are small compared to l. Note that the length of the vibrating fluid mass is known in all cases except for the

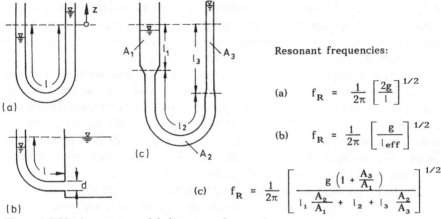

Resonant frequencies:

(a) $f_R = \dfrac{1}{2\pi} \left[\dfrac{2g}{l} \right]^{1/2}$

(b) $f_R = \dfrac{1}{2\pi} \left[\dfrac{g}{l_{eff}} \right]^{1/2}$

(c) $f_R = \dfrac{1}{2\pi} \left[\dfrac{g\left(1 + \dfrac{A_3}{A_1}\right)}{l_1 \dfrac{A_2}{A_1} + l_2 + l_3 \dfrac{A_2}{A_3}} \right]^{1/2}$

Figure 4.5. U-tube systems and their resonant frequencies.

one depicted in Figure 4.5b. In that case, $l_{eff} = l + l'$ where $l' \simeq 0.3d$ for a circular cylinder tube of diameter d and for low frequencies, in analogy to the acoustic counterpart described with Equation 4.11. The possibilities of coupling between U-tube and sloshing modes are discussed by Yeh (1966). Information on sloshing modes is presented in Section 4.3.2.

4.3 DISTRIBUTED-MASS FLUID OSCILLATORS

4.3.1 *Basic relationships*

As was pointed out in Section 4.1, the distributed-mass type of fluid oscillator is characterized by oscillations within the entire expanse of fluid at different amplitudes and velocities. This oscillatory motion can be viewed as the result of two or more wave trains, one originating at a location of disturbance or excitation and the other(s) being reflected from the boundaries surrounding the fluid mass considered. As the waves move through the fluid, the fluid particles are cyclically displaced from their equilibrium positions and the fluid pressure rises and falls periodically. If reductions in amplitude due to various sources of damping and the effect of fluid flow are disregarded (reflection coefficient $K_r = 1.0$), the reflected waves will be similar to the original direction of wave propagation. At a rigid boundary (i.e., a 'closed end') the fluid velocity normal to the wall must be zero at all times and the wave becomes positively reflected; at the juncture between the fluid oscillator and a large reservoir (i.e., an 'open end'), the pressure must remain approximately equal to the outside pressure and the wave becomes negatively reflected.

Standing waves are formed when waves return on themselves. In such cases, closed ends coincide with pressure antinodes (i.e., sections of maximum pressure amplitude) and open ends coincide with pressure nodes (i.e., sections of zero pressure amplitude). The reason that the boundary condition at the open end is only approximately achieved is described in Figure 4.2.

One may distinguish between one-, two-, and three-dimensional standing waves. Figure 4.6 illustrates a one-dimensional wave in a narrow duct (acoustic wave) and a narrow open channel (gravity wave) for the closed/closed end condition (narrow means $B < \lambda/2$). Between the two ends, there may be any number of nodes, $N = 1, 2, 3, ...$; only one of the corresponding wave modes, the second with two nodes between the ends, is shown in this figure. With respect to symbols, $T = 1/f_R$ is the period of oscillation and $\tilde{v}, \tilde{p}, \tilde{h}$ denote the fluctuating fluid velocity, pressure, and free-surface displacement, respectively.

An overview of all possible end conditions and the corresponding one-dimensional standing-wave patterns is given in Figure 4.7. Clearly, the distance of

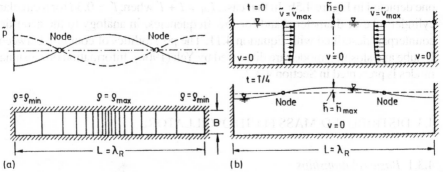

Figure 4.6. (a) Characteristics of a standing pressure wave in a narrow duct, $N = 2$, and (b) successive phases of a standing gravity wave in a narrow open channel, $N = 2$.

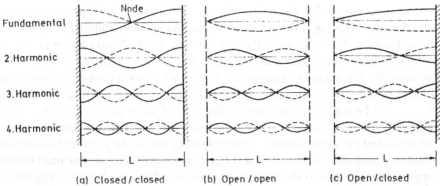

Figure 4.7. One-dimensional standing wave patterns showing longitudinal pressure variation for different end conditions.

length L between the two ends may contain any integer-number multiple of one-half of a resonant wavelength, $\lambda_R/2$. In general, therefore,

$$\lambda_R = \frac{2L_{eff}}{N},$$

$N = 1, 2, 3, \ldots$ for closed/closed, or open/open end condition $\qquad$ (4.18)

and

$$\lambda_R = \frac{4L_{eff}}{2N - 1},$$

$N = 1, 2, 3, \ldots$ for open/closed end condition $\qquad$ (4.19)

Again, as in Section 4.2, $L_{eff} \geq L$ is used in these equations to account for added-mass effects at the open ends. In the case of closed/closed end conditions (Figure 4.7a), $L_{eff} = L$. For a circular duct of diameter d with one end open and one closed, Alster (1972) found

$$\frac{L_{eff}}{L} = \left(1 + 0.48\frac{d}{L}\right)^{1/2} \qquad (4.20)$$

for the fundamental acoustic mode ($N = 1$).

With these expressions for λ_R, the all-important natural or resonant frequency f_R of a one-dimensional distributed-mass type of fluid oscillator can now be computed with the aid of Equation 4.2, $f_R = c/\lambda_R$. One needs to simply substitute the relevant celerity c according to Equations 4.3 through 4.9 and the resonant wavelength λ_R according to Equation 4.18 or 4.19. In distinction to the discrete-mass fluid oscillator, there exist, theoretically, an infinite number of resonant frequencies corresponding to the mode of vibration ($N = 1$ for the fundamental mode, $N = 2$ for the second harmonic, etc., Figure 4.7). One-dimensional systems with junctions are discussed, e.g., by Merkli (1978).

If fluid is contained within boundaries extending over more than one-half wavelength in the three orthogonal directions, standing waves, when excited, may take a three-dimensional form. For an *enclosure with orthogonal boundaries*, the general expression for the resonant frequency is

$$f_R = c\left[\frac{1}{\lambda_x^2} + \frac{1}{\lambda_y^2} + \frac{1}{\lambda_z^2}\right]^{1/2} \qquad (4.21)$$

If these boundaries are closed walls, all resonant wavelengths are related to the corresponding cavity dimensions L_x, L_y, and L_z by way of Equation 4.18, and one obtains

$$f_R = \frac{c}{2}\left[\left(\frac{N_x}{L_x}\right)^2 + \left(\frac{N_y}{L_y}\right)^2 + \left(\frac{N_z}{L_z}\right)^2\right]^{1/2} \qquad (4.22)$$

where $N_x = 1, 2, \ldots, N_y = 1, 2, \ldots, N_z = 1, 2, \ldots$ are the normal modes in the x, y, z

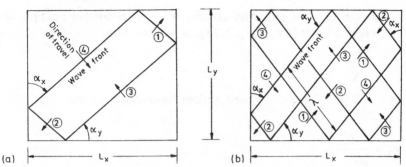

Figure 4.8. Wave fronts and directions of wave travel for two two-dimensional standing-wave modes (N_x, N_y): (a) Mode (1.1) and (b) Mode (3.2) (after Beranek, 1954).

directions. For a gravity-governed standing wave in a rectangular basin of horizontal dimensions L_x and L_y, of course, the counterpart to Equation 4.22 is

$$f_R = \frac{c}{2} \left[\left(\frac{N_x}{L_x} \right)^2 + \left(\frac{N_y}{L_y} \right)^2 \right]^{1/2} \quad \text{where } N_x = 0, 1, 2, ..., N_y = 0, 1, 2, ... \quad (4.23)$$

The relationships of Equations 4.22 and 4.23 can be easily verified by kinematic reasoning (Figure 4.8). Starting from the premise that waves can travel in the enclosure forward and backward between any two opposing walls as well as around the enclosed space involving the walls at various angles of incidence, standing waves will be seen to emerge from the superposition whenever the waves return on themselves. In the examples in Figure 4.8, the numbers 1 and 3 denote forward-travelling waves, and the numbers 2 and 4 denote backward-travelling waves. The angles α_x and α_y at which these waves are incident upon and reflect from the walls are

$$\alpha_x = \tan^{-1} \frac{N_y / L_y}{N_x / L_x}; \qquad \alpha_y = \tan^{-1} \frac{N_x / L_x}{N_y / L_y}$$

Some of the standing-wave patterns corresponding to the infinite number of combinations (N_x, N_y, N_z) are schematically represented in Figure 4.9 for $N_z = 0$. The corresponding mode shapes of the pressure amplitudes (for pressure waves) or free-surface displacement amplitudes (for gravity waves) are given by

$$\cos \frac{N_x \pi x}{L_x} \cos \frac{N_y \pi y}{L_y}$$

In case of a *right cylindrical enclosure* of radius R and height L, the mode shapes are described in terms of the cylindrical coordinates r (radius), θ (circumferential angle), and z (height). The resonant frequency in this case is

$$f_R = \frac{c}{2\pi} \left[\frac{C_{r\theta}^2}{R^2} + \left(\frac{N_z \pi}{L} \right)^2 \right]^{1/2} \tag{4.24}$$

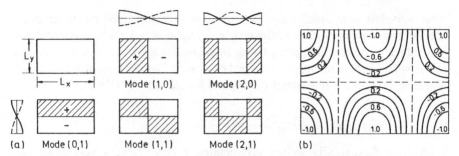

Figure 4.9. (a) Schematic of standing-wave patterns in a rectangular cavity or basin. (b) Contours of constant pressure or surface-displacement amplitudes for mode (2, 1) (after Beranek, 1954).

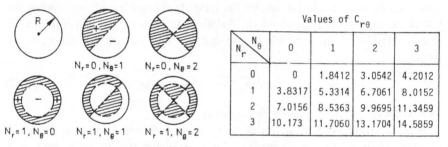

Figure 4.10. Schematic of standing-wave patterns in a right cylindrical cavity or basin and values of the constant $C_{r\theta}$ in Equation 4.24.

where $C_{r\theta}$ is a constant depending on the number N_r of nodal circles and the number N_θ of nodal diameters (Figure 4.10). Additional values of $C_{r\theta}$ are tabulated in the books by Blevins (1986) and Abramowitz & Stegun (1970).

In an *elliptical cylinder*, the fundamental mode $(0, N_\theta = 1, 0)$ has the frequency, according to Lamb (1945),

$$f_R = \frac{c}{2\pi a} \left[\frac{18 + 6(b/a)^2}{5 + 2(b/a)^2} \right]^{1/2} \tag{4.25}$$

where a = one-half of the major axis and b = one-half of the minor axis.

When the enclosure of a fluid is *irregularly shaped*, standing waves can also be excited. Solutions in cases like this have to be obtained numerically except for the fundamental resonant frequency which is:

$$(f_R)_{min} = \frac{c}{2L_m} \tag{4.26}$$

with L_m = maximum linear dimension of the enclosure. The corresponding higher harmonics in this case are so large in number and so complex in pressure or surface-displacement distribution that waves will be produced travelling in all directions and involving each wall at nearly all angles of reflection, even if excited by a source of narrow frequency band. In acoustics, therefore, one speaks of a diffuse sound field in an irregular-shaped enclosure. Further information on such oscillators is available in Beranek (1954) and Kinsler & Frey (1962).

4.3.2 *Specific compressibility-governed systems*

A group of fluid oscillators not yet mentioned consists of a volume of *gas enclosed by a liquid* (Figure 4.11). The simplest oscillator belonging to this group is a spherical gas bubble in a liquid, which can oscillate in a symmetric radial mode analogous to a discrete spring-mass system (Plesset & Prosperetti, 1977). For larger bubbles ($R > 1$ cm), surface-tension effects are negligible and one can formulate the equation of motion for the (added) mass of liquid surrounding the bubble in analogy to Equation 4.10. Relating the restoring spring force to the compressibility of the gas within the bubble, one obtains for the resonant frequency:

$$f_R = \sqrt{3}\ c_a/R, \qquad c_a = \sqrt{kp_*/\rho}, \qquad R > 1\ \text{cm} \qquad (4.27)$$

The symbols used are defined with Equation 4.5.

When a larger volume of gas is enclosed by a flowing liquid as shown in Figures 4.11b, c, the corresponding fluid oscillator has to be treated as a distributed-mass type oscillator with an infinite number of modes. Problems involving the flow-induced excitation of this kind of fluid oscillator may be treated in two alternative ways: The fluid oscillations may be considered as brought about by various types of excitation of this fluid oscillator; or they may be viewed as part of a distinct excitation mechanism belonging to the group of IIE (Figures 6.33 and 6.35). There are advantages for each of these viewpoints. In this monograph, the oscillators in Figures 4.11b, c are treated as IIE mechanisms (or

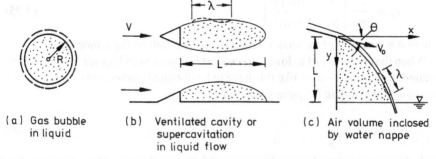

(a) Gas bubble   (b) Ventilated cavity or   (c) Air volume inclosed
    in liquid   supercavitation   by water nappe
                          in liquid flow

Figure 4.11. Gas volumes enclosed by liquids. (a) Gas bubble in liquid. (b) Ventilated cavity or supercavitation in liquid flow. (c) Air volume enclosed by water nappe.

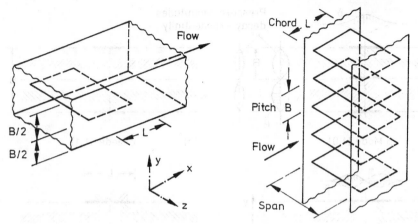

Figure 4.12. Sketch of plate in conduit and cascade of plates.

'flow-oscillators', as explained in Section 3.3.2), since they cannot exist without flow.

A configuration often encountered in engineering is shown in Figure 4.12. The *plate* in this figure represents a simplified guide vane, flow divider, or pier in a conduit, and the *cascade of plates* is an idealization of stator or other vanes in flow machines or of guide vanes in bends. When fluid flows past such structures, alternating vortices are shed from the plates (Section 6.3) introducing a source of excitation for standing waves between them. As pointed out by Parker (1966), the amplitude of pressure fluctuations so excited may be of the same order of magnitude as the dynamic pressure $\rho V^2/2$ of the free stream approaching the plates.

The four standing-wave patterns of this type which occur most frequently are sketched in Figure 4.13. Depending on the chord length L, one or more antinodes (regions of high pressure amplitudes) may exist along each plate. With larger pitches B between the plates, moreover, a node may develop in the x, z planes midway between the plates, as seen in Figures 4.13c, d. In all cases, pressure amplitudes also occur in the duct downstream and upstream of the plates, which, according to Parker (1967), decay exponentially in the x-direction as

$$\bar{p}_x = \bar{p}_{ref} \exp\left[-\frac{2\pi \sqrt{f'^2 + f_R^2}}{c_a} x \right] \tag{4.28}$$

Here, f_R is the resonant frequency of the standing wave and f' is known as the cut-off frequency, i.e., the resonant frequency of a standing transversal wave well downstream and upstream of the plates. [For Parker modes α and β, $f' = c_a/2B$; for Parker modes γ and δ, $f' = c_a/B$.]

A more precise picture of the standing-wave pattern for one of the possible modes is given in Figure 4.14. The contours of pressure amplitude shown were

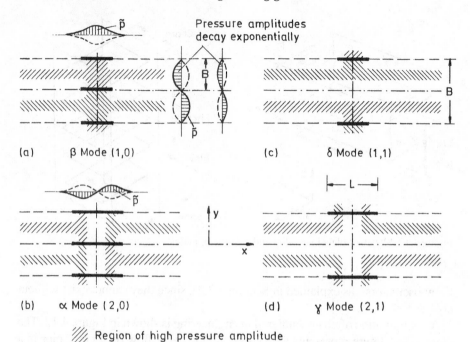

(a)   β Mode (1,0)

(c)   δ Mode (1,1)

(b)   α Mode (2,0)

(d)   γ Mode (2,1)

▨ Region of high pressure amplitude

Figure 4.13. Schematic of standing waves in a cascade of plates; (a, b) modes $(N_x, 0)$; (c, d) modes (N_x, N_y).

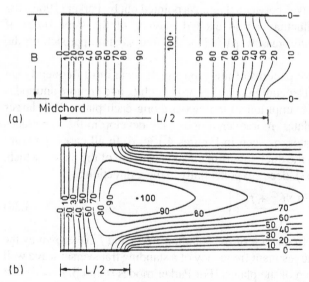

(a)

(b)

Figure 4.14. Contours of constant pressure amplitude for α-mode standing waves $(N_x = 2,$ $N_y = 0)$ between parallel plates. (a) $L/B = 5.1$ (Point P_2 in Figure 4.15a) and (b) $L/B = 1.35$ (Point P_1 in Figure 4.15a) (after Parker, 1967).

derived by Parker (1967) by solving the transformed version of the two-dimensional wave equation in conjunction with the appropriate boundary conditions. As seen from Figure 4.14, the wave pattern becomes more nearly one-dimensional as L becomes larger in relation to B; in the limit it resembles the pattern of a standing wave in an open-ended organ pipe.

An interesting interpretation of these standing waves, or Parker modes as they are called, is obtained if the space between the plates is viewed as an enclosure bounded on two sides by solid boundaries (the plates) and on two sides by 'open' boundaries. Of course, this viewpoint is acceptable only for cases in which f_R is far removed from f', so that the pressure amplitudes are negligible outside the plate region (e.g., the case in Figure 4.14a). Nevertheless, it permits estimates of f_R in all cases provided that the ratio L/B is large enough (say > order 1). In other words, certain Parker modes can be treated, in an approximate way, as 'distorted' versions of resonant waves in an 'open' box with dimensions $L_x = L_{eff} \geq L$ and $L_y = B$ (the dimension L_z is disregarded in this two-dimensional treatment). After substitution of $\lambda_x = 2L_{eff}/N_x$ and $\lambda_y = 2B/N_y$ (from Equation 4.18) into Equation 4.21,

$$ f_R = \frac{c}{2}\left[\left(\frac{N_x}{L_{eff}}\right)^2 + \left(\frac{N_y}{B}\right)^2\right]^{1/2} \text{ where } N_x = 0, 1, 2, ..., N_y = 0, 1, 2, ... \quad (4.29) $$

According to this equation for two-dimensional standing waves, the Parker modes α and β correspond to modes (2, 0) and (1, 0), and the Parker modes γ and δ to modes (2, 1) and (1, 1), respectively. As seen from the plot of f_R in Figure 4.15a,

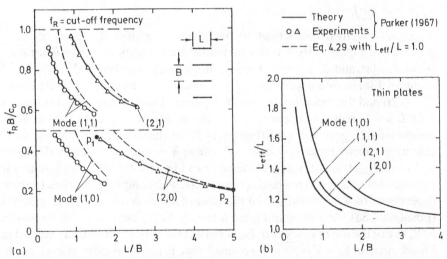

Figure 4.15. (a) Resonant frequencies f_R of standing pressure waves in a cascade of plates (Figure 4.12b) for low-Mach-number flow. (b) Corresponding values of L_{eff}/L according to Equation 4.29.

the experimental data for moderately high L/B values are surprisingly well predicted by this simple formula even without accounting for added-mass effects ($L_{eff}/L = 1.0$). For lower values of L/B, of course, the added mass discussed in conjunction with Figure 4.2c cannot be neglected, and the ratio L_{eff}/L assumes increasing magnitudes as shown in Figure 4.15b.

The results presented for plate cascades relate to the four simplest modes. Higher two-dimensional modes are also possible, of course, and all may be combined with pressure variations in the z-direction as well (Parker, 1966); in terms of the open-box counterpart:

$$f_R = \frac{c}{2} \left[\left(\frac{N_x}{L_{eff}} \right)^2 + \left(\frac{N_y}{B} \right)^2 + \left(\frac{N_z}{L_z} \right)^2 \right]^{1/2} \tag{4.30}$$

Note that oscillations in the y direction with frequencies f_R larger than the cut-off frequency f' cannot occur in cascades (more information on the cut-off frequency was presented by Tyler & Sofrin, 1962).

Figure 4.15 may be used also to determine f_R for a single plate in a rectangular duct if the plate is placed in the plane of symmetry (Figure 4.12a). The side walls in this case take the place of the midplanes in the multiple-plate case. The modes indicated in Figures 4.13c, d are impossible here as they would be incompatible with the side-wall boundary condition.

None of the material presented, of course, shows the effects of mean flow on the standing-wave characteristics. Such effects are particularly important for high Mach numbers $Ma = V/c_a$ as can be inferred from Figures 4.21 and 4.22.

4.3.3   Specific free-surface systems

A gravity-governed counterpart to the plates in a duct discussed above is presented in Figure 4.16. For this particular arrangement of a *pier in an open channel*, Marvaud & Ramette (1964) observed the fundamental mode (1, 0) (Figure 4.13a) in their experiments. Again one can see that the cut-off frequency $f' = c/2B$ and the 'open-box' mode (1, 0) without added-mass effect (Equation 4.29, $L_{eff} = L$) represent upper bounds to the measured values of f_R just as in the compressibility-governed case (Figure 4.15). In the case considered, the wave celerity c is approximately equal to $\sqrt{gh}$, where h = mean water depth (Equation 4.8b), and L_{eff}/L is larger than obtained from Figure 4.15b due to differences in plate geometry (d/L was not held constant). Another complication arises from the dependence of wave celerity c on the ratio of water depth to wavelength, h/λ (Equation 4.8a). According to Figure 4.16b, $\lambda = c/f_R$ increases with increasing L/B, so that the use of $c = \sqrt{gh}$ in Equation 4.29 is only approximately valid. The Froude number $Fr = V/\sqrt{gh}$ was so small that it had negligible effects on the resonant frequency (Equation 4.33).

A fluid oscillator of great importance in civil engineering is a *harbor*. Harbors differ from the basins depicted in Figures 4.9 and 4.10 because of the channel

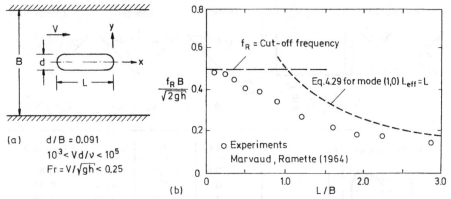

Figure 4.16. Resonant frequency f_R of the $(1,0)$ mode standing wave in an open channel with pier (h = depth of flow).

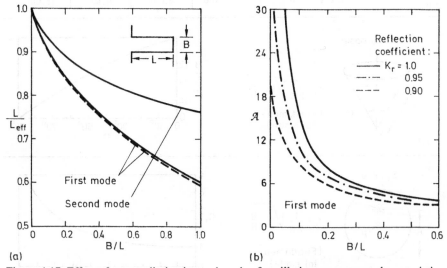

Figure 4.17. Effect of energy dissipation and mode of oscillation on resonant characteristics of a fully open harbor (after Ippen & Goda, 1963). (a) Values of L/L_{eff} according to Equation 4.29; (b) Amplification factor at resonance. (K_r = measure of energy dissipation.)

connecting the harbor to the open sea. This channel gives rise to at least one more mode of fluid oscillation, called the pumping or harbor mode: a mode corresponding to the discrete-mass oscillation in the Helmholtz resonator schematically represented in Figure 4.1. Depending on its width d, moreover, the channel changes the basin-boundary condition from that of a close end for small d to that of an open end for large d. In the two extremes, therefore, Equations 4.18, 4.19

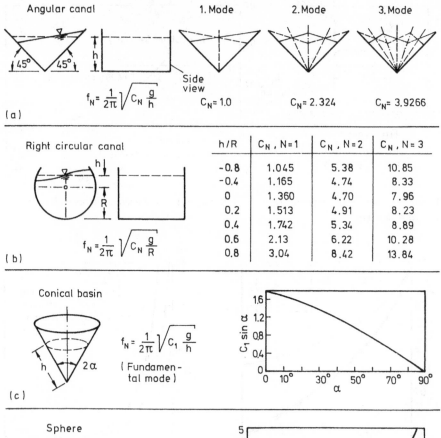

Figure 4.18. Resonant frequencies f_R of standing waves in basins of variable depth (from Blevins, 1986).

and 4.21 lead to the following expressions for the distributed-mass oscillation within the harbor basin:

$$f_R = \frac{c}{2}\left[\left(\frac{N_x}{L_x}\right)^2 + \left(\frac{N_y}{L_y}\right)^2\right]^{1/2}$$

(4.31)

for $d/L_y = 0$ (closed/closed end condition), and

$$f_R = \frac{c}{2} \left[\left(\frac{2N_x - 1}{2L_{x\,\text{eff}}} \right)^2 + \left(\frac{N_y}{L_y} \right)^2 \right]^{1/2} \tag{4.32}$$

for $d/L_y = 1$ (open/closed end condition).

Harbor-mode resonant frequencies f_R for a rectangular harbor as a function of the opening ratio d/L_y are depicted in Figure 4.4b. The added-mass or end effect for a fully open harbor is shown in Figure 4.17a in terms of the ratio L_{eff}/L. This effect is different for the various standing-wave modes. The energy dissipation, expressed in terms of a reflection coefficient, K_r, has almost no effect on L_{eff} and, hence, on f_R, but it definitely reduces the amplification factor at resonance, $\mathcal{A}_R$. More information on harbors is found in Section 8.2. Numerical techniques for determining f_R of an arbitrary-shaped harbor are described by Raichlen (1966), Lee (1971), and Lee & Raichlen (1972).

For basins or *tanks of variable depth*, approximate or numerical solutions are usually required (e.g., Abramson, 1966; Moiseev & Petrov, 1966). Some data on the resonant frequency f_R of sloshing, as the standing-wave action is called in these cases, are presented in Figure 4.18. In general, the depth variations have little influence on f_R (Blevins, 1986), so that a good estimate can be obtained from a solution for a similarly shaped basin of mean depth. For basins with compliant walls, flexing of the walls can have a significant influence on the resonant frequency (Bauer & Siekmann, 1971).

4.4 EFFECTS OF MEAN FLOW ON RESONATOR CHARACTERISTICS

An important effect on resonator characteristics in view of flow-induced oscillations (Chapter 8) is the effect of mean flow past or through a resonator. A summary of flow effects on Helmholtz resonators is given by Worraker (1980). In general, *tangential flow* past the resonator mouth increases the resonant frequency f_R, even though decreases in f_R have also been observed. The problem in investigations of this nature is that it is difficult to eliminate the influence of the dominant frequency of the shear-layer instability, which drives the resonant oscillations, on f_R. According to Anderson (1977), the mean flow reduces the added mass and, hence, l_{eff} in Equation 4.13 beyond a critical value of the reduced velocity $V/(f_R d)$ of about 10. The corresponding increase in f_R obtained with a cylindrical side-branch resonator in a circular duct of diameter 22.9 mm is shown in Figure 4.19.

With mean *throughflow* of velocity $V < c$ through a fluid oscillator such as a duct or a channel, a periodic excitation will produce wave trains which will travel with propagation velocities $c - V$ upstream and $c + V$ downstream, where c denotes the wave celerity (Blevins, 1984; Quinn & Howe, 1984). Even though a standing wave cannot form under this condition, there are frequencies f_R at which resonance is also possible for this type of a wave system. Modification of

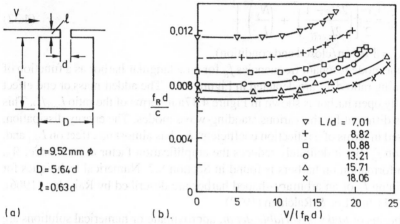

Figure 4.19. Helmholtz resonance frequency f_R as a function of mean-flow velocity V (after Anderson, 1977).

Equation 4.2 by the factor $(1 + V/c)(1 - V/c)$ yields

$$f_R = \frac{c}{\lambda_R}\left[1 - \left(\frac{V}{c}\right)^2\right] \tag{4.33}$$

Here, λ_R is the resonant wavelength in accordance with either Equation 4.18 or 4.19, depending on end conditions, and V/c represents either the Mach number Ma or the Froude number Fr, depending on whether the fluid oscillator is dominated by compressibility (c = speed of sound, Equations 4.3 through 4.7) or by gravity (c = celerity of gravity waves, Equation 4.8 or 4.9).

An illustration of the effect of mean through flow on the response of an open-ended duct to random noise excitation is given in Figure 4.20. Without flow (Ma = 0), there are sharp peaks in the pressure spectrum corresponding to the fundamental and higher harmonic resonance modes described by Equations 4.33 and 4.18, or

$$f_R = \frac{Nc}{2L_{\text{eff}}}(1 - \text{Ma}^2) \qquad \text{where } N = 1, 2, 3, ... \tag{4.34}$$

With air flow at Ma $\equiv V/c = 0.25$ through the duct, the peaks not only shift toward lower frequencies in agreement with Equation 4.34, but they also attenuate substantially. At Ma = 0.5 there is little resonant response left. Of course, Figure 4.20 represents an extraneously induced excitation of a fluid oscillator; in cases of instability- or movement-induced excitations, this damping effect of the mean flow may not be so drastic (Chapter 8).

Further insight into this damping effect is obtained if one considers the influence of mean throughflow on the pressure reflection coefficient K_{pr} (defined as the ratio of the amplitude of the pressure wave reflected from a boundary to its

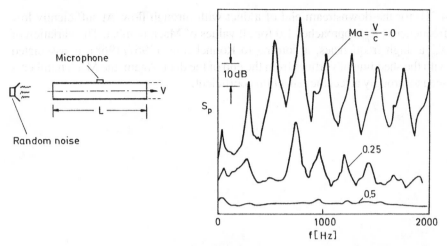

Figure 4.20. Effect of air flow of mean velocity V on the pressure spectra of the resonant response of an open duct subjected to random excitation by a loudspeaker (after Ingard & Singal, 1975).

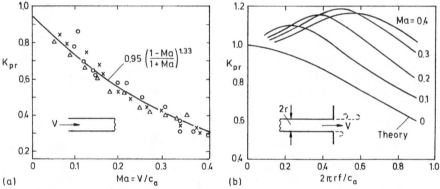

Figure 4.21. Pressure reflection coefficient K_{pr} (a) at the upstream open end of a duct as a function of Mach number (after Ingard & Singal, 1975), and (b) at the downstream open end of a duct terminating in an infinitely large wall as a function of the frequency f of the acoustic signal and the Mach number (after Ronneberger, 1967/1968).

incident amplitude). Figure 4.21a shows that the value of K_{pr} at an upstream open end of a duct with through flow decreases from about 0.95 for Ma = 0 to a low value of only 0.3 for Ma = 0.4. This drastic change of K_{pr} contrasts with a relatively weak variation of the pressure reflection coefficient at the downstream end of the duct, studied by Ingard & Singal (1975).

The effect of the frequency f of an acoustic signal on K_{pr} is illustrated in Figure

4.21b for the downstream end of a duct with through flow. At sufficiently low frequencies, K_{pr} approaches 1.0 for all values of Mach number. The variation of K_{pr} at high frequencies, according to Ronneberger (1967/1968), is associated with the shedding of vortices from the end of the duct. Again, the Mach number is seen to strongly affect the reflection coefficient.

CHAPTER 5

Vibrations due to extraneously induced excitation

5.1 SOURCES OF EXTRANEOUSLY INDUCED EXCITATION (EIE)

Excitation that is independent of any flow instability and independent of any structural motion except for added-mass and fluid-damping effects is designated in this monograph as extraneously induced (EIE). In contrast to instability- and movement-induced excitation (IIE and MIE, Chapters 6 and 7), EIE gives rise to forced vibrations as defined in Section 2.3. Among the most common sources of EIE are:
1. Turbulence;
2. Cavitation and some aspects of two-phase flows;
3. Oscillating flows including waves;
4. Machines and machine parts;
5. Earthquakes.
A brief survey of these EIE sources is given in Section 2.2 of the IAHR Monograph on Hydrodynamic Forces (Naudascher, 1991). In the following, only turbulence buffeting is presented in some detail, and only the excitation of body oscillators (structures and structural parts) is considered. Extraneously induced excitation of fluid oscillators (piping systems, resonating basins, etc.) is treated in Section 8.2 of this monograph.

5.2 EXCITATION DUE TO TURBULENCE

5.2.1 *Overview and definitions*

In cases of turbulence buffeting, the fluctuating structural load is caused by turbulent eddies flowing past or impinging upon the structure. The turbulence may be produced by one of several types of free-shear flow such as a wake, a mixing layer, or a jet (Figure 5.1a); by a wall-shear flow such as a boundary-layer or pipe flow (Figure 5.1b); or by flow through a grid or lattice (Figure 5.1c). From the many parameters available to describe turbulence (Tennekes & Lumley,

91

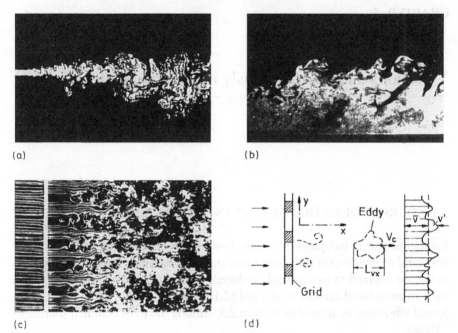

Figure 5.1. Turbulence generated (a) by a jet (photo by Dimotakis, Lye & Papantoniou in Van Dyke, 1981), (b) by a wall boundary (photo by Falco in Van Dyke, 1981), and (c, d) by a grid (photo by Corke & Nagib in Van Dyke, 1981).

1972), at least a few must be mentioned here for the sake of order-of-magnitude engineering considerations. The two most basic parameters are the turbulence level or intensity, Tu, and the integral scale or macroscale of turbulence, L_{vx}, which is a measure of the size of the eddies containing most of the turbuence energy:

$$Tu = \frac{v'_{rms}}{\bar{v}}, \qquad v'_{rms} = \sqrt{\overline{v'^2}} = \lim_{T \to \infty} \left[\frac{1}{T} \int_o^T v'^2 (t)dt \right]^{1/2} \qquad (5.1)$$

$$L_{vx} = \bar{v} \int_o^\infty R(\tau)d\tau, \qquad R(\tau) = \frac{\overline{v'(t) \, v'(t + \tau)}}{\overline{v'^2}}$$

Here, $\bar{v}$ = characteristic mean velocity; v' = fluctuating part of the velocity in the direction of $\bar{v}$; t = time, τ = delay time; $R(\tau)$ = velocity autocorrelation function; a bar over a symbol denotes time-averaging. As an eddy of size L_{vx} moves along with convection velocity V_c ($V_c = \bar{v}$ in the case of grid turbulence, Figure 5.1d), it generates a fluctuation in velocity and pressure associated with a frequency

$$f^* = V_c/L_{vx} \qquad (5.3)$$

(cf., Figure 5.5). Strictly speaking, one should account for different streamwise and transverse velocity fluctuations and length scales, but in most engineering applications, the assumption of homogeneous turbulence with just one pair of values (v', L_v) is acceptable.

In all cases presented in Figure 5.1, the length scale L_{vx} increases with distance x downstream. In shear flows (Figures 5.1a, b), turbulence continues to be produced as long as there is shear (or a velocity gradient) in the cross-stream direction in the flow. In grid flows (Figures 5.1c, d), shear is active only near the grid; far downstream of the grid, the turbulence intensity Tu decreases in proportion to $x^{-\frac{1}{2}}$ while L_{vx} increases proportional to $x^{\frac{1}{2}}$ (Naudascher & Farell, 1970; Tennekes & Lumley, 1972). Both L_{vx} and Tu, of course, are time-averaged quantities. Since the turbulent eddies change continuously as they are convected by the mean flow, the instantaneous velocity and pressure fields induced by them change continuously as well. Figure 5.2 depicts, as an example, the pressure field on a wall underneath a turbulent boundary layer. Clearly, additional variables must be defined before the turbulence-induced exciting force can be formulated.

To describe the frequency content of either the velocity fluctuations v' or the pressure fluctuations p', one may employ the power spectral densities

$$S_v(f) = \frac{\overline{v_f'^2}}{df}, \quad S_p(f) = \frac{\overline{p_f'^2}}{df} \tag{5.4}$$

where $\overline{v_f'^2}$ and $\overline{p_f'^2}$ are the mean-square values of those parts of v' and p', respectively, which lie within a frequency band df centered around the frequency f. In other words, S represents the magnitude per unit bandwidth of frequency. It

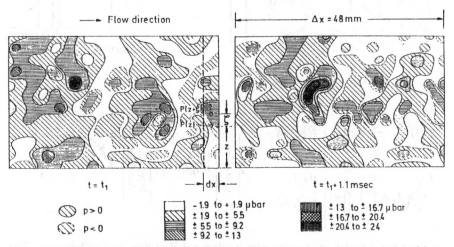

Figure 5.2. Instantaneous contours of wall pressure along part of a flat plate induced by a turbulent boundary layer (Figure 5.1b) after Emmerling, 1973 (V_∞ = 8.5 m/s, boundary-layer thickness δ = 3.7 cm, $V_\infty \delta/\nu \approx 2.1 \times 10^4$).

follows that the areas under the power spectra $S(f)$ are equal to the total mean-square values:

$$\overline{v'^2} = \int_o^\infty S_v(f)df,$$

$$\overline{p'^2} = \int_o^\infty S_p(f)df \qquad (5.5)$$

The variables defined in Equations 5.4 and 5.5 relate to a given point in the flow. If one wishes to determine the fluctuations of the overall load acting on a structure in the flow, one needs information on how effectively the pressure fluctuations at various locations on the surface of the structure act together to produce a resultant fluctuating force F'. An indication in this regard is given by the correlation $\overline{p'_1 p'_2}$ between the pressure fluctuations p'_1 and p'_2 on two incremental areas dA_1 and dA_2 on the face of the structure (Figure 5.3a). The larger the scale of the eddy in relation to a characteristic structural dimension, $\sqrt{A}$, the more nearly identical will be the time variations of p'_1 and p'_2 over the surface of the structure, in general, and the greater will be $\overline{p'_1 p'_2}$ and the mean-square of the force fluctuations, $\overline{F'^2}$, acting on the area of loading, A.

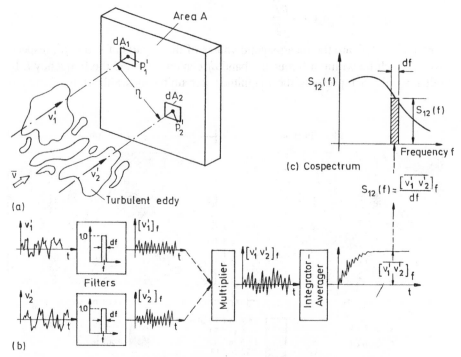

Figure 5.3. (a) Sketch of structure with incident turbulent flow. (b, c) Illustration of co-spectrum on the basis of the analogue technique of evaluation (in modern practice, this spectrum is computed by digital methods).

The incident turbulence is usually made up of eddies with a broad range of scales and, hence, frequencies. To obtain information on the frequency characteristics of F' in such a case, it is necessary to deduce the correlation $\overline{p_1' p_2'}$ as a function not only of distance η between the points of measurement but also of frequency f and to find the corresponding cospectrum of the pressure fluctuations. The significance of a cospectrum can best be illustrated with the aid of the analogue method of evaluation. As seen from Figure 5.3b, the cospectral density of the longitudinal velocity fluctuations

$$S_{12}(f) = \frac{[\overline{v_1' v_2'}]_f}{df} \tag{5.6}$$

is defined as the time average of that part of the normalized product $v_1' v_2'$ which is associated with the frequency f within a given frequency band df. In contrast to $S_{12}(f)$ the lateral velocity correlation coefficient R_{vy} is defined as

$$R_{vy}(\eta) = \frac{\overline{v_1' v_2'}}{\overline{v'^2}}, \qquad L_{vy} = \int_o^\infty R_{vy}(\eta) d\eta \tag{5.7}$$

In many cases, $R_{vy}(\eta)$ may be approximated by exp $(-\eta/L_{vy})$. Here, η = distance between points 1 and 2; L_{vy} = lateral scale of turbulence which is of the same order of magnitude as L_{vx} (Equation 5.2). For oscillating or gust flow, the velocity and pressure fluctuations are perfectly correlated over the whole structure, $L_{vy} \to \infty$, and the cospectrum $S_{12}(f)$ becomes identical to the spectrum $S_v(f)$ determined at a point in the flow (Equation 5.4).

5.2.2 *Flow-induced loading*

The prediction of the response of a structure to EIE such as turbulence buffeting may be conveniently considered in two parts. The first of these concerns the fluid-dynamic problem of predicting the loading on the structure from the known mean and fluctuating components of the flow. The second concerns the structural problem of predicting the response of the structure (i.e., the deflections, stresses, etc.) from the known fluid-dynamic loading and structural characteristics. A brief introduction to these two steps of predicting a typical flow-induced vibration of the EIE category is given in the following and in Section 5.2.3.

The problem of relating the pressure and force fluctuations on a bluff body to the velocity fluctuations in a turbulent approach flow is complicated by the distortion that the turbulence eddies undergo as they are swept past the body (Bearman, 1972; Hunt, 1973). In a simplified approach it is assumed that the instantaneous load in the direction of flow (i.e., the drag) dF acting on an incremental area dA normal to the mean flow, is related to the instantaneous velocity $v = \bar{v} + v'$ incident on that area in the form

$$\frac{dF}{dA} = C_D \frac{\rho v^2}{2} + C_m \rho \frac{\partial v}{\partial t}, \tag{5.8}$$

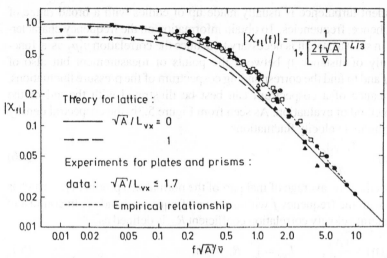

Figure 5.4. Fluid-dynamic admittance for plates and prisms normal to a turbulent flow $(Re \simeq 2 \times 10^4)$ (after Vickery & Davenport, 1967).

This equation was first formulated by Morison et al. and employed in the field of wind engineering by Vickery & Davenport (1967). Here, C_D and C_m are the drag and inertia coefficients (Sarpkaya & Garrison, 1963). The approach is justifiable for open-lattice structures which influence the velocity field on a scale much smaller than that of the smallest significant eddies, but it can also be used as an approximation for bluff bodies of relatively small aspect ratios (Figure 5.4).

If the inertial force component $C_m \rho \partial v / \partial t$ and the terms of order $(v'/\bar{v})^2$ are ignored and v and F are expressed in terms of mean and fluctuating components,

$$v = \bar{v} + v', \qquad F = \bar{F} + F',$$

Equation 5.8 yields the quasi-steady simplification:

$$d\bar{F} \simeq C_D \frac{\rho \bar{v}^2}{2} dA, \qquad dF' \simeq C_D \rho \bar{v} v' dA \qquad (5.9)$$

An approximation to the root-mean-square value of the fluctuating drag acting on the complete structure, F'_{rms}, can now be obtained through a process of integration in which the correlation between the contributions from all incremental areas dA (Figure 5.3a) is properly accounted for:

$$F'_{rms} = \sqrt{\overline{F'^2}} \simeq C_D \rho \bar{v} \sqrt{\overline{v'^2}} \left[\iint \frac{\overline{v'_1 v'_2}}{\overline{v'^2}} dA_1 \, dA_2 \right]^{1/2} \qquad (5.10)$$

Hence, the root-mean-square drag coefficient due to turbulence excitation is

$$C_{D\,\mathrm{rms}} = \frac{F'_{\mathrm{rms}}}{A\rho\bar{v}^2/2} \simeq 2\,C_D\,\mathrm{Tu}\left[\frac{1}{A^2}\iint\frac{\overline{v'_1 v'_2}}{\bar{v}^2}\,\mathrm{d}A_1\,\mathrm{d}A_2\right]^{1/2} \tag{5.11}$$

A relationship similar to these expressions can also be derived for the power spectral density of the drag fluctuation F'

$$S_F(f) = \frac{\overline{F_f'^2}}{\mathrm{d}f} \tag{5.12}$$

defined, by analogy to the expressions in Equation 5.4, as the normalized mean-square contribution to F' centered at frequency f and lying within a frequency band $\mathrm{d}f$. From Equations 5.6, 5.9b and 5.10, one obtains an expression for F_f' which, if substituted in Equation 5.12 along with $\bar{F} = C_D A\rho\bar{v}^2/2$, yields

$$S_F(f) = 4\left(\frac{\bar{F}}{\bar{v}}\right)^2 |\chi_{fl}(f)|^2\,S_v(f) \tag{5.13}$$

with

$$|\chi_{fl}(f)|^2 = \frac{1}{A^2}\iint\frac{S_{12}(f)}{S_v(f)}\,\mathrm{d}A_1\,\mathrm{d}A_2 \tag{5.14}$$

Here, $\chi_{fl}(f)$ is the fluid-dynamic admittance. Experimental data on the absolute value of $\chi_{fl}(f)$, obtained for various plates and prisms in turbulent flow with $\mathrm{Tu} = 10.5\%$ and $L_{vx}/\sqrt{A} = 0.6$ and larger, are presented in Figure 5.4 together with theoretical predictions for lattice structures. According to these results, the simplified approach described in the foregoing works surprisingly well for bluff bodies of relatively small aspect ratios L/d (for the data in Figure 5.4, the ratio of length to diameter, L/d, was equal to or smaller than 3). There are two possible reasons for this success: First, a significant part of F' for such bodies is caused by pressure fluctuations on the front face rather than in the wake; and, second, the neglected inertial or 'accelerative' and turbulence-distortion effects partially compensate each other.

In accordance with Equation 5.13, the spectrum of the drag fluctuation, $S_F(f)$, can be constructed by multiplying each ordinate value of the velocity spectrum $S_v(f)$, point by point, with $4(\bar{F}/\bar{v})^2$ times the corresponding value of $|\chi_{fl}(f)|^2$. The root-mean-square of the drag fluctuation is then obtained from the area under the $S_F(f)$ curve, as

$$\sqrt{\overline{F'^2}} = C_{D\,\mathrm{rms}}\,A\,\frac{\rho\bar{v}^2}{2} = \left[\int_o^\infty S_F(f)\mathrm{d}f\right]^{1/2} \tag{5.15}$$

Examples of this procedure are presented in Figure 5.5 for two cases with identical flow conditions (maximum turbulent energy near $f^* \simeq \bar{v}/L_{vx}$, Equation 5.3) but different body sizes $\sqrt{A}$. The range of correlated flow fluctuations has an upper limit of about $f_{\mathrm{cor}} \simeq \bar{v}/\sqrt{A}$ according to Figure 5.4. It follows that f_{cor}/f^* is approximately equal to $L_{xv}/\sqrt{A}$, and that the admittance curves in Figure 5.5 shift

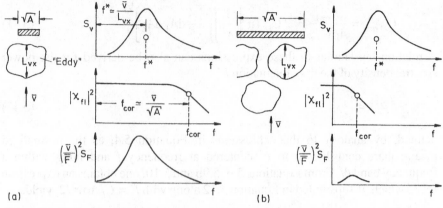

Figure 5.5. Evaluation of the drag spectrum $S_F(f)$ according to Equation 5.14 for two bluff bodies placed normal in the same turbulent flow: (a) $L_{vx}/\sqrt{A} > 1$; (b) $L_{vx}/\sqrt{A} < 1$.

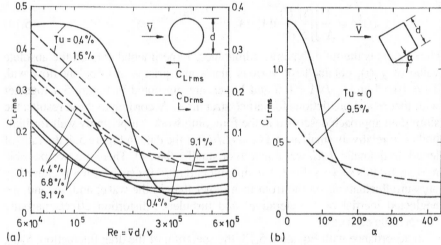

Figure 5.6. Root-mean-square lift and drag coefficients for stationary circular and square cylinders as functions of the background-turbulence level Tu. (a) $L_{vx}/d \leq 0.75$ (after Cheung & Melbourne, 1983); (b) $L_{vx}/d = 1.33$, Re $= 10^5$ (after Vickery, 1966).

in proportion to the difference in $\sqrt{A}$. As a result, the turbulence-induced excitation almost disappears for $L_{vx}/\sqrt{A} < 1$ (Figure 5.5b).

$S_F(f)$ and C_{Drms} referred to in this section do not contain contributions from vortex shedding in the wake. Since these are caused by an instability-induced phenomenon (IIE), they are treated in Chapter 6; suffice it to outline here some of their features in distinction to the force fluctuations due to turbulence buffeting (EIE):

– For bluff bodies in cross-flow, the buffeting forces act primarily in the direction of flow, whereas vortex-induced forces act mainly normal to the flow, corresponding to fluctuations of lift (Figure 5.6a).

– Since background turbulence disturbs vortex shedding in general, it usually reduces the vortex-related lift and drag fluctuations. For a square cylinder at zero angle of incidence, $\alpha = 0$, the root-mean-square lift coefficient C_{Lrms} may drop by 50% due to turbulence (Figure 5.6b). According to Vickery (1966), this change is accompanied by a decrease in the length of spanwise pressure correlation from $5.6d$ to $3.3d$ and a decrease in mean-drag coefficient from $C_D = 1.32$ to 0.68.

– A similar reduction in C_{Lrms} due to small-scale turbulence has been observed in the critical and the high subcritical regime of circular cylinders (Figure 5.6a). As shown by So & Savkar (1981), this drastic reduction is associated with a shift in the transition from laminar to turbulent separation towards a lower (critical) Reynolds number. Only in the supercritical flow regime (Figure 6.12), where organized vortex shedding disappears, do the fluctuating forces grow with increased turbulence intensity (Figure 5.6a).

– In contrast to extraneously induced excitation (EIE), the force fluctuations due to vortex shedding (IIE) are highly dependent on the response of either a body or fluid oscillator (Figure 8.12 and Naudascher, 1991, Figure 2.39). Data like those presented in Figure 5.6, therefore, cannot be applied to problems of oscillating cylinders or cylinders in an oscillating fluid.

To be sure, the division of fluctuating forces into front-face (EIE) and wake-related (IIE) contributions can only be a gross simplification of what should actually be treated as a mixed EIE/IIE case. Moreover, the above observations pertain to bluff cylindrical bodies and cannot be generalized. If turbulence, or gusts, act on a streamlined body or a flat plate aligned with the mean-flow direction, for example, EIE induces lift rather than drag fluctuations (Sears, 1941; Rockwell & Naudascher, 1990).

In the foregoing predictions of turbulence buffeting, the body was considered to be a discrete-mass type oscillator. If one deals with a distributed-mass system such as a flexible cylinder of length L, $|\chi_{fl}(f)|$ in Equation 5.13 must be replaced by $|J_N(f)| / \int_o^L \tilde{y}_N(z)dz$, where $J_N(f)$ is the joint acceptance and $\tilde{y}_N(z)$ is the mode shape along the axis z of the body. In $J_N(f)$, the correlation between body deflections y is also taken into account as follows:

$$|J_N(f)|^2 = \frac{1}{A^2} \iint \left[\tilde{y}_N(z_1)\tilde{y}_N(z_2) \frac{S_{12}(f)}{S_v(f)} \right] dA_1 \, dA_2 \tag{5.16}$$

An example is the simply-supported cylinder in an *axial* turbulent flow sketched in Figure 5.7. Here, the body axis coincides with the direction of flow, x, and one is interested in the excitation of mode shapes $\tilde{y}_N(x)$ (Figure 2.18). Clearly, eddies with scales of the order of L are best suited to excite mode $N = 1$ as they bring about force increases and decreases well correlated over the entire length L of the body. With regard to mode $N = 2$, such large eddies are completely ineffective.

Evidently, this second mode of vibration is excited more effectively by eddies of an intermediate scale for which the probability of opposite-force actions along the two half spans becomes a maximum. (This simplified argument, of course, does not account for the circumferential correlation of the eddy-induced forces.)

In Figure 5.8, the joint acceptance for this case is normalized by a constant K representing the circumferential correlation of the turbulence at low dimensionless frequencies fL/V_c (V_c = convection velocity). In analogy to Equation 5.3, V_c/f can be interpreted as approximately equal to the correlation length or scale L_c of those turbulence eddies which generate velocity and pressure fluctuations of frequency f (Figure 5.7a). The large-scale eddies of the total turbulence are thus represented on the left-hand side of the diagram in Figure 5.8 and the small-scale eddies on the right-hand side. As expected, $J_{N=2}$ tends to zero for small values of $fL/V_c \simeq L/L_c$ and it assumes a maximum for intermediate values. The small eddies ($L_c \ll L$) affect all modes alike, and the corresponding J_N^2 for large values

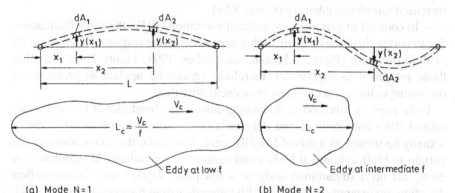

Figure 5.7. Schematic of turbulence excitation of a simply supported cylinder of length L in axial flow. (a) Mode $N = 1$: $J_N = 1$ for $L_c > L$; (b) Mode $N = 2$: $J_N = 0$ for $L_c > L$.

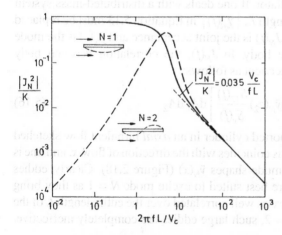

Figure 5.8. Joint acceptance for the first two mode shapes of a simply-supported flexible cylinder in axial turbulent flow (after Chen & Wambsganss, 1972).

of fL/V_c becomes proportional to L_c/L. More details are contained in the book by Blevins (1977, p. 148ff). The method for calculating the force spectrum $S_F(f)$ in the case considered is given by Chen & Wambsganss (1972).

5.2.3 Flow-induced vibration

If the fluid loading due to extraneously induced excitation is known, the next step is to determine the corresponding response of the structure as affected by the structural dynamics. For a structure that can be treated as a discrete-mass type, linear, lightly damped system with one degree of freedom in the x direction (Equation 2.8), the relation between the spectrum of the exciting force, $S_F(f)$, and the spectrum of the structural response, $S_x(f)$, is given by

$$S_x(f) = |\chi_m(f)|^2 S_F(f) \tag{5.17}$$

in which $\chi_m(f)$ is the mechanical admittance of the structure. In analogy to the fluid-dynamic admittance (Equation 5.13), it represents a transfer function between a known and an unknown quantity. For a system which is discrete, linear, and lightly damped, the absolute value of $\chi_m(f)$ follows from Equations 2.6 and 2.28 as

$$|\chi_m(f)| = [2\pi m f_n]^{-1} \left\{ \left[1 - \left(\frac{f}{f_n}\right)^2\right]^2 + \left[2\zeta\frac{f}{f_n}\right]^2 \right\}^{-\frac{1}{2}} \tag{5.18}$$

Here, m = structural mass including added mass (Section 3.2); f_n = natural frequency of the structure including added-mass effect (Equation 3.3b); and ζ = structural damping ratio including structural and fluid damping (Equations 2.10 and 5.21). The response spectrum is defined as

$$S_x(f) = \frac{\overline{x_f^2}}{df} \tag{5.19}$$

In analogy to the expressions in Equations 5.4 and 5.12, $\overline{x_f^2}$ is the mean-square value of structural deflections x' falling within a frequency band df centered around the frequency f. The root-mean-square of the structural deflection, or mean response amplitude, is obtained from an integration of the response spectrum as

$$x'_{rms} = \sqrt{\overline{x'^2}} = \left[\int_o^\infty S_x(f)df\right]^{\frac{1}{2}} \tag{5.20}$$

The probability that this value is exceeded follows from the amplitude-distribution function. If the latter is Gaussian, for example, amplitudes of $3x'_{rms}$ will be exceeded 0.28% of the time.

Figure 5.9 shows how a response spectrum $S_x(f)$ can be constructed point by point by multiplying each ordinate value of the spectrum of the exciting force $S_f(f)$ with the corresponding value of $|\chi_m(f)|^2$ in accordance with Equation 5.17. The

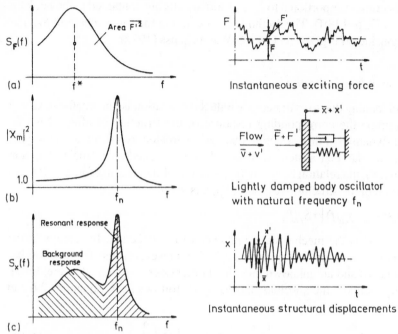

Figure 5.9. Evaluation of response spectrum $S_x(f)$ from drag spectrum $S_F(f)$ according to Equation 5.17.

response spectrum contains a predominant peak near the natural frequency f_n in spite of the fact that f_n does not coincide with the dominant frequency f^* of the excitation. The more random the excitation, of course, the weaker will be the broadband background response as compared to the relatively narrow-band resonant response around f_n (Figure 5.9c). Thus, a highly periodic structural response with large amplitudes may be produced by a completely random exciting force if the structure exposed to this random force is lightly damped. If one keeps this in mind, one is not misled to classify such typical extraneously induced excitation (EIE) either as self-excitation (MIE) or as the result of resonance between f_n and a frequency of flow instability (IIE). The so-called 'resonant response' in Figure 5.9c has nothing to do with coincidence of two frequencies!

For illustration, the next figure depicts four structural responses (Figures 5.10b to e) to the same instantaneous load due to turbulence (Figure 5.10a). The data were obtained from experiments with a 3.66 m square board in a gusty wind; however, similar results are possible also with trashracks, especially when they are partially plugged by trash (Section 9.1.3). Note that a change in natural frequency has little effect as compared to a change in damping, an important conclusion regarding means of attenuation (Section 5.3). According to Davenport

(1963), the fluid-dynamic part of the damping caused by drag may be described by the ratio

$$\zeta_{fl} = \frac{\overline{F}}{2\pi f_n m\bar{v}}, \qquad \overline{F} = C_D A \rho \overline{v^2}/2 = \text{mean drag} \tag{5.21}$$

For distributed-mass type structures such as flexible cylinders, the approach is more involved, requiring the use of a joint acceptance $J_N(f)$ plus a mechanical admittance in the derivation of the response spectrum for each vibration mode N (Blevins, 1977). If the structure is lightly damped so that the damping is predominantly fluid-dynamic, and if the spanwise scale of the turbulent eddies associated with the natural frequency, $L_N \simeq \bar{v}/f_N$, is small compared to the span of the structure L, then $J_N^2(f_N)$ becomes proportional to L_N/L [or $\bar{v}/(f_N L)$] independent of the mode shape. The situation is comparable to that described in Figure 5.8. Davenport (1962) has shown that the root-mean-square response amplitude x'_{rms} in this case is of the form

$$\frac{x'_{rms}}{\bar{x}} \simeq \text{constant} \left[\frac{\bar{v}}{f_N L} \frac{f_N m\bar{v}}{\overline{F}} \frac{f_N}{\overline{F'^2}} S_F(f_N) \right]^{1/2} \tag{5.22}$$

Here, $\bar{x}$ = mean deflection due to the mean drag $\overline{F}$; f_N = natural frequency of mode N (Section 2.6); $2\pi f_N m\bar{v}/\overline{F}$ = inverse of fluid-dynamic damping ratio (Equation 5.21); and $S_F(f_N)$ = ordinate of the exciting-force spectrum at frequency f_N. Although the application of this expression is restricted, it clarifies several features which help in a general understanding of the response to turbulence

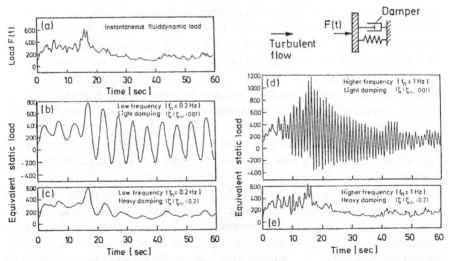

Figure 5.10. Response of structures with different dynamic characteristics to the *same* instantaneous fluid-dynamic load exerted by turbulent wind on a 3.66 m square board (after Davenport, 1963). (Response in terms of equivalent static load; scale relative only.)

excitation (Davenport, 1963). From the first factor on the right-hand side, it follows that the normalized response amplitude will be smaller for longer structures due to smaller spanwise correlation. The second factor indicates that the normalized response will be smaller for structures with high drag-to-weight ratio because of the relative increase in fluid-dynamic damping. And the third factor implies that the response can be reduced by an increase in natural frequency because of the decrease in $S_F(f_N)$ which this usually brings about.

5.3 CONTROL OF EXTRANEOUSLY INDUCED EXCITATION

For the control or attenuation of vibrations due to extraneously induced excitation, the first consideration, naturally, should be attenuating or even eliminating the excitation itself. For example, water-hammer-like pressure pulsations exerted on control gates during two-phase flows in conduits can be prevented by measures which let the entrained air escape before it reaches the gate (Naudascher, 1991, Figures 2.26 and 2.27). Or an oscillating flow produced by the excitation of a fluid oscillator in the system can be changed into a steady flow by preventing resonating oscillations of that oscillator (Figure 8.34). Or the pulsations of pressure in a penstock generated by the turbine runner buckets passing by the stationary guide vanes can be attenuated by changing the number of buckets in the runner (Naudascher, 1991, Figure 2.10).

When measures of eliminating or limiting the excitation of a structure are not feasible, one may control its vibrations by means of either an increase in structural damping or an increase (or possibly a decrease) in the natural frequency of the structure. The effect of increased damping is illustrated in Figure 5.10. Less obvious are the practical means by which greater structural damping can be achieved. The possibilities range from the use of materials of higher material damping (e.g., concrete as opposed to steel) or the use of laminated material or of vibration dampers (e.g., Blevins, 1977 p. 230; Lazan, 1968; Derby et al., 1969) to small changes in construction causing large damping effects. Especially important with respect to damping are the details of structural support. Examples are the loose support of heat-exchanger tubes in perforated plates, in which a substantial amount of energy is dissipated by impact and scraping (Blevins, 1975); or the addition of a damping element as shown in Figure 5.11a; or the applications of bracing cables with dampers, which are particularly useful as preliminary devices during construction periods (Figures 5.11b, c, d). Even though the devices in Figure 5.11 were originally developed for wind-engineering applications, their principle can also be applied to attenuate flow-induced vibrations in hydraulic structures. The difficulty is that the damping produced by such devices, while significant compared to air-induced forces, is usually dwarfed by water-induced forces.

An illustration of how the structural response can be attenuated through

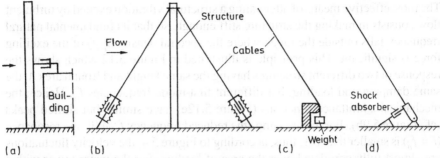

Figure 5.11. Attenuation by (a) Providing additional damping for a smoke stack near a building; (b, c) Bracing cables with (b) spring-damper elements and (c) friction dampers (after Ruscheweyh, 1978). (d) Employing shock absorber as a measure against vibration of the bracing cable.

increased stiffness is given in the next section. As a last resort, one may apply tuned mass dampers to reduce excessive vibration amplitudes of structures (Section 7.7.3) or filters to avoid excessive excitation of fluid oscillators (Section 8.5.2).

5.4 PRACTICAL EXAMPLES

A great number of practical examples on extraneously induced excitation involving structures are given in the IAHR Monograph on Hydrodynamic Forces (Naudascher, 1991). They include:
 1. Dams during earthquakes (Figure 2.9);
 2. Turbine runner vanes passing guide vanes (Figure 2.10a);
 3. Axial compressor with rotating stall (Figure 2.10b);
 4. Centrifugal compressors and pumps with rotating stall (Section 2.2.1.2);
 5. Power-house roof under spillway flow (Figure 2.11);
 6. Stilling-basin slab under hydraulic jump (Figures 2.12-2.14);
 7. Baffle wall in stilling basin (Figures 2.21-2.23);
 8. Baffle blocks in stilling basin (Figure 2.23);
 9. Sluice-gate wall (Figure 2.24);
 10. Tandem gates during emergency closure (Figure 2.25);
 11. Tainter gate with two-phase flow (Figures 2.26, 2.27);
 12. Pipe bend with two-phase flow (Figure 2.28);
 13. Navigation-lock miter gate with oscillating flow (Figures 2.57, 2.58).
Examples involving fluid oscillators are presented in Section 8.6. In the following, two more types of structures are to be covered; these are:
 14. Wind-excited structures (Figure 5.12); and
 15. Heat-exchanger tube bundles (Figures 5.13-5.15).

The most effective means of attenuating a structural vibration excited by turbulent flow consists in making the structure stiff enough, so that its fundamental natural frequency falls outside the range where the spectral density $S_F(f)$ of the exciting force is significant. This principle is illustrated in Figure 5.12 which shows the response of two different structures having the same height and frontal area A, the same damping and loading, but different first-mode frequencies f_N. Hence, the mechanical admittance functions (Figure 5.12e) have similar but shifted peaks (cf., Figure 5.9b). For the arc lamp, the reduced frequency $f\sqrt{A}/\bar{v}$ at resonance ($f = f_n$) is smaller than 0.1; thus, according to Figure 5.4, the velocity fluctuations are almost fully correlated over the area of loading. For the water tower with a reduced frequency at resonance of 1.0, in contrasts, these same fluctuations are poorly correlated and almost ineffective. The response spectrum of the water tower, consequently, shows no resonant peak, contrary to that of the arc lamp (Figure 5.12f). The resulting root-mean-square deflections, expressed in terms of the equivalent static wind pressure, are 2200 N/m² for the arc lamp and 950 N/m² (or less than half!) for the water tower. Further examples are described by Scruton (1981).

Turbulence buffeting may lead to exciting forces of subtantial magnitude in

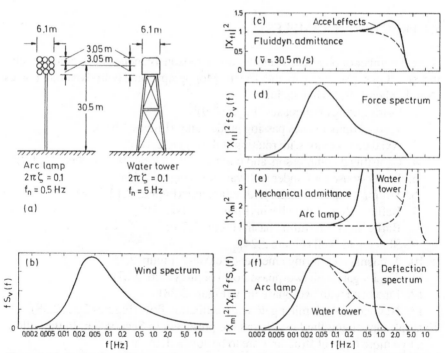

Figure 5.12. Evaluation of force and response spectra for two structures of equal height and area of loading but different stiffness (after Davenport, 1961).

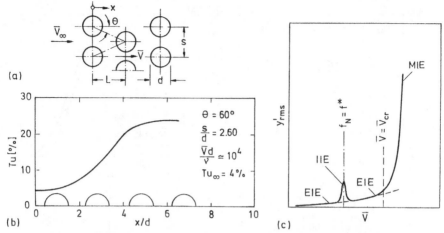

Figure 5.13. Tube bundle in cross-flow. (a) Definition; (b) Turbulence level as a function of depth into the bundle (after Sandifer & Bailey, 1984); (c) Typical response of a tube in the bundle to extraneously (EIE), instability-(IIE) and movement-induced excitation (MIE).

cases of groups of structures where one structure is placed in the wake of another. Examples of this situation include trashracks with large inclinations to the flow (Section 9.1.3) and heat-exchanger tube bundles in cross-flow. In the latter case, turbulence intensities were found to increase drastically from the first to the third or fourth row of tubes (Figure 5.13b), and levels in the range of 20% < Tu < 40% were observed in the interstitial flow and over 50% in the near wakes behind tubes in a closely packed triangular array (Zdravkovich & Namork, 1979).

Figure 5.13c shows a typical response of an individual cylinder in a tube bundle. The EIE component, which follows roughly the proportionality $y_{rms} \propto \bar{V}^2$ (Equations 5.15, 5.17, 5.20), is present at all values of the flow velocity $\bar{V}$. In general, the upstream tube rows are excited most. Amplitude levels are usually small except at resonance with an IIE mechanism ($f_N = f^*$) and past the critical velocity $\bar{V}_{cr}$ where MIE dominates the response (Section 7.6.2). Even though there are different conceptual models available on the IIE-resonance mechanism (Paidoussis, 1983; Weaver & Fitzpatrick, 1988), the important points to realize are that the velocity and pressure spectra within the bundle show an identifiable peak near a dominant frequency f^* (Figures 5.14a, b), and that a resonant response with maximum amplitude occurs if a natural frequency f_N of the system comes close to f^* (Figure 5.13c). From the many design guidelines available on the magnitude of f^*, Figures 5.14c, d depict those most frequently cited. Note that $\bar{V}$ in this figure denotes the gap velocity rather than the velocity of approach V_∞ (Figure 5.13a), $\bar{V} = V_\infty /(1 - d/s)$.

The response to turbulence buffeting of a cylinder in a tube bundle with cross-flow follows from the procedure outlined in Section 5.2. To make the

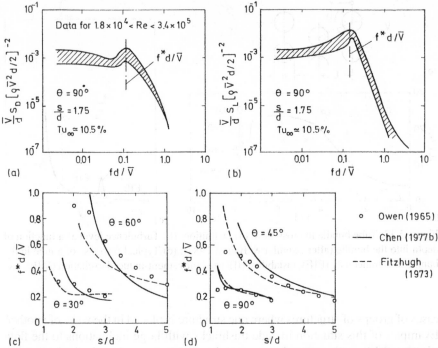

Figure 5.14. Dominant flow periodicity in stationary tube bundles with cross-flow (Figure 5.13a). (a, b) Normalized power spectra of drag and lift fluctuations on second-row cylinder (after Chen & Jendrzejczyk, 1987). (c, d) Strouhal number for dominant flow periodicity (from Paidoussis, 1980).

solution tractable for simple use, Blevins et al. (1981) arrive at the following expression for the root-mean-square amplitude at midspan:

$$y'_{rms}\left(\frac{L}{2}\right) = J_N \tilde{y}_N\left(\frac{L}{2}\right)\sqrt{S_F(f_N)}\,[64\,\pi^3 f_N^3 m^2 \zeta_N]^{-\frac{1}{2}} \qquad (5.23)$$

Here, J_N = joint acceptance which is equal to unity for a perfectly correlated flow field; $\tilde{y}_N(L/2)$ = eigenfunction at midspan; $S_F(f_N)$ = power spectral density of the exciting (lift) force per unit length at $f = f_N$; m = cylinder mass per unit length including added mass (Section 3.2); and f_N and ζ_N are the cylinder Nth mode frequency and damping ratio, respectively, both evaluated in stationary fluid rather than in vacuum. Typical values of $S_F(f)$ are presented in Figure 5.15. Depending on the intensity and scale of the incident turbulence, however, the buffeting response of the first-row cylinders may well be higher than that of the interior ones.

The presented approach yields the 'background' buffeting response rather than the resonance peak at $f_N = f^*$ (Figure 5.13c). Prediction of the latter is compli-

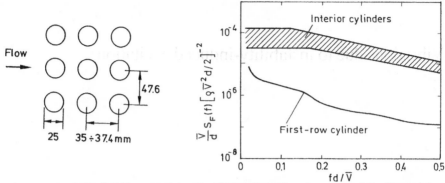

Figure 5.15. Normalized power spectral density $S_F(f)$ yielding, in conjunction with Equation 5.23, the background buffeting response of a cylinder in a tube bundle with cross-flow according to wind-tunnel measurements of Blevins et al. (1981) (from Paidoussis, 1983).

cated by the typical feature of all instability-induced excitations (IIE): the exciting force is strongly dependent on the vibration amplitude (Pettigrew, 1981; Paidoussis, 1983). This condition applies also to cases of acoustic resonance, $f_R = f^*$, and especially to cases of 'triple coincidence', $f_N = f_R = f^*$ (f_R = resonance frequency of acoustic standing wave, Section 4.3 and Figure 8.7). The simplest solution regarding these cases consists in avoiding resonance altogether by designing the system so that the reduced velocities $\overline{V}/(f_N d)$ and $\overline{V}/(f_R d)$ are some ±40 to ±50% away from the inverse of the Strouhal number, $(f^* d/\overline{V})^{-1}$. A means of suppressing Strouhal resonance, at least as far as tightly spaced tube bundles are concerned, is the use of upstream turbulators (Gorman, 1980).

Taylor et al. (1986) conducted a comprehensive study on the buffeting response of tube bundles in liquid and two-phase cross-flows. Using the same bounding spectra approach as Blevins et al. (1981), they found good collapse of data for water flow but not for air-water mixtures where void fraction plays an important role. Au-Yang (1986) has extended Blevins' approach to multi-span tubes where the variable tube mass from span to span becomes a factor. For the prediction of buffeting of clusters of tubes in axial flow, Paidoussis & Curling (1985) have developed an analytical model which agrees satisfactorily with experimental results.

CHAPTER 6

Vibrations due to instability-induced excitation

6.1 SOURCES OF INSTABILITY-INDUCED EXCITATION (IIE)

The basic source of instability-induced excitation (IIE) is an instability of flow which gives rise to flow fluctuations if a certain threshold value of flow velocity is exceeded, irrespective of any extraneously or movement-induced excitation (EIE or MIE). Depending on the control or amplification mechanism affecting the instability, these flow fluctuations and the exciting forces they generate may become so intense, well correlated, and concentrated near a dominant frequency that they can lead to large-amplitude vibrations of body or fluid oscillators. One can broadly distinguish between extraneous control and feedback control, and the latter can be fluid-dynamic, fluid-elastic, or fluid-resonant in nature (Figure 6.1). The most dangerous IIEs are those of the fluid-elastic or fluid-resonant variety. Their main feature is a resonating body or fluid oscillator which not only amplifies the exciting force within a given velocity range but also 'locks-in' its frequency to that of the oscillator. An important design consideration regarding IIE, therefore, is the assessment of possible resonances between the dominant frequencies (f_o) of IIE and the natural frequencies (f_N, f_R) of the body and fluid oscillators in the system. To be avoided under all circumstances are 'triple coincidences', i.e., $f_o = f_N = f_R$.

The variety of flow instabilities is large. In order to provide a minimum framework for an engineering design guide, this chapter addresses the following basic groups of instabilities:

1. Vortex shedding;
2. Impinging shear layers;
3. Interface instabilities;
4. Bistable-flow instabilities;
5. Swirling-flow instabilities.

The response of body oscillators to IIE (vibration of structures or structural parts) is covered in Section 3.3 (common methods of prediction) and Chapter 9 (examples involving various types of structures). The response of fluid oscillators to IIE (oscillation of compressibility-governed or free-surface systems) is presented in Section 8.3.

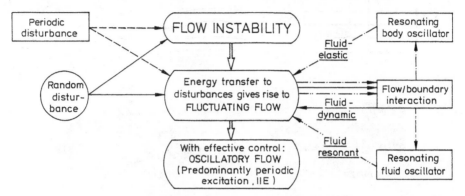

Figure 6.1. Schematic of control (or amplification) mechanisms by means of which flow instabilities may lead to oscillatory flow and instability-induced excitation (after Naudascher, 1967).

6.2 AMPLIFICATION MECHANISMS

For the sake of clarification, the basic mechanisms by which a flow instability may be controlled or amplified (Figure 6.1) are discussed here for a typical unstable flow: a planar jet with an efflux velocity V emerging from a nozzle of width B. In a simplified way, this jet can be said to be unstable because of shear layers commencing in the nozzle. Figure 6.2a shows an idealized model of a shear layer: a surface of abrupt velocity discontinuity. As was first point out by Prandtl (1904), a slight lateral perturbation of such a surface will produce a change in the velocity and pressure field which, according to the Bernoulli theorem, acts to modify the streamline pattern further in the direction of perturbation. As the disturbance is thus amplified, it becomes asymmetric on account of the mean-velocity distribution and develops ultimately into a region of concentrated vorticity. In a viscous fluid, this process of shear-layer roll-up is either damped or leads to the formation of discrete and gradually decaying vortices or eddies (Figure 6.2b). In summary, then, flow instability initiates a transfer of energy from the flow to the disturbance giving rise to flow fluctuations.

A feature which all shear flows have in common and which is essential with respect to possible excitation of structural vibrations is the principle of selective amplification (Naudascher, 1967). As exemplified for a jet and a free shear layer in Figure 6.3, the effectiveness of the initial energy transfer from the mean flow to flow disturbances depends not only on geometry and on flow parameters but also on the frequency of the disturbances, f_d. Above a critical Reynolds number, only disturbances within an intermediate frequency range are amplified, and the rate of amplification increases toward the middle of this range (dash-dotted line in Figure

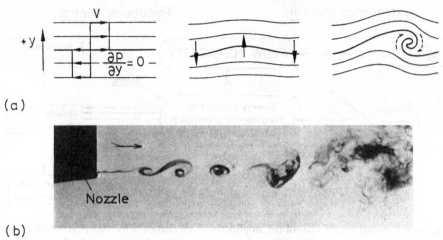

(a)

(b)

Figure 6.2. (a) Development of vorticity concentrations from a disturbed thin shear layer (after Prandtl, 1904). (b) Development of discrete vortices from the free shear layer of a jet (from a motion picture by Rouse & O'Loughlin).

6.3a). As a consequence, even the *random* disturbances present in any flow environment may be selectively amplified to yield almost periodic flow oscillations near a frequency for which the overall growth rate is a maximum. This tendency toward periodicity, however, is weak unless it is enhanced through additional control mechanisms in the system. One such mechanism involves fluid-dynamic feedback from a downstream edge of impingement (Figure 6.3b) as explained in the next section. For most practical applications, the Reynolds number is sufficiently large that one need not consider its value in predicting the most unstable frequency. Figure 6.3c depicts the amplification conditions for a free shear layer, i.e., a layer of velocity nonuniformity between uniform streams with velocities V_1 and V_2.

6.2.1 *Amplification involving extraneous control*

The most obvious mechanism by which the selective amplification inherent to shear-flow instabilities is augmented is extraneous control by *periodic* disturbances of finite amplitudes (Figure 6.1). In engineering practice, such disturbances may be sound or pressure pulsations emitted by engines, pumps, or turbines. Figure 6.4 shows a typical photograph of the periodic eddy formation generated by an antisymmetric transverse disturbance of amplitude 0.09 B (B = width of nozzle). Of course, generation and maintenance of such oscillatory flow are possible only if the Strouhal number of the disturbance, $\mathrm{Sh}_d = f_d B / V$, lies within a range of sufficient amplification (Figure 6.3a). The larger eddies downstream of eddies 1 to 4 are the result of vortex coalescence.

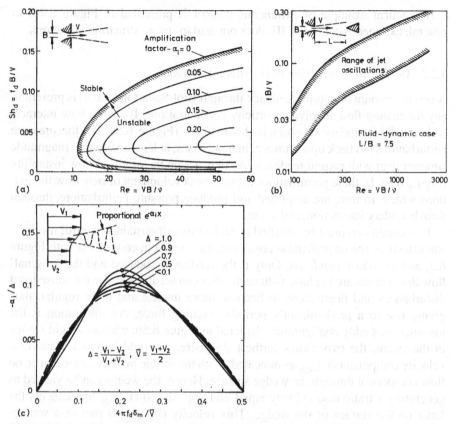

Figure 6.3. Amplification conditions for (a, b) a planar, submerged jet and (c) a planar free shear layer. (a) Effect of Reynolds number and Strouhal number of planar, antisymmetric disturbances on initial amplification (after Bajaj & Garg, 1977). (b) Range of jet oscillations for a jet of parabolic velocity distribution with a rigid 30° wedge placed in its centerplane (after Powell, 1961). (c) Effect of frequency of disturbance f_d on amplification factor for a free shear layer (after Monkewitz & Huerre, 1982). (δ_m = momentum thickness at shear-layer origin.)

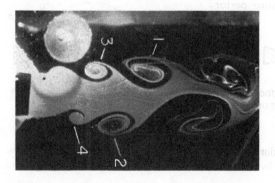

Figure 6.4. Extraneous excitation of planar jet by antisymmetric transverse disturbances at frequency close to natural breakdown frequency which, for the particular nozzle used, corresponds to $f_d B/V = 0.51$ at $VB/\nu = 1860$ (after Rockwell, 1972).

A practical example of extraneous control is presented in Figure 6.40. In general, extraneous control of IIE does not lead to severe structural vibrations.

6.2.2 *Amplification involving fluid-dynamic feedback*

A certain amount of natural feedback through Biot-Savart induction is present in any fluctuating flow involving vorticity concentrations. If such a flow interacts with a solid boundary or obstacle downstream (Figure 6.5a, II), the pressure perturbations fed back upstream are, in most cases, at least an order of magnitude stronger than with natural feedback. As they act on the origin of flow instability (Figure 6.5a, I), these perturbations trigger the development of new flow fluctuations which, in turn, are amplified and produce pressure perturbations through flow-boundary interaction, and so on.

Two conditions must be satisfied to make this self-sustained control mechanism effective: the *amplification condition*, mentioned in connection with Figure 6.3, and the *phase condition*. Only if the feedback-produced and the 'original' flow fluctuations are in phase with each other can feedback cause the subsequent disturbances and fluctuations to become more intense and more regular, thus giving rise to a predominantly periodic exciting force. An illustration is the instantaneous eddy configuration depicted in Figure 6.5b; without a rigid wedge in the steam, the two eddies furthest downstream would induce a transverse velocity component $(V_i)_{eddy}$ as indicated. With the wedge in place, there can be no flow component through the wedge surface. Hence, the wedge can be viewed as generating a transverse velocity equal and opposite to $(V_i)_{eddy}$ that cancels the latter on the surface of the wedge. This velocity $(V_i)_{wedge}$ is part of a wedge-induced flow field which gives rise to an induced transverse velocity component V_i near the nozzle as shown schematically in Figure 6.5b. The disturbance created by the latter, finally, initiates an eddy formation on the lower side of the jet; it thus happens to be exactly in phase with the disturbances associated with the 'original' eddy configuration.

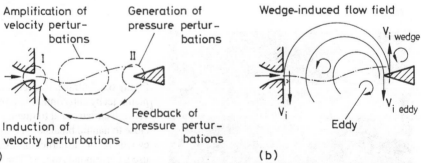

(a)   (b)

Figure 6.5. (a) Overview of events during fluid-dynamic feedback for impinging shear flow. (b) Schematic of flow field induced by interaction between disturbed flow and impingement edge.

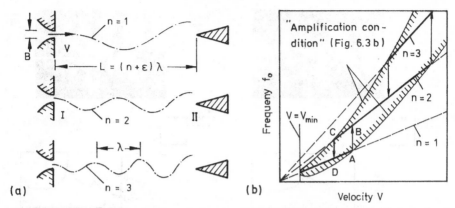

Figure 6.6. Phase condition for planar, submerged jets (fluid-dynamic case). (a) Schematic of first three modes of jet oscillation. (b) Corresponding relationship between dominant frequency of jet oscillation f_o and jet velocity V for constant L/B (after Powell, 1961).

In Figure 6.6a, the optimal phase condition is represented for the first three modes $n = 1, 2, 3$ of jet oscillation. Evidently, the number of wavelengths λ included between the region of greatest sensitivity to disturbances (near I) and the region at which new disturbances are generated by flow-boundary interaction (near II) must be equal to an integer n plus (or minus) a fraction $\varepsilon < 1$, provided the speed of sound c transmitting the disturbances from II to I is much larger than the mean convection velocity V_c at which a perturbation or an eddy is transported downstream $(V/c \equiv \text{Ma} \ll 1)$. The frequency f_o of the flow oscillation that is compatible with this optimal phase condition can therefore be expressed as

$$f_o = \frac{V_c}{\lambda} = (n + \varepsilon)\frac{V_c}{L}, \qquad n = 1, 2, 3, \dots$$

or in terms of a Strouhal number as

$$\text{Sh}_L \equiv \frac{f_o L}{V} = (n + \varepsilon)\frac{V_c}{V}, \qquad n = 1, 2, 3, \dots \qquad (6.1)$$

The variables ε and V_c/V depend on the particular flow conditions and on the mode n $(0 < \varepsilon < \frac{1}{2})$. In general, however, the variation in ε and V_c/V is small, at least in the absence of body and fluid oscillators, so that the phase criterion of Equation 6.1 can be represented approximately as a set of straight lines on a plot of f_o versus V (Figure 6.6b).

The typical characteristics of flow oscillations (and accompanying sound production) depicted in Figure 6.6b can now be interpreted in terms of compatibility with the amplification and phase conditions. For a given flow configuration, these conditions can be simultaneously satisfied only for velocities above the critical velocity V_{min}, and only with a dominant frequency of oscillation equal to

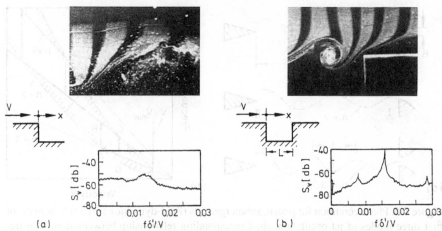

Figure 6.7. Comparison of flow pictures and velocity spectra without and with impingement edge at same streamwise location $x = 125\ \delta'$ ($L = 144\ \delta'$; $V\delta'/\nu = 243$; $\delta' =$ displacement thickness at separation) after Rockwell & Knisely (1979).

that of the first mode ($n = 1$). If V increases past point A, the flow is no longer sufficiently unstable to maintain first-mode oscillations. At the same time, the rate of amplification for the next potential mode has increased, so that a jump to this mode is likely to occur (point B in Figure 6.6b). For similar reasons, a jump back from C to D is obtained if V is reduced, i.e., as soon as the degree of instability has dropped to the smallest value required for maintaining second-mode oscillations. (Hysteretic jumps much like those in Figure 6.6b have also been observed with changes in L for constant velocity: Powell, 1961; Ziada & Rockwell, 1982.)

The pronounced effect of fluid-dynamic feedback on the degree of organization and the periodicity of flow fluctuations is illustrated in Figure 6.7 with the example of a mixing layer. In much the same way as in the case of jets, an impingement edge is seen to amplify those parts in the spectra which are compatible with the amplification and phase conditions relevant to that flow. Note that more than one peak, each corresponding to an oscillation mode, can be present at any given flow velocity V and impingement length L (Figure 6.7b). As V or L is changed, the amplitudes of the modes are altered in accordance with these conditions, and jumps occur from one to another value of predominant frequency somewhat as indicated in Figure 6.6b. (More detailed information concerning impinging shear layers is contained in Section 6.4.)

6.2.3 *Amplification involving fluid-elastic or fluid-resonant feedback*

As illustrated in Figure 6.1, feedback control may also occur with a body oscillator or a fluid oscillator affecting the events in the feedback loop shown in

Figure 6.5a. The corresponding flow oscillations are referred to as fluid-elastic and fluid-resonant, respectively. If, for example, the rigid wedge at the center plane of a planar jet is replaced by an elastic and hence dynamically active flow boundary such as a flexible blade, then the flow field induced by the interaction between disturbed flow and impingement edge (Figure 6.5b) may be controlled by the dynamic characteristics of the flexible blade (i.e., the body oscillator). Similarly, if the feedback loop contains a fluid oscillator such as in the free-surface system sketched in Figure 6.8c, the properties of that oscillator may control the interaction-induced flow field. One may view the situations in Figures 6.8b, c as cases in which two feedback mechanisms compete for control: the fluid-dynamic feedback discussed in the preceding section, and the fluid-elastic or fluid-resonant feedback characterized by an overriding influence from a body or fluid oscillator. In a simplified way, the switch of control can be interpreted in terms of optimal kinematic compatibility: whichever feedback gives rise to a motion near the location of maximum sensitivity (near I) to which the flow instability is most receptive, and whichever satisfies the optimal phase condition best, that feedback is likely to take over the control of energy transfer from the mean to the oscillating flow.

A possible mechanism for this takeover of control is schematically illustrated in Figure 6.8. The vibration of the flexible blade in Figure 6.8b or the standing-wave motion in Figure 6.8c both amplify the induced transverse velocities V_i near the jet nozzle and, hence, the generation of eddies (Figure 6.5b). In both cases, the oscillator movements can be viewed as either changing the mean-velocity profile of the jet or forcing the eddies to move along paths that are further removed from the jet centerline than in the case unaffected by body or fluid oscillator (Figure 6.8a). As a consequence, the eddy convection velocity V_c decreases, and the frequency f_o of eddy formation related to the optimal phase condition is reduced accordingly (Equation 6.1). By 'locking-in' f_o to the natural frequency of the

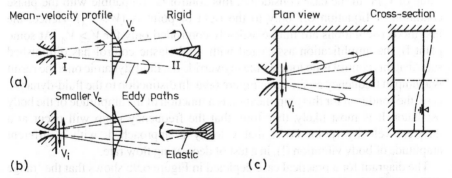

Figure 6.8. Schematic of the effects of a body oscillator (sketch b) and a fluid oscillator (sketch c) on the induced transverse velocities near the planar-jet nozzle and, possibly, on the eddy paths. (a) Fluid-dynamic case; (b) Fluid-elastic case; (c) Fluid-resonant case.

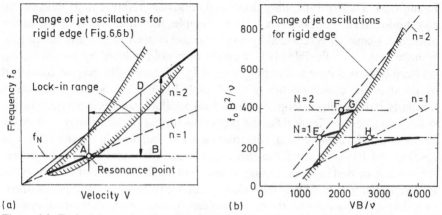

Figure 6.9. Typical frequency diagrams for instability-induced excitation, involving fluid-elastic feedback. (a) Schematic representation of dominant frequency f_o of flow oscillation as a function of velocity V. (b) Dominant frequency of flow oscillation for a jet-wedge system with flexible blade (Figure 6.8b, $L/B = 3.33$; $B = 0.51$ mm) after Brackenridge & Nyborg (1957).

controlling oscillator in this (or a comparable) way, fluid-elastic or fluid-resonant feedback control can prevail over large ranges of velocity.

Figure 6.9a shows a typical IIE frequency diagram that can be interpreted as follows: With fluid-dynamic control, the dominant frequency of flow oscillation, f_o, follows the dashed lines, which correspond to the phase condition given by Equation 6.1 with a more or less constant V_c/V. Fluid-elastic control, on the other hand, implies $f_o \simeq f_N$, where f_N = natural frequency of the body oscillator, and is therefore represented by the dash-dotted line in Figure 6.9a. For the first mode of jet oscillations, A marks the point of resonance for which f_N coincides with the fluid-dynamic value of f_o. Since fluid-elastic control was shown to reduce the value of V_c/V in the case considered, this control is compatible with the phase condition of Equation 6.1 only to the right of point A ($V > V_A$). Hence, the first-mode oscillations are fluid-elastically controlled only for $V > V_A$. At some point B, the amplification associated with fluid-elastic control has diminished such that this control is no longer able to override the fluid-dynamic one: the result is a jump in frequency as shown in Figure 6.9a. In distinction to the fluid-dynamic case, the condition for this amplification is a function of the amplitude of the body oscillator. It is most likely, therefore, that the frequency jump will occur at a different velocity (point D) if that velocity is approached, with a different amplitude of body vibration (!), in a test of decreasing flow rate.

The diagram for a practical case depicted in Figure 6.9b shows that the 'range of jet oscillations' for a flexible blade differs from that for a rigid edge with respect to the low-velocity limit as well. Corresponding to the two possible vibration modes $N = 1, 2$, peculiar to the flexible blade support investigated, there are three resonance points E, F, H in this case. Again, lock-in occurs only to the right of these points in agreement with the kinematic-compatibility considerations given above. In contrast to the case in Figure 6.9a, however, the lock-in ranges at E and F involve second-mode jet oscillations at frequencies above the amplification limit for a rigid edge. The fluid-elastic control in these modes is evidently stronger than the fluid-dynamic control involving first-mode oscillations, $n = 1$, at least up to point E.

Frequency diagrams similar to that in Figure 6.9b have been observed also with systems containing a fluid oscillator (Figure 6.8c) in place of the body oscillator (e.g., Nyborg et al., 1952). Essential with respect to the location and extent of the lock-in ranges in those cases are compatibility considerations much as those presented in connection with Figure 6.8.

6.2.4 Amplification involving mixed control

The various amplification mechanisms discussed so far can combine in a number of ways. Of particular interest are cases in which coupling exists between body oscillators and/or fluid oscillators, each of which can be excited to vibrate on its own. In the pipe of a reed organ (Figure 6.10a) containing both a body and fluid oscillator, for example, flow oscillations can presumably be controlled by fluid-elastic *and* fluid-resonant feedback. Actually, however, coupling between the two oscillators, the blade and the resonator pipe, produces control conditions which are quite different from those of the purely fluid-elastic or the purely fluid-resonant case. Remarkably, the coupled system is characterized by an eigenfre-

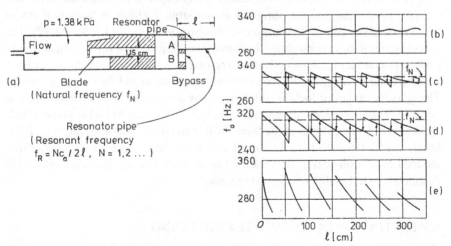

Figure 6.10. Dominant frequency f_o of flow oscillations in a reed pipe of variable length for various degrees of coupling between reed and resonator oscillations for constant flow velocity (after Vogel, 1920). (b) Bypass 'A' partially closed; (c) 'A' and 'B' open; (d) 'B' partially closed; (e) 'B' closed.

quency f_C that exceeds the smaller and falls short of the larger of the two eigenfrequencies f_N and f_R of blade and resonator pipe, respectively. For the special case of negligible damping, e.g.,

$$f_C^2 = \frac{1}{2}(f_N + f_R)^2 \pm \frac{1}{2}\left[(f_N^2 - f_R^2)^2 + 4K^2 f_N^2 f_R^2\right]^{1/2} \qquad (6.2)$$

where K is a coefficient ($K \le 1$) indicating the degree of coupling. Vogel (1920) found that the dominant frequency of flow oscillations f_o in the reed pipe follows the trend indicated by this equation very closely. The variation of f_o with pipe length l is characterized by subsequent jumps near the resonance points, $f_N = f_R$, as well as by hysteretic shifts of these jumps depending on whether the pipe length is increased or decreased. In Figures 6.10b-e, the degree of coupling (and thus the value of K) was gradually increased by first removing a restriction from opening A while the bypass at B was open and then gradually closing the bypass.

The increased susceptibility to instability-induced excitation due to multiple-body coupling is treated in Section 7.6.1.

6.2.5 *Summary*

Section 6.2 focusses attention on the basic mechanisms by which the energy transfer associated with flow instabilities may be amplified and controlled to yield, by way of an oscillating flow, instability-induced excitation (IIE) for body oscillators (Section 3.3) or fluid oscillators (Section 8.3). The phenomenology of these mechanisms has been greatly simplified in order to emphasize their main features.

Several types of amplification mechanisms often exist simultaneously and compete for control. Strong vibrations are possible if a body or fluid oscillator in the system gives rise to fluid-elastic or fluid-resonant control, respectively. The main feature of these controls is a lock-in of the frequency of flow oscillation f_o to a natural frequency f_N of the body or to a resonant frequency f_R of a fluid oscillator, accompanied by an intensification of flow organization and flow-induced forces. The velocity ranges of lock-in and their extent are greatly affected by amplification and phase conditions, the latter of which are intimately linked to a compatibility between the kinematics of the flow and that of the oscillator. At any rate, IIE is highly correlated with the vibration it excites, quite in contrast to extraneously induced excitation (EIE). Because of this distinct feature, the response to IIE cannot be treated as that of a forced vibration.

6.3 EXCITATION DUE TO VORTEX SHEDDING

6.3.1 *Characteristics of vortex-induced excitation*

Of all flow instabilities, the one associated with wakes past bluff cylindrical bodies has received by far the greatest research attention. Comprehensive reviews have been written by Morkovin (1964), Wille (1974), Parkinson (1974, 1989), Sarpkaya (1979), and Bearman (1984), to name only a few. The instability mechanism giving rise to vortex formation behind a bluff body is analogous to that described for jets and mixing layers in Section 6.1. If a bluff-body wake is present, the difference between the high velocity in the free stream and the low (or even negative) velocity in the base region promotes what has become known as a global instability. This global instability grows in place and leads to highly robust self-sustained oscillations of the near-wake. In contrast to the flow oscillations induced in jets and mixing layers, these oscillations are insensitive to small perturbations. Descriptions of the global instabilities for bluff-body flows are given by Monkewitz (1988), Oertel (1990), and Huerre & Monkewitz (1990). In the following, the physical consequences of this type of instability are emphasized in terms of the vortex-shedding process.

For any bluff cylinder in a cross-flow, near-wake oscillations accompanied by more or less well organized shedding of vortices commence if a certain Reynolds number is exceeded; the predominant mode is the alternate shedding shown in Figure 6.11. Mutual interaction between the two separating shear layers gives rise to the alternate formation of vortices which grow, fed by circulation from their connected shear layers, until they are strong enough to draw the opposing shear layer across the near wake. Eventually, further supply of circulation is cut off through entrainment of counter-rotating vorticity (Figure 6.11a); as a consequence, the vortices are shed and convected downstream. Directly connected with this vortex shedding is an initial flow around the body and a fishtail-like flapping of the early wake associated with the feedback control from the near-wake to the body. The strength of this feedback depends strongly on the shape of the afterbody. The major fluctuating forces on the cylinder act normal to the free-stream direction with a frequency equal to that of vortex shedding from one side of the cylinder, f_o; however, minor force fluctuations are also induced at twice the

(a)   (b)

Figure 6.11. (a) Model of vortex formation from a bluff cylinder showing entrainment flows in the wake (after Gerrard, 1966). (b) Visualization of the flow in the wake of a bluff body (after Bearman, 1984).

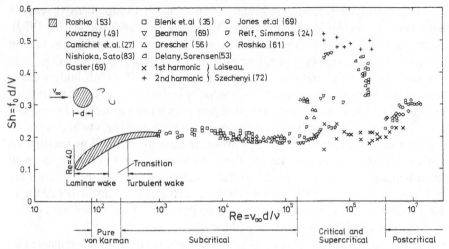

Figure 6.12. Strouhal number of vortex shedding from a stationary, smooth circular cylinder in low-turbulence cross-flow. (Note that lift spectra in the critical and supercritical range are broadband.)

frequency in the direction of flow, since *each* vortex produces a drop in base pressure during its formation. The most common excitation of vibrations is hence in the cross-flow direction, even though streamwise vibrations and vibrations at an incident angle to the flow are also possible, depending on the cross-sectional shape of the cylinder (Section 9.1.2).

Studies over the past 60 years have shown that vortex shedding is affected by a number of factors, in addition to cross-sectional shape and Reynolds number (Section 6.3.2). Information on the vortex-shedding frequency f_o as a function of these two variables only, such as that in Figure 6.12, must therefore be used with caution. Moreover, even perfectly two-dimensional bodies are unlikely to shed two-dimensional vortices when stationary. As a consequence, there is only limited spanwise correlation of the vortex-induced sectional load acting on the cylindrical body. Typical correlation lengths (Equation 5.7) within the range $10^4 < \text{Re} < 10^5$ are 3 and 6 cylinder diameters for a stationary circular cylinder (King, 1977), $5.5d$ for a stationary square prism (Vickery, 1966), and $10d$ for a stationary flat plate normal to the stream (Bearman & Trueman, 1972). Thus, if the cylinder is long compared with the correlation length, not all vortices cause forces in phase with each other, and the net exciting force is smaller. Drastic increases of correlation length (Figure 8.11) and exciting force (Figure 6.16b) occur with increasing amplitude of cylinder vibration.

If a cylinder is free to vibrate in any direction perpendicular to its axis (Figure 6.13a), the vortex shedding may be controlled in many ways including a change from alternate to symmetric shedding within a relatively small range of approach-

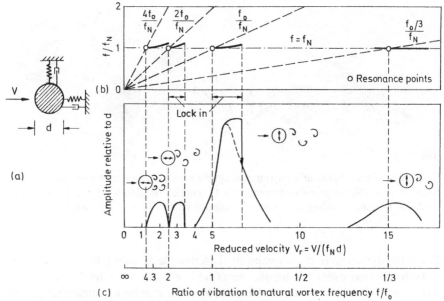

Figure 6.13. Schematic of the response of a lightly damped circular cylinder in cross-flow with the same natural frequency f_N for both streamwise and cross-flow vibrations.

flow velocities. As shown by Aguirre (Naudascher, 1987b), *streamwise vibrations* may set in for lightly damped cylinders when the ratio of vibration to natural vortex frequency, f/f_o, is equal to 4 (Figure 6.13c). Near this superharmonic resonance point, the cylinder starts shedding symmetric vortices such that four vortices of the same rotational sense coalesce to form one dominant vortex in the far wake. If f/f_o is less than 4, movement-induced excitation takes over the control (Section 7.3.5) until, with f/f_o approaching 2, the mode of vortex shedding changes from predominantly symmetric (Figure 6.14a) to clearly antisymmetric (Figure 6.14b). For these two modes, there are two excitation regions for streamwise vibrations, one below and one above the limit $f/f_o \simeq 2$. Since the frequency of vibration f is usually close to the natural frequency of the cylinder f_N, f/f_o can be expressed as

$$f/f_o \simeq f_N/f_o = (\text{Sh}\,V_r)^{-1} \tag{6.3}$$

where

$$V_r \equiv \frac{V}{f_N d} \quad \text{and} \quad \text{Sh} \equiv \frac{f_o d}{V} \tag{6.4}$$

are the reduced velocity and the Strouhal number of alternate vortex shedding from the stationary cylinder, respectively. By correlating the shear-layer kinema-

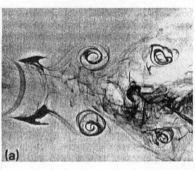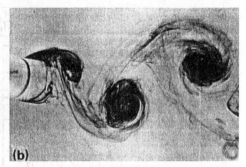

Figure 6.14. Vortex shedding from a circular cylinder vibrating in streamwise direction (after King, 1974). (a) Symmetric shedding in the first excitation range ($4 > f/f_o > 2$, Figure 6.13c). (b) Antisymmetric shedding in the second excitation range ($f/f_o < 2$, Figure 6.13c).

tics to the dynamics of the pressure field around the vibrating body, Naudascher (1987b) derived estimates for the onset and self-limitation tendency of these streamwise vibrations. For circular cylinders, the maximum attainable streamwise amplitude is about $0.2d$ (see also King et al., 1973), as compared to a maximum cross-flow amplitude of roughly $1.5d$ (Figure 6.16a).

The most important excitation regarding *cross-flow vibrations* is that involving resonance near $f/f_o = 1$. However, excitation can also involve superharmonic and subharmonic resonance near $f/f_o = 3$ and $1/3$, respectively. While superharmonic resonance for cross-flow vibration (Stansby, 1976) is seldom observed, subharmonic excitation (Durgin et al., 1980; Shirakashi et al., 1985) may occur in practice if a system is operated at the unusually high value of V_r corresponding to $f/f_o = 1/3$ (Figure 6.13c). The reason that resonance for the higher and lower modes occurs at frequency ratios of 3 and $1/3$, rather than 2 and $1/2$, is compatibility between vortex- and movement-induced forces on the cylinder as explained by Durgin et al. (1980) and Ericsson (1980). At subharmonic resonance, for example, this compatibility demands three vortex-shedding events for each cycle of body vibration. In all cases represented in Figure 6.13, the vortex street in the far wake has either the Strouhal frequency $f_o = $ Sh V/d or, within the lock-in ranges, a frequency deviating only little from it. If the cylinder possesses higher-mode frequencies f_N ($N = 2, 3, ...$) of course, there will be additional resonance points in Figure 6.13b, much as illustrated in Figure 3.29 (for an explanation of lock-in, the reader is referred to the discussion of Figure 6.9a).

Typical diagrams for the resonance response of cross-flow vibrations near $f/f_o = 1$ (or $V_r \simeq 1/$Sh, Equation 6.3) are shown in Figure 6.15. As seen from Figure 6.15a, this response is strongly affected by the 'after-body shape' of the cylindrical body. In the design of flexible vanes and blades, therefore, the proper shaping of the trailing edge is important. With a notched edge, for example, the maximum attainable amplitude $(y_o)_{max}$ is almost two orders of magnitude smaller

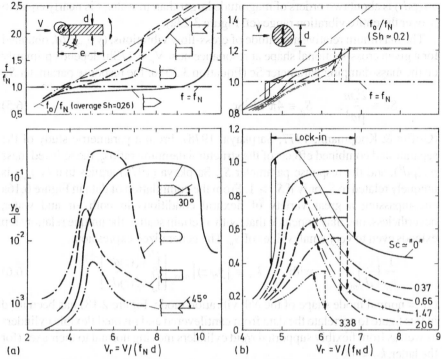

Figure 6.15. Response diagrams for cylinders in cross-flow. (a) Effect of trailing-edge geometry on torsional vibration of plates in water ($L/d = 30$; $e/d = 8$; $d = 6.35$ mm; Sc $\approx$ constant) after Toebes & Eagleson (1961); (b) Effect of mass-damping parameter Sc $\equiv 2m\zeta/\rho d^2$ on transverse vibration of a circular cylinder in water ($L/d = 1.8$; $m/\rho d^2 = 3.675$) after Meier-Windhorst (1939).

than for a well-rounded or pointed trailing edge. Similar reductions in the severity of vibrations are obtained with sharply beveled trailing edges (Figure 6.46). For the range of V_r represented in Figure 6.15a, the Reynolds number varied between $VL/\nu = 8\times10^4$ and 3×10^5. Information on the vortex-induced response of other cylindrical bodies such as square prisms, D-sections, and normal plates is available (e.g., Bearman & Obasaju, 1982; Bearman & Davies, 1977; see also Section 9.1).

A decisive factor in controlling vibration amplitude and lock-in range, in addition to geometry, is the damping. By increasing the damping ratio ζ in the supporting device of his cylindrical model, Meier-Windhorst (1939) found the reductions in maximum amplitude and width of the lock-in range depicted in Figure 6.15b. Even though his data include unusual end effects, because of the relatively small width of his test section ($L = 1.8d$), the variation in vibration frequency f between 80% and 120% of the natural frequency f_N (measured in still water) is rather typical for water flow. In air flow for which the reduced mass

$m/(\rho d^2)$ is about two orders of magnitude larger than in water, f is nearly equal to f_N over the entire vibration range (cf. Figure 3.23a).

The maximum relative amplitude of cross-flow vibrations, $(y_o)_{max}/d$, obtained for a given cross-sectional shape at resonance near $V_r = 1/\text{Sh}$, depends primarily on the mass-damping parameter Sc (Equation 3.36) or the response parameter S_g

$$\text{Sc} \equiv \frac{2\zeta m}{\rho d^2}; \qquad S_g \equiv 4\pi^2 \text{Sh}^2 \text{Sc} \tag{6.5}$$

(Griffin & Koopmann, 1977; Sarpkaya, 1978). From a parametric study of the separate and combined effects of the structural damping ratio ζ, the reduced mass $m/(\rho d^2)$, and the response parameter S_g, Sarpkaya (1978) argues that $(y_o)_{max}$ is uniquely related to S_g only if $S_g > 1$. From the compilation of data in Figure 6.16a encompassing a great variety of flexural conditions in both air and water, nevertheless, one may conclude that, with a certain scatter, the unique relationship extends even to very small values of S_g. The generalized expression

$$\frac{1}{k_N}\left(\frac{y_o}{d}\right)_{max} \qquad \text{with} \qquad k_N \equiv \left|\bar{y}_N(z)\right|_{max} \left[\frac{\int_o^L \bar{y}_N^4(z)dz}{\int_o^L \bar{y}_N^2(z)dz}\right]^{-1/2} \tag{6.6}$$

($\bar{y}_N$ = normal mode shape of the Nth vibration mode, Figure 2.18) has been used in this figure to correlate the data from cantilevered and pivoted flexible cylinders as well as from flexibly supported rigid cylinders moving normal to their axes (for the latter, $k_N = 1$).

As already stated, a distinct feature of instability-induced excitation is the dependence of the exciting fluid force on the vibration. In Figure 6.16b, one finds quantitative data on the lift component of this force acting on freely-vibrating

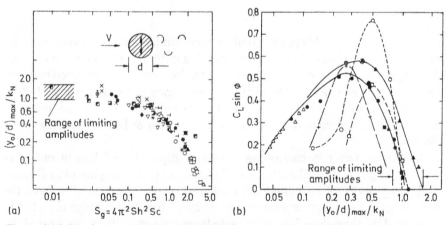

Figure 6.16. Maximum cross-flow amplitude for resonantly vibrating circular cylinder at subcritical Reynolds numbers and corresponding coefficient of the excitation-force component, $C_L \sin \phi$ (after Griffin & Koopman, 1977). The legend of the data points is given in: (a) Griffin & Ramberg (1982); (b) Griffin (1980).

circular cylinders during resonance or on circular cylinders forced to vibrate at corresponding amplitudes. The diagram shows that this force component reaches a maximum at amplitudes between $0.3d$ and $0.5d$ for all cases; beyond a limiting amplitude of roughly $1.5d$, the exciting force is zero. [One may view this increase in exciting force during vibration also as due to a superposition of instability- and movement-induced excitation mechanisms. In fact, as first suggested by Marris (1964), vibration of a cylinder with continuous surface curvature does bring about movement-induced angular displacements of the separation lines which are associated with transverse Magnus forces such that the system experiences negative damping (see also Ericsson, 1980).]

However, body vibrations do not affect the exciting force only. As reported by Griffin (1984), the mean drag on a circular cylinder, C_D, is amplified by as much as 250% if the cylinder vibrates freely in the transverse direction at amplitudes near the limiting value, and similar changes in C_D are possible with streamwise vibrations (Naudascher, 1991, Figure 2.98; Torum & Anand, 1985). The practical implication of this finding is that large-amplitude vibrations due to vortex shedding can cause both fatigue-related stresses and steady deflections simultaneously.

The excitation of gates due to vortex shedding is discussed in Section 9.2.5.

6.3.2 *Factors affecting vortex shedding*

An important design parameter is the vortex-shedding frequency f_o for the stationary body, commonly given in terms of a Strouhal number Sh (Equation 6.4). Unfortunately, Sh is usually presented as a function of only cross-sectional shape in spite of the fact that it depends on a great number of factors. One such factor is *Reynolds number*, $Re = Vd/v$, as seen from Figure 6.12. For cylinders of angular shape such as wedges, plates, and prisms, for which the lines of separation are uniquely determined by sharp edges, Sh is a constant for $Re \geq 10^4$, in contrast to cylinders with curved shapes. Based on the spacing d_s between the separation streamlines and the velocity v_s along them, Roshko (1954) developed a universal Strouhal number

$$Sh_s \equiv f_o d_s / v_s \qquad (6.7)$$

which is about 0.165 for a variety of cross-sectional shapes (Naudascher, 1991, Figure 2.36). Results of more recent experiments show that the concept of a universal wake Strouhal number is more general than was originally assumed (Griffin, 1989).

The vortex shedding is also strongly affected by *end effects* as illustrated in Figure 6.17. Both the junction between cylinder and wall (or ground surface) and the free end or tip are seen to introduce vortices with axes parallel to the flow (Figure 6.17a). These secondary vortices disturb the shedding of vortices parallel to the cylinder axis and weaken the excitation generated by the latter. Wootton

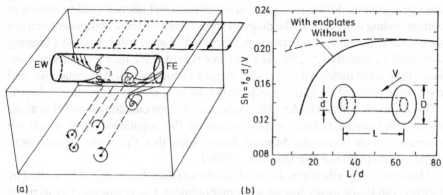

(a)   (b)

Figure 6.17. (a) Effects of an end wall (EW) and a free end (FE) on vortex shedding (after Etzold & Fiedler, 1976). (b) Effect of aspect ratio L/d and endplates ($D/d = 10$) on the Strouhal number of stationary circular cylinders in low-turbulence cross-flow; $10^3 < Re < 10^4$ (after Gowda, 1975).

(1969) observed that significant vortex-induced vibrations are unlikely to occur for circular stacks with heights smaller than 5 to 6 diameters. The three-dimensional effects shown in Figure 6.17a are minimized by thin end plates parallel to the flow. The effect of installing or omitting such end plates on the Strouhal number for different *aspect ratios L/d* is depicted in Figure 6.17b.

Sometimes even small geometric changes bring about large changes in flow conditions. Figure 6.18a shows the effect of the *rounding of corners* on the Strouhal number of square prisms at different angles of flow incidence, α. One of the consequences of such rounding is a shift of the region of vortex shedding formation away from the body. The effect of blockage or *confinement* on Sh is depicted in Figure 6.18b. All data reported in this section have been corrected for blockage effects unless stated otherwise.

An important factor, which is often disregarded, is free-stream *turbulence* (Figure 5.6). As was pointed out by Gerrard (1966), an increase in turbulence level $Tu = v'_{rms}/V$ (Equation 5.1) decreases f_o for a circular cylinder at subcritical Reynolds numbers, in contrast to Roshko's universal law (Equation 6.7). Similar decreases in Sh with increasing Tu have been observed for prismatic bodies, as evident from the work of Gartshore (1984) and from Figure 6.19a. This figure also illustrates the effect of *angle of incidence*. The Strouhal number is defined in terms of the dimension d' normal to the velocity vector V.

The effect of angle of incidence α on the Stouhal number for cylinders with rectangular cross-sections is given in Figures 6.20 and 9.11. Remarkable are the jumps and the double values occurring with side ratios smaller than roughly $e/d = 16$ for $\alpha = 0°$. One can distinguish four flow regimes in Figure 6.20a. Regime (1) is associated with non-reattaching separation streamlines in the range of approximately $0 < e/d < 3$ (or $0 < e/d < 2$ for turbulent flow). Within this

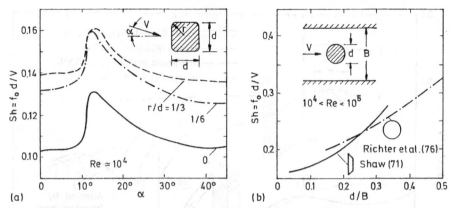

Figure 6.18. (a) Effect of rounded corners and angle of incidence on the Strouhal number in turbulent cross-flow (after Soliman, 1986). (b) Effect of flow confinement on the Strouhal number of stationary circular cylinders and flat plates placed normal to low-turbulence cross-flow (cf. Figure 9.21).

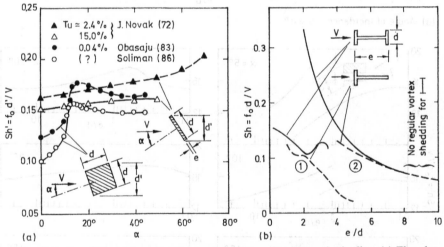

Figure 6.19. Strouhal number of stationary, two-dimensional, angular bodies. (a) Flat plate ($e/d = 0.1$) and square prism (Re $\equiv Vd/\nu = 2 \times 10^5$, 4.7×10^4, and 10^4, respectively, in the experiments of J. Novak (1972), Obasaju (1983), and Soliman (1986)). (b) H and T section prisms in low-turbulence cross-flow (after Nakamura & Nakashima, 1986); $Ve/\nu = 5 \times 10^4$ to 30×10^4.

range of leading-edge vortex shedding (LEVS, Figure 9.1), vortex formation is strongly controlled by the dynamics in the near-wake and the feedback from the vortex street downstream. In regime (2a), which extends roughly over the range $2 < e/d < 8$, the separated shear layers reattach intermittently. In an adjacent regime (2b), $8 < e/d < 16$, hardly any regular vortex shedding occurs for a

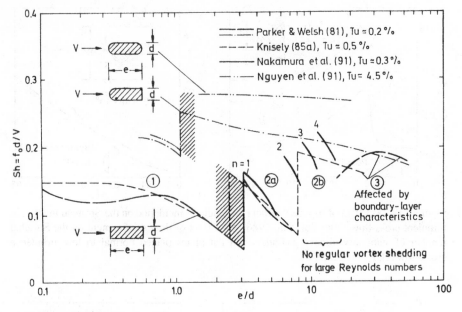

(a) Angle of incidence $\alpha = 0°$

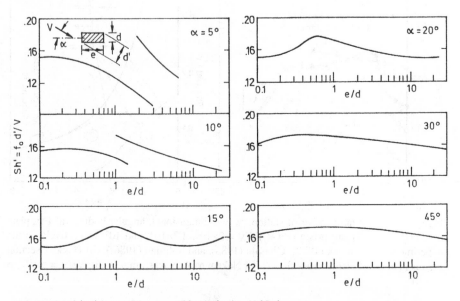

(b) Angle of incidence $5° < \alpha < 45°$ (Knisely, 1985a).

Figure 6.20. Strouhal number for stationary rectangular cylinders at various angles of incidence as a function of side ratio e/d. Data by Parker & Welsh (1981) for $17000 < Vd/v < 35000$; by Knisely (1985a) for $720 < Vd/v < 31000$; by Nakamura, Ohya & Tsuruta (1991) for $Vd/v = 1000$; and by Nguyen & Naudasher (1991) for $Vd/v = 1300$.

stationary cylinder; and in regime (3) beyond $e/d \simeq 16$, boundary layers develop past reattachment and combine to form a stable vortex street a short distance downstream of the trailing edge. The corresponding mode of trailing-edge vortex shedding (TEVS, Figure 9.1) is strongly affected by the specific boundary-layer characteristics. (The scaling of f_o with the thickness of the boundary layer at separation has been described by Monkewitz & Nguyen, 1987.) For cylinders with semi-circular leading edge, Parker & Welsh (1981) and Nguyen & Naudascher (1991) found a transition from regime (1) directly to regime (3) for values of e/d of about 1.2.

The vortex formation in regime (2a) is controlled by an IIE mode equivalent to that of impinging shear layers (Section 6.4) or impinging leading-edge vortices (ILEV, Figure 9.1). The same applies to regime (2) in Figure 6.19b. The Strouhal number in these regimes can be described by an expression analogous to that of Equation 6.1, i.e.,

$$\mathrm{Sh}_n \equiv \frac{f_o d}{V} = (n + \varepsilon)\,\frac{d}{e}\,\frac{V_c}{V}\,; \qquad \mathrm{Sh}_1 \simeq 0.6\,\frac{d}{e} \quad \text{for } n = 1 \tag{6.8}$$

where e is equivalent to the length between shear-layer origin and impingement edge. As shown in Figure 6.20a, two subsequent harmonics (e.g., $n = 2$ and 3 or 3 and 4) may coexist for certain side ratios e/d; actually, they occur one after another in random time intervals.

One point warrants special emphasis regarding regimes (2a) and (2b). Whereas for high Reynolds numbers only the fundamental ILEV mode ($n = 1$) was found in regime (2a) and none in regime (2b), all ILEV modes up to $n = 4$ were found in succesion over the full regime (2) for a low Reynolds number, $Vd/v = 1000$. As suggested by Nakamura, Ohya & Tsuruta (1991), the disappearance of regular vortex formation in regime (2b) means merely that the ILEV mechanism is too weak to materialize in high-Reynolds-number flow without additional control of the leading-edge vortex formation. Such control can be obtained either externally by means of a sound field or internally by leading-edge movements. In fact, Stokes & Welsh (1986) have shown that periodic vortices with Strouhal numbers close to those of Nakamura et al. can indeed be excited in regime (2b) by the use of sound. An even stronger augmentation of the periodicity and strength of the leading-edge vortices can thus be expected due to vibrating leading edges as the vibration frequency approaches one of the dominant ILEV frequencies given by Equation 6.8. In other words, although Nakamura et al. obtained their regime (2) data for flow at a low Reynolds number, these data are more relevant to practical cases of flow-induced vibration than the others because even minute vibrations of a cylinder cause enough control on the leading-edge vortex generation to revive the ILEV mechanism even at high Reynolds numbers (cf. Section 9.1.1).

Similar coexistent modes of vortex shedding as those in regime (2) have been observed in connection with *interference* between neighboring cylinders. The insert in Figure 6.21a depicts a typical bistable situation, in which the jet emerging

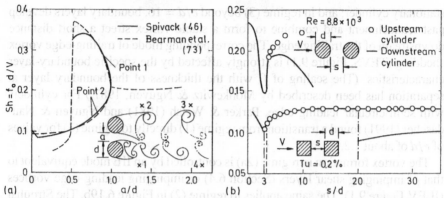

Figure 6.21. Interference effects on the Strouhal number for two stationary cylinders in different arrangements. (a) Two circular cylinders side by side (Quadflieg, 1977); (b) Tandem circular cylinders (Oka et al., 1972) and tandem prisms (Sakomoto et al., 1987).

between the two cylinders is deflected towards either the upper or the lower wake centerline. The vortex-shedding frequency in this case depends on the point of measurement, as high- and low-frequency shedding coexist in the range of roughly $0.5 < a/d < 1.0$. Quadflieg (1977) found that two cross-flow vibrations are possible for flexibly supported, lightly damped cylinders in the range $0.2 < a/d < 1.0$: in-phase vibration at low frequency [or high reduced velocity $V_r \equiv V/(f_n d)$] and vibration that is 180° out-of-phase at high frequency [or low reduced velocity]. The Coanda jet switches back and forth in the former case, frequently in step with the vibration, and it clings to one side in the latter. (Interference effects are also discussed in Sections 5.4, 7.3.4, and 7.6.)

For two *cylinders in tandem*, no distinct vortex shedding behind the upstream cylinder is detectable for a relative spacing s/d of less than about 3.8 for circular cylinders and 3.0 for square prisms (Figure 6.21b). As the spacings increase, the upstream shedding appears suddenly and its frequency approaches gradually the shedding frequency for a single cylinder. For square prisms, Sakomoto et al. (1987) found synchronized vortex shedding from the two prisms for $3 < s/d < 20$ (stable) and $20 < s/d < 27$ (unstable) and individual lower-frequency shedding from the downstream prism for $s/d > 20$ (unstable) and 27 (stable). Further information on tandem prisms is presented with Figures 9.9c-g. Reviews of the great variety of other interference effects have been given by Zdravkovich (1977, 1983, 1984a, 1985, 1987) and Chen (1987); effects of normal and parallel *splitter plates* have been discussed by Gerrard (1966) (also in Section 6.8.2d). Instability-induced excitation in cylinder arrays (tube bundles) is mentioned in Section 5.4.

Experiments with circular, rectangular, and D-section cylinders in *shear flow* have demonstrated that, (1) with a velocity gradient in the direction of the cylinder axis, vortices are being shed in cells of constant shedding frequency f_o, and (2) the cell-boundary regions are marked by two characteristic frequencies, one from each of the adjacent cells (Griffin, 1985). A typical example is shown in Figures 6.22a, b. The number of cells depends, mainly, upon aspect ratio L/D and end conditions. When forced to oscillate at amplitude $y_o = 0.06d$ and frequency $0.198 V_M/d$ (V_M = centerline velocity), the cylinder was found to shed vortices in four cells with much greater regularity than in the stationary case (Figure 6.22c), the main cell being locked to the vibration frequency. Similar evidence proves that cylinders can be excited to vibrate at large amplitudes in the presence of moderate shear levels if the mass damping is sufficiently small and the critical reduced velocity is exceeded [$V_M/(f_N d) > (0.198)^{-1} \simeq 5$ in the case considered].

A shedding of vortices in cells has also been observed with slender cones of small *taper* (Gaster, 1969; Vickery & Clark, 1972; Walker & King, 1988). Figure

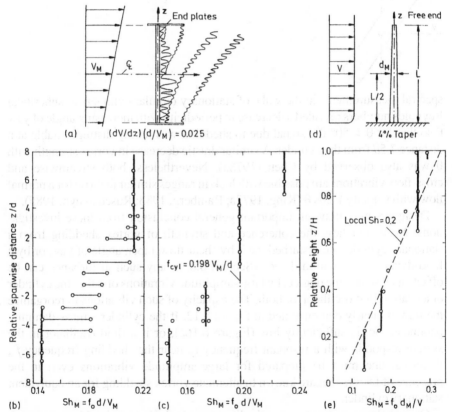

Figure 6.22. (a) Schematic of vortex shedding from stationary circular cylinder in shear flow (Stansby, 1976; Griffin, 1985). (b) Corresponding local Strouhal number Sh_M ($L/d = 16$; $V_M d/\nu \simeq 4000$). (c) Local Strouhal number for same cylinder during forced vibration at frequency $0.198 V_M/d$ and amplitude $0.06d$. (d, e) Local Strouhal number Sh_M for stationary slender cone ($L/d_M = 19.7$, $V d_M/\nu \simeq 2.4 \times 10^4$) after Vickery & Clark (1972).

6.22e depicts a typical variation of local Strouhal number Sh_M for a uniform flow with a very low turbulence level. Double-peaked spectra were measured at a number of positions in this case. In turbulent shear flow, the variation of Sh_M is broadly consistent with a constant value of local Strouhal number near 0.18, although some. variation was found near the tip. Significant, with respect to vortex-induced vibration, is Vickery and Clark's finding that the maximum response in a given mode will occur if the shedding frequency, from the region at which d^4 times the modal deflection is a maximum, equals the natural frequency of that mode.

For *yawed flow*, Surry & Surry (1967) found the Strouhal number to remain approximately constant if based on the normal velocity component, $V \cos \theta$. Their

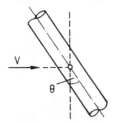

spectral measurements in the wake of stationary circular cylinders at subcritical Reynolds numbers revealed a decrease in periodicity with increasing angle of yaw θ so that for $\theta > 50°$ the signal due to shedding was hardly distinguishable at a distance $7.5d$ from the cylinder. A similar drastic decrease in vortex strength with θ was also observed by Chen (1978a). Nevertheless, both streamwise and cross-flow vibrations are possible with lock-in ranges similar to those for a normal flow with velocity $V \cos \theta$ (King, 1977a; Ramberg, 1983; Ruscheweyh, 1985).

Designers may draw an important general conclusion from these investigations. No matter how the coherence and strength of vortex shedding from a stationary cylinder are disturbed, be it by shear flow, taper, angle of yaw, or by a fluid-dynamic attenuation device (Section 6.8.2), any such disturbance can be offset, at least to some degree, by large-amplitude vibrations of either the cylinder or a nearby fluid oscillator, or both. The capacity of such vibrations to reorganize the flow is clearly demonstrated in Figure 8.12. If the cylinder's mass-damping parameter, Sc, is sufficiently low (Figure 6.16a), or if a fluid oscillator in the system responds with a resonant frequency f_R near the shedding frequency f_o, therefore, one must be prepared for large-amplitude vibrations even in the presence of factors that are known to attenuate vortex shedding for an equivalent stationary cylinder.

Overwhelmingly extensive as the literature on the subject is, this monograph can only attempt to sensitize the reader to the great variety of influencing factors. For example, nothing has yet been mentioned on the effect of surface roughness (Achenbach et al., 1981; Sarpkaya, 1982; Okajima, 1977), or cavitation (Young & Holl, 1966; Arndt, 1981), or two-phase composition of the fluid (Hara, 1982,

1984), to name just two of the remaining factors. An exhaustive coverage of such effects is obviously impossible.

6.4 EXCITATION DUE TO IMPINGING SHEAR LAYERS

In the event that a shear layer impinges on an obstacle, or edge, a process of feedback control (Figure 6.5) may alter the unstable flow substantially, giving rise to a more or less periodically oscillating flow. The significance of the various types of feedback control (Figure 6.1) and the conditions under which they may lead to structural vibrations (Figure 6.8b) or oscillations in a fluid (Figure 6.8c) are discussed in Section 6.2. The following is a brief review of the characteristics of this excitation for some configurations of interest to the hydraulic engineer.

Figure 6.23 depicts impingement configurations for which strong flow oscillations may occur (Rockwell & Naudascher, 1978, 1979; Blake, 1986, Vol. 1, p. 134ff). For a given configuration and length L, the frequency of these oscillations, f_o, is a function of conditions at separation, which is specified by the velocity distribution $\bar{v}(y)$ and the Reynolds number Re (based on the boundary-layer thickness δ or momentum thickness δ_m at separation). The variation of Strouhal number, $\mathrm{Sh} \equiv f_o L/V$, with dimensionless length shows a remarkable similarity for the jet/edge, cavity, and mixing-layer/edge configurations (Figure 6.24). In all cases, there are well-defined frequency jumps in correspondence with the different modes, n, of flow oscillations as explained in conjunction with Figure 6.6. Qualitatively similar variations in frequency occur for other values of Re (Figure 6.26).

In most practical applications, the boundary layer at separation is turbulent. As

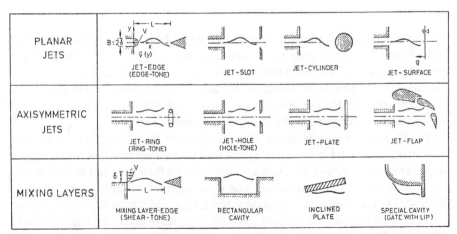

Figure 6.23. Basic configurations of shear-layer and impingement-edge conditions producing self-sustained flow oscillations (after Rockwell & Naudascher, 1979).

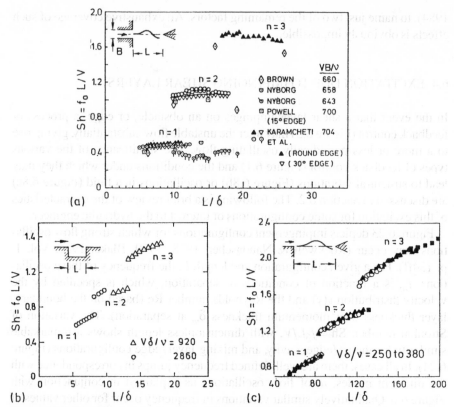

Figure 6.24. Strouhal number for (a) jet-edge oscillations (from Karamcheti et al., 1969); (b) Cavity oscillations (from Sarohia, 1977); and (c) Mixing-layer/edge oscillations (from Hussain & Zaman, 1978). The boundary layer at separation is laminar in all cases.

seen from Figure 6.25, the fluctuations of flow may be organized in these cases as well. When the cavity length L is increased for a given cavity width W, the pressure spectra are altered so that the predominant Strouhal number jumps from the value for the first mode to that for the second mode. According to that figure, one must consider the possibility of multiple-mode excitation when evaluating the structural loading excited by shear-layer impingement.

The frequency jumps obtained when the jet velocity is changed for a jet-edge system are shown in Figure 6.26a. The jump locations are different for increasing and decreasing values of Re. Although the basic jet flow becomes turbulent near Re = 50 in this system, jet oscillations (and edge tones) were observed up to Re $\simeq$ 4000. The transverse fluctuating force F'_{rms} exerted on the wedge by the oscillating jet is seen from Figure 6.26b first to increase and then to decrease with increasing Reynolds number. Similar variations in F'_{rms} have been obtained with changes in relative distance of the wedge, L/B (Powell, 1961; Naudascher, 1967).

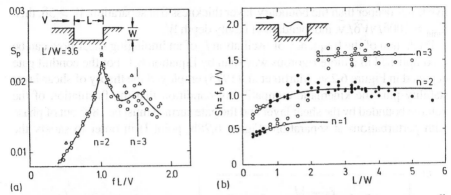

Figure 6.25. (a) Typical spectrum of pressure fluctuations measured on the downstream wall of a two-dimensional cavity for a turbulent boundary layer at separation ($\delta_m/W = 0.025$; $V\delta/\nu = 2.4\times10^5$) after Ethembabaoglu (1973, 1978). (b) Strouhal number at spectral peaks according to Figure 6.25a. Dark points represent predominant mode; solid lines represent theory of Rockwell (1977a).

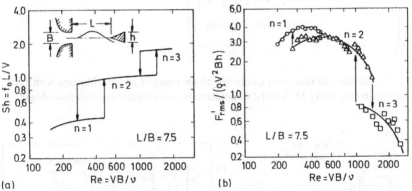

Figure 6.26. (a) Dominant Strouhal number of jet oscillation for a jet-edge system of constant $L/B = 7.5$. (b) Corresponding root-mean-square transverse force acting on rigid 30° wedge (after Powell & Unfried, 1964, and Powell, 1961).

These trends are typical for the instability behavior of free-shear flows. In fact, there is a direct link between the fluid-dynamically induced force F'_{rms} as a function of L and the change in disturbance-growth rate with streamwise distance (Miksad, 1972). Design of impinging-flow systems should, therefore, include consideration of the Reynolds number at separation, $V\delta/\nu$, which generally controls the distance required for the saturation amplitude of the disturbance to be reached (in Figure 6.26, $\delta = B/2$). As seen from Figure 6.27 for the example of an axisymmetric cavity with laminar separation, $V\delta/\nu$ also controls the minimum impingement distance, L_{min}, for onset of flow oscillation. If the cavity is more than five times deeper than the boundary-layer thickness δ at separation, $W > 5\delta$, then $L_{min} \simeq 300\delta/\sqrt{V\delta/}\ \nu$, independent of cavity depth W.

Prediction of the frequency of oscillation f_o of an impinging shear flow can be accomplished in a more rigorous way than by Equation 6.1. For the conduit gate depicted in Figure 6.28a, Martin et al. (1975) employed the theory of shear-layer stability plus the kinematic compatibility condition that the fluctuation of the volume bounded by the shear layer and the gate surface must be 180° out of phase with perturbations at separation (Figure 6.28b, point I) in order to satisfy the

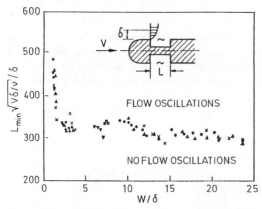

Figure 6.27. Variation of minimum cavity length for onset of flow oscillations with cavity depth (after Sarohia, 1977). Different data points represent various upstream nose shapes.

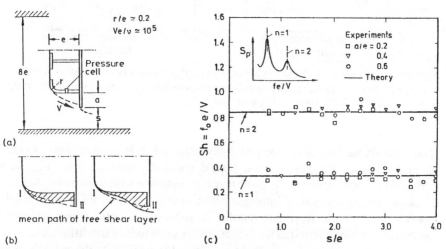

Figure 6.28. (a) Stationary conduit gate with extended lip. (b) Shear-layer deflections for the critical condition that its mean free path passes through the lip edge. (c) Strouhal number at peaks of pressure spectra measured on bottom of stationaty gate (after Martin et al., 1975).

condition of mass conservation. For the conditions of the tests reported in Figure 6.28c, they predicted Sh $\simeq 0.33$ and 0.84 for $n = 1$ and 2, respectively. The satisfactory agreement with the peaks in the measured pressure spectra (Figure 6.28c) proves that the possible flow oscillations in impinging-flow systems are not necessarily harmonically related to each other. Note that critical flow conditions (Figure 6.28b) can be avoided completely by designing the gate bottom with a sufficiently long lip extension, $(a + r)/e \geq 1$ or with an inclination of about $45°$ or more (Section 9.2.2).

Drastic amplification of the exciting force due to impinging shear layers is possible if the feedback is controlled fluid-elastically or fluid-resonantly rather than fluid-dynamically (Figure 6.1). Of special interest in this regard are vibratory systems with sharp edges such as, for example, cylinders with rectangular cross-section free to vibrate in a translatory or torsional mode (cf. discussion of Figures 6.20 and 9.10) and fexible skinplates with flat bottoms. Movements of the leading edge, in all these cases, give rise to a strongly enhanced vortex formation in the shear layer (sometimes referred to as movement-induced leading-edge vortices) which, in turn, leads to a substantial amplification of the vortex-induced pressure fluctuations along adjacent structural boundaries. Vibrations of rectangular cylindrical structures, gates, and skinplates, which can be excited in this way, are discussed in Sections 9.1 and 9.2. A few examples of vertical gate vibrations are given in the following.

For the systems presented in Figures 6.29 and 6.30, in contrast to that in Figure 6.28, the flow separates from a sharp leading edge, and the Strouhal number Sh is based on peaks in the spectra of load fluctuations acting on the entire gate bottom. In the spectra for $a/e = 0.35$, only the first-mode oscillations ($n = 1$) were clearly detectable (Figures 6.29b, d). Vertical gate vibrations were observed within a limited range of relative gate openings, s/e, in which the shear layer was neither permanently reattached to the trailing edge ($s/e > 1.5$ in Figure 6.29a) nor too far removed from it ($s/e < 3$ in Figure 6.29a). Maximum amplitudes, $(y_o)_{max}$, consistently occurred close to the resonance condition $V_r = 1/Sh$, and, in the case considered here, their magnitude depended strongly on the presence of a free surface in the submerged-flow region downstream of the gate. The implication of this dependency (fluid-resonant amplification) and other details are discussed in Section 9.2.2. With $a/e = 0.6$ (or larger), the shear layer is so close (or reattached) to the gate bottom that the excitation remains small even with a free surface downstream (Figure 6.29e).

Figure 6.30 shows how a flat-bottom gate vibrates in the case of impinging shear-layer instability. The typical lock-in behavior, comparable to that in Figures 3.23 and 6.15b, is depicted in Figure 6.30c; here, f = frequency associated with the peak in the load spectra. The influence of the Scruton number $Sc \equiv 4\zeta m_r$ (in which ζ = damping ratio in air; $m_r \equiv m/(\rho e^2)$ = reduced mass, and m = gate mass per unit width) is shown in Figure 6.30d. The significance of Sc is evident from the fact that cases with different values of ζ (3.85%; 1.32%; 0.9%) and m_r

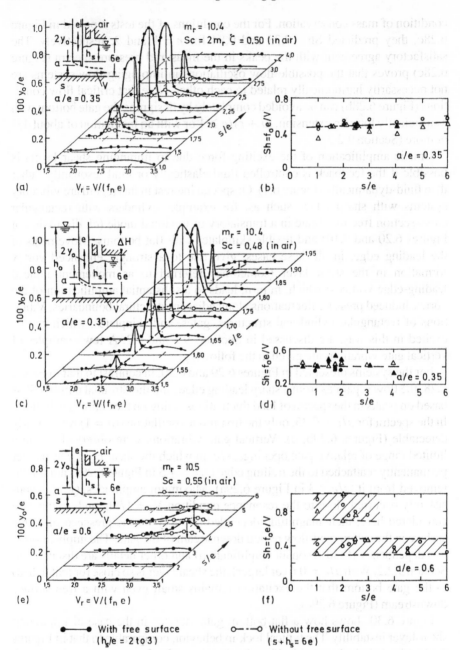

Figure 6.29. (a, c, e) Response diagrams for gates, elastically suspended in the vertical direction, with and without free tailwater surface ($f_n e^2/\nu = 51000$; $T_u \simeq 1\%$; $\Delta H/e \simeq 1.06\ V_r^2$). (b, d, f) Strouhal number for stationary gate. (a, b) Extended lip, $a/e = 0.35$; (c, d) Inclined bottom, $a/e = 0.35$; (e, f) Inclined bottom, $a/e = 0.6$ (after Nguyen, 1984).

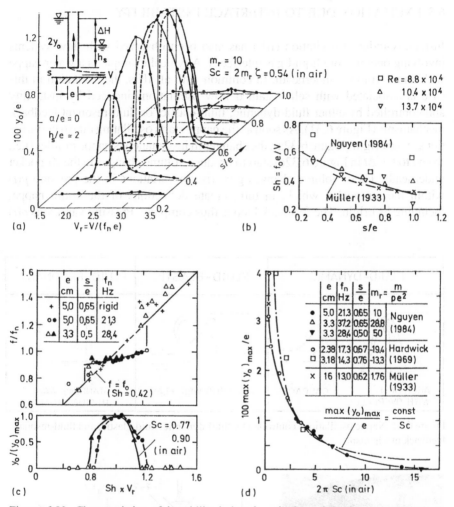

Figure 6.30. Characteristics of instability-induced excitation of flat-bottom gate, elastically suspended in the vertical direction ($f_n e^2/\nu = 53000$; $T_u \simeq 1\%$; $\Delta H/e \simeq 1.15\ V_r^2$). (a) Response diagrams; (b) Strouhal number for stationary gate; (c) Lock-in behavior; (d) Maximum attainable amplitude as a function of mass-damping parameter, Sc (after Nguyen et al., 1984, 1986a).

(10; 28.8; 50) but similar values of Sc (4.84; 4.78; 5.65) yield the same relative maximum amplitude $(y_o)_{max}/e$ (the extension of this finding to cases with $m_r < 10$ is not permissible according to Sarpkaya, cf. Zdravkovich, 1990). The inverse proportionality between the maximum attainable amplitude $(y_o)_{max}$ and Sc corresponds, approximately, to the prediction obtained from the energy relationship, Equation 3.31.

6.5 EXCITATION DUE TO INTERFACE INSTABILITY

Instability-induced excitation (IIE) has also been observed with flow systems involving one or more liquid-gas interfaces. A typical example is a water nappe with an air volume underneath. In common with all IIEs, the excitation in this case is associated with self-sustained flow oscillations induced by instability and controlled by either fluid-dynamic, fluid-elastic, or fluid-resonant feedback mechanisms (Figure 6.31). As soon as the nappe starts oscillating due to perturbations at the origin (Figure 6.32a), the changes in instantaneous nappe profiles (i.e., from t to $t + \Delta t$ in Figure 6.32c) start to produce volume changes in the air pocket underneath. These volume changes give rise to variations in the pressure $p(t)$ within the air pocket which, in turn, create deflections of the entire nappe, including its origin. The feedback loop is thus complete. Pressure variations $p(t)$

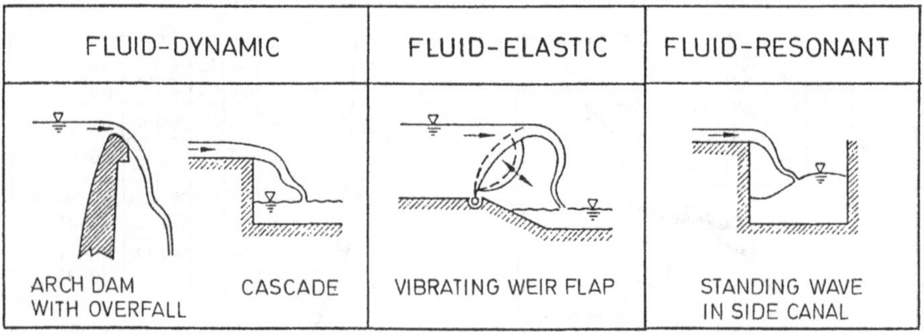

FLUID-DYNAMIC
FLUID-ELASTIC
FLUID-RESONANT
ARCH DAM WITH OVERFALL
CASCADE
VIBRATING WEIR FLAP
STANDING WAVE IN SIDE CANAL

Figure 6.31. Nappe oscillations controlled by fluid-dynamic, fluid-elastic, and fluid-resonant feedback mechanisms.

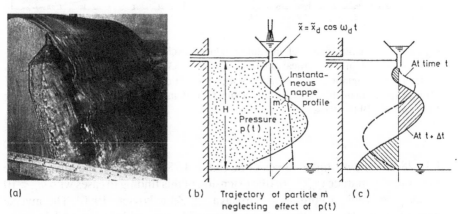

x̃ = x̃d cos ωd t
Instanta-neous nappe profile
m
H
Pressure p(t)
At time t
At t + Δt
(a)
(b) Trajectory of particle m neglecting effect of p(t)
(c)

Figure 6.32. (a) Oscillating water nappe falling from stationary crest (photo by Schwartz). (b, c) Schematic of nappe oscillation produced by disturbance $\tilde{x}_d \cos \omega_d t$ (after Müller, 1937).

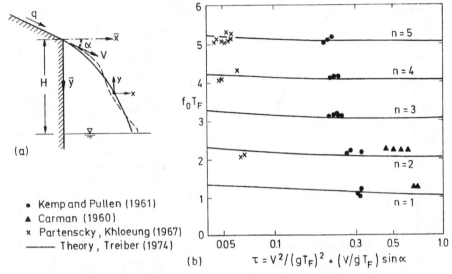

Figure 6.33. Frequency f_o of nappe oscillation as a function of initial velocity V and fall time T_F (Equation 6.10) after Treiber (1974).

are most intense if the air pocket is closed, but nappe oscillations have been observed with aerated air pockets as well (e.g., Falvey, 1980, p. 392).

There are a number of striking similarities between the oscillating nappe and the impinging shear-layer flows treated in Section 6.4, including the 'multiple-mode' behavior (Figure 6.33), the frequency jumps and hysteresis effects, and the existence of a critical phase relationship between the deflections of the two ends (I and II) of the undulating nappe and shear-layer configuration, respectively. For feedback to be effective, a number of n plus a fraction wavelengths must be included between these ends. In analogy to Equation 6.1, hence, the various possible modes of nappe oscillations can be approximated by the relation

$$f_o T_F = n + \varepsilon, \qquad n = 1, 2, 3, \ldots \tag{6.9}$$

where f_o = frequency of nappe oscillation; $\varepsilon < 1$; and T_F = fall time of a particle of the liquid jet:

$$T_F = \sqrt{(V \sin \alpha / g)^2 + 2H/g} - V \sin \alpha / g \tag{6.10}$$

For the fluid-dynamic case (Figure 6.33a), Treiber (1974) derived a more rigorous relation for f_o. As seen from the satisfactory agreement between his theory and the various experimental data in Figure 6.33b, the 'multiple-mode' behavior for any given value of the dimensionless fall-time parameter τ is well predicted. For $\tau \to 0$, ε tends toward $1/2$ and f_o toward $(n + 1/2)\sqrt{g/(2H)}$. Further analysis is required, however, to determine the 'preferred' modes of oscillation, indicated by

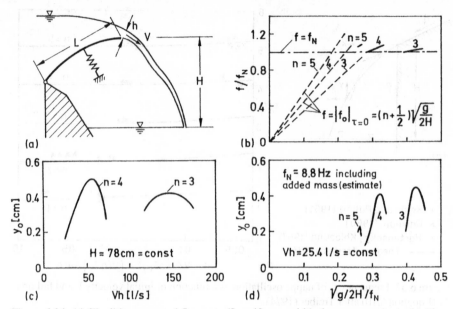

Figure 6.34. (a) Flexibly supported flap gate ($L = 40$ cm, width 6 m, no aeration). (b) Flap vibration frequency f; (c, d) Variation of flap-tip amplitude y_o, with: (c) discharge $q = Vh$ at constant H and (d) fall height H at constant Vh (after Müller, 1937).

the location of the data points. (The added-mass effect with flaps has been investigated by Gal, 1971.)

Some typical results for the fluid-elastic case of a flexibly supported flap with overflow are shown in Figure 6.34. The dependence of the flap-tip amplitude y_o on the nappe thickness h is shown in Figure 6.34c. As h increases for a given fall height H and a given number of wavelengths n, the sensitivity of the nappe to pressure changes underneath decreases, and so does its potential for the excitation of flap vibrations. If h decreases below a certain limit, on the other hand, the nappe oscillations become increasingly disturbed by air leakage through the nappe, by possible onset of nappe instabilities, and by breakup associated with surface-tension effects (Binnie, 1972). Whether a given overflow gate will experience nappe oscillations cannot be predicted with certainty. However, a number of devices are available today for attenuating nappe oscillations (Figures 6.68, 6.69). The nappe-induced noise has been investigated by Knisely (1989).

Interface instabilities also exist in flow configurations involving large cavities produced at high velocities in liquids by supercavitation or by ventilation (Figure 6.35a). The key to the excitation of cavity oscillations, according to Song (1962) and Woods (1966), lies in the way such cavities part with the gas inside. Instead of by steady entrainment into the bounding stream, the gas leaves the cavity intermittently by a fission process in which the last wave of the cavity surface is periodically pinched off at times when the cavity pressure p_c is too low to sustain a cavity length greater than the average l_* (Figure 6.35b). Hsu & Chen (1962) have modeled some features of the self-sustained oscillations using potential-flow theory. According to their solution in Figure 6.36a, two families of possible oscillation modes n occur for a given flow velocity V, one for large and one for small ratios of cavity length to instability wavelength, l_*/λ. Since the large values of l_*/λ are physically unreal, only the lower family of curves can occur in practice.

Another interesting result of Hsu's & Chen's analysis is the existence of a fixed

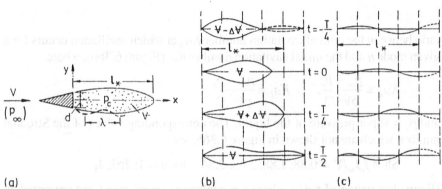

Figure 6.35. (a) Cavity past a body due to supercavitation or ventilation: steady profile just before pulsation. (b, c) First-state pulsating cavity profile ($n = 1$) and corresponding non-uniform travelling wave (after Song, 1962).

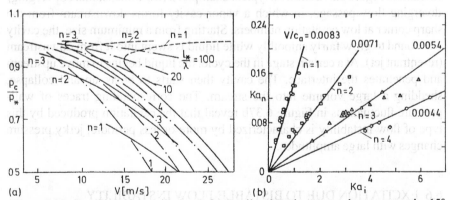

Figure 6.36. Conditions for oscillations of a ventilated cavity past a plane-symmetric, 15° wedge (Figure 6.35a) at 1.52 m depth under the free surface. (a) Instability wavelength λ as a function of geometric and flow parameters (after Hsu & Chen, 1962); (b) Critical caviation number Ka_n at which oscillation mode n sets in as a function of initial cavitation number Ka_i. (Lines: Hsu, 1975; data: Silberman & Song, 1959. Mach number V/c_a based on $c_a = 1520$ m/s.)

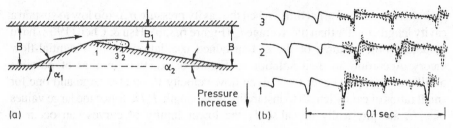

Figure 6.37. (a) Two dimensional converging/diverging flow passage with supercavitating flow. (b) Pressure history during the cavity growth/collapse cycle (after Furness & Hutton, 1975).

ratio between the critical cavitation number Ka_n at which oscillation occurs for a given mode n and the initial cavitation number Ka_i (Figure 6.36b), where

$$Ka_n \equiv \frac{p_\infty - \bar{p}_c}{\rho V^2/2}, \qquad Ka_i \equiv \frac{p_\infty - p_v}{\rho V^2/2},$$

and p_v = vapor pressure of the liquid. The corresponding values of the Strouhal number, which are not shown in Figure 6.36b, are:

$$\mathrm{Sh} \equiv f_o l_*/V = 0.88;\ 1.88;\ 2.68;\ 3.84 \quad \text{for } n = 1;\ 2;\ 3;\ 4.$$

(Equivalent values of $f_o d/c$, where c = analogous sound speed, are presented by Song, 1962.) Since V/f_o can be interpreted as the wavelength λ, these values of Sh correspond to the l_*/λ ratios within the lower family of modes in Figure 6.36a.

A frequent source of undesirable pressure pulsations in hydraulic systems is cavitation. Figure 6.37a shows a typical example: a two-dimensional converging/ diverging flow passage, in which a vapor cavity forms downstream from the sharp corner at low cavitation numbers. Starting from a minimum size, the cavity was found to grow fairly smoothly while liquid is swept into it from downstream (re-entrant jet). At a certain stage in the cycle, the liquid begins to move upstream and penetrates the interface. The cavity then starts to break up and collapse, shedding a large volume into the stream. The corresponding traces of wall-pressure fluctuations in Figure 6.37b reveal that the excitation produced by this type of flow instability is characterized by more or less periodic, jerky pressure changes with large amplitudes.

6.6 EXCITATION DUE TO BISTABLE-FLOW INSTABILITY

For certain types of separated flow, intermittent fluctuations and even oscillations can arise that are not traceable to classical fluid-dynamic instability. A possible means of identifying these oscillations is a test of 'bistability', but similar flow oscillations have also been observed with recirculation zones at backward facing steps (Eaton & Johnston, 1980, 1981), at leaf gates (Narayanan & Reynolds, 1968), and a variety of other configurations in hydraulic structures (Nikitina & Kulik, 1980). When a flow configuration lies between two configurations corresponding to two different stable states of flow as shown in Figure 6.38, it is likely to be unstable, or at least prone to intermittency. Common to all of the more regular oscillatory flows of the bistable category are, first, movements of the *whole* shear layer relative to the adjacent structural boundary (rather than undulations within that layer as in the case of impinging shear-layer instability); and, second, global variations of a volume V enclosed within a separation pocket (Figure 6.39). The variations in V are associated with an imbalance between the mass flow ρQ_E entrained by the separated shear layer and the mass flow ρQ_R

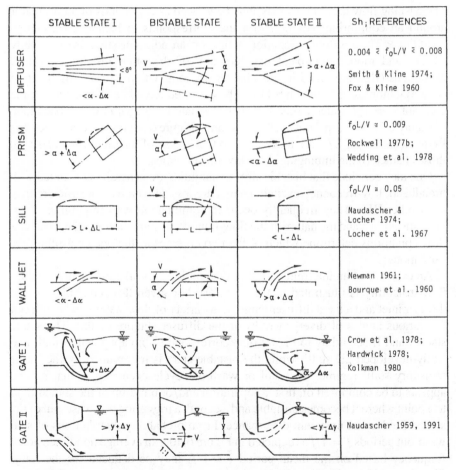

	STABLE STATE I	BISTABLE STATE	STABLE STATE II	Sh ; REFERENCES
DIFFUSER				$0.004 \gtrless f_0 L/V \gtrless 0.008$ Smith & Kline 1974; Fox & Kline 1960
PRISM				$f_0 L/V \approx 0.009$ Rockwell 1977b; Wedding et al. 1978
SILL				$f_0 L/V \approx 0.05$ Naudascher & Locher 1974; Locher et al. 1967
WALL JET				Newman 1961; Bourque et al. 1960
GATE I				Crow et al. 1978; Hardwick 1978; Kolkman 1980
GATE II				Naudascher 1959, 1991

Figure 6.38. Illustration of some bistable states of flow which may give rise to low-frequency oscillations.

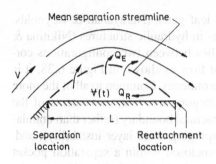

Figure 6.39. Schematic illustrating the rate of volume change $d\Psi/dt$ associated with flow-rate imbalance $(Q_R - Q_E)$.

returned to the pocket near impingement:

$$d\Psi/dt = Q_R - Q_E \tag{6.11}$$

The ensuing flow oscillations may thus be viewed as deflections of an oscillator around an equilibrium configuration that corresponds to balanced mass-flow conditions. The resulting excitation acting on an adjacent structure is usually weaker and more broad-band than are any of the other instability-induced excitations (IIE) presented in this chapter. In conjunction with the momentum equation, moreover, Equation 6.11 can be expected to yield a characteristic time constant $T_o \simeq 1/f_o$ much larger than that associated, e.g., with the feedback loop of Figure 6.5a. As seen from the last column in Figure 6.38, typical dimensionless frequencies Sh $\equiv f_o L/V$ are indeed one to two orders of magnitude smaller than those characterizing impinging shear-layer instabilities.

One may conclude that bistable-flow instabilities are generally associated with broadband, low-frequency excitations that are less dangerous to a structure than most other IIEs. They frequently occur in combination with impinging shear-layer instability, and the nature of the flow-boundary interaction and the particular flow conditions determine which of the two emerges as the one controlling the excitation.

An example of an intermittent rather than periodic oscillation due to bistable-flow instability is illustrated in Figure 6.40a. For given flow conditions at the diffuser inlet and a fixed diffuser length L, a variety of flow states can be induced for various angles of divergence 2α of the diffuser (Figure 6.40b). The stable states I and II according to Figure 6.38 correspond to 'no appreciable stall' and 'fully-developed 2D stall', while the bistable state is represented by the 'large transitory stall' regime. The stall or wash-out cycle occurring in that regime appears to be composed of, first, a separation and/or build-up of the stalled fluid to a point where it becomes unstable and, second, a triggering of a wash-out of the stalled fluid by an upstream disturbance. From probability distributions of stall wash-out periods $T = 1/f$ as depicted in Figure 6.40c, an average nondimensional frequency based on the mean period $\bar{T} = 1/f_o$ may be deduced for the range $5° < 2\alpha < 30°$ as $f_o L/V_1 \simeq 0.006$ with a scatter reaching values of 0.004 and 0.008. An interesting result of this study by Smith & Kline (1974) relates to the

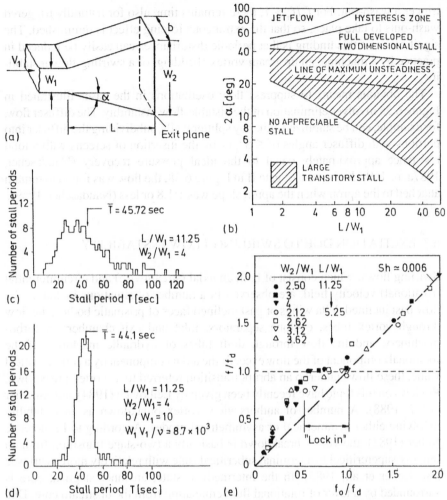

Figure 6.40. (a) Definition sketch of diffuser. (b) Regime chart (after Fox & Kline, 1960). (c, d) Probability distribution of stall wash-out period T: (c) without and (d) with extraneous control by pulsed inlet flow at $f_d = 1/45$ Hz (amplitude $\pm 0.06 V_1$); (e) Effect of pulsed-inlet-flow frequency f_d on frequency f of stall oscillations; f_o = undisturbed oscillation frequency (after Smith & Kline, 1974).

sensitivity of transitory diffuser stall to extraneous control by periodic pulsations of the inlet flow ($\pm 6\%$). As seen from a comparison of Figures 6.40c and d, such control may enhance the periodicity of the flow oscillations substantially. According to Figure 6.40e, the disturbance 'drives' the stall periods within a 'lock-in' range of roughly $0.5 < f_o/f_d < 1.0$, where f_d = disturbance frequency. Only within this range is the time sufficient for a stall to develop before a disturbance can trigger a wash-out. For $f_o/f_d > 1$, there remains time also for naturally triggered wash-outs to take place, so that the extraneous forcing effect is diminished. The significance of this finding is that periodic disturbances can easily be induced in practice by a machine, an upstream vortex shedding, or a swirling-flow instability.

The obvious means to suppress flow oscillations in the cases illustrated in Figure 6.38 involves elimination of the bistable-flow instability. The diffuser flow, for example, can be stabilized either by splitter walls, subdividing the diffuser into portions with diffuser angles of $\leq 8°$, or by the insertion of screens with a total resistance approximately equal to the ideal pressure recovery (Naudascher, 1987a, p. 169). In the case of gate II in Figure 6.38, the flow was found to remain attached to the apron when the apron slope was 1:1.8 or less (Naudascher, 1959).

6.7 EXCITATION DUE TO SWIRLING-FLOW INSTABILITY

Swirling flows, which consist of both an axial (streamwise) and circumferential (rotational) velocity field, are observed in a number of situations including the flow past inclined delta wings or past inclined faces of prismatic bodies, the flow through vortex tubes, cyclone separators, inlet and exit chambers of turbo-machinery, and the flow through draft tubes of hydraulic machines. If the rotational component of the flow exceeds the axial component by a certain critical value, these flows undergo an abrupt transition referred to as vortex breakdown. Reviews on this topic have recently been given by Leibovich (1984) and Escudier (1987, 1988). A number of authors view vortex breakdown as an instability involving either symmetrical or asymmetrical modes. According to Escudier & Keller (1983), the vortex breakdown is basically a two-stage transition from an initially supercritical to a terminal subcritical state with relatively more swirl (see also Keller et al., 1985). In the intermediate state, an inner stagnant core is surrounded by a layer of rotational fluid emanating from the upstream core. Due to instability, this layer of intense shear rolls up and, at the higher Reynolds numbers typical of most practical situations, degenerates into intense turbulence fluctuations. The final transition to a subcritical flow state is in many ways analogous to the transition from supercritical to subcritical open-channel flow, known as hydraulic jump.

Irrespective of the theoretical interpretation given, vortex breakdown can be uniquely described by parameters including the Reynolds number and the swirl parameter (or reciprocal Rossby number), $\Omega \equiv 2M_\theta /(M_x D)$, defined as the ratio of the fluxes of angular and axial momenta, M_θ and M_x, normalized by the radius $D/2$ of the tube. One may distinguish two types of vortex breakdown (Sarpkaya, 1971), as illustrated in Figure 6.41 for a slit-tube arrangement: the symmetrical-mode and the spiral- or helical-mode breakdown. Both modes have been observed, e.g., in external flows past delta wings, in constant-diameter tubes, and in

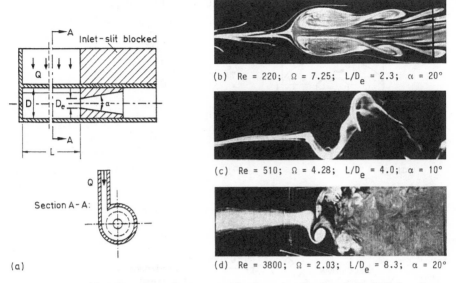

Figure 6.41. (a) Slit-tube vortex generator used for flow visualization. (b, c, d) Various forms of vortex breakdown (after Escudier, 1988).

complex tube-diffuser configurations. Which of the two modes will occur for a given system and geometric condition depends on the Reynolds and swirl numbers.

The characteristics of the oscillatory flow resulting from such breakdown processes can best be elucidated with the example of a simple vortex whistle (Figure 6.42a). In fact, as was shown by Escudier & Merkli (1979), Escudier (1980), and Merkli & Escudier (1979), these characteristics are entirely consistent with those of ring-type exit and inlet chambers of turbomachines. When gas or liquid enters a cylindrical cavity tangentially and exits through a tube as sketched in Figure 6.42a, pronounced oscillations can occur, giving rise to whistle-like sounds in the case of gas flow. Figure 6.42b shows that the Strouhal number Sh of the oscillation in air is strongly dependent on the dimensionless tube length L/D for a given Reynolds number, $Re \equiv VD/v$. Comparable experiments in water produced data which agreed with those in air within 10% over the range $4000 < Re < 13500$. In all cases, Reynolds-number effects diminished with increasing values of Re. As to the influence of cavity dimensions, Chanaud (1963) found that Sh increased with decreasing values of D_1/D and increasing values of D_1/d. Depending on the type of swirl generator employed, moreover, Chanaud (1965) observed changes in Strouhal number of up to 240%. The lower limit of Re at which periodic flow ceases to exist for a given value of swirl parameter $\omega_T D/(2\pi V)$ is shown in the 'stability diagram' in Figure 6.42c, in which $\omega_T/2\pi$ = frequency associated with the angular velocity of flow ω_T and

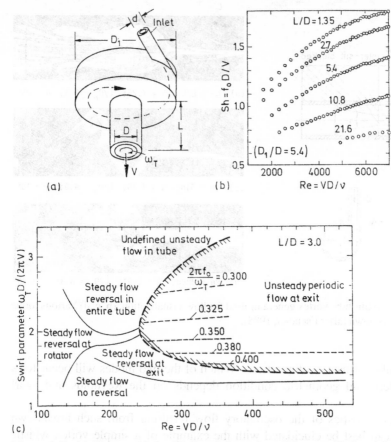

Figure 6.42. (a) Vortex whistle; (b) Strouhal number of flow oscillation (after Chanaud, 1963); (c) Regimes of vortex-whistle operation (after Chanaud, 1965).

f_o = frequency of oscillation, or precession, of the fluid within the tube. The ratio $f_o/(\omega_T/2\pi)$ tends toward constant values which increase with decreasing swirl parameter.

The phenomenon referred to as vortex breakdown above is known as draft-tube surging in the hydraulic-machinery literature (Falvey, 1971). It has long been recognized that the low-frequency vibrations (typically a few Hz) experienced with Francis and Kaplan turbines operating at partial load or overload are associated with the swirl remaining in the water as it leaves the turbine runner (Figure 6.43). The problem is complicated by many factors including geometry of the turbine and the draft tube, cavitation, and, because of possible coupling effects such as power swings and penstock resonance, by the dynamic characteristics of the electrical network and the penstock. With cavitating flow (Figure 6.43b) or with entrained air in the fluid, moreover, the compliance of the fluid is increased so that resonance may also be induced with fluid-oscillator vibrations in the draft tube (Dörfler, 1980; Nishi et al., 1984).

To clarify the basic relationships involved in draft-tube surging, Falvey & Cassidy (1970) carried out systematic experiments with the idealized draft-tube model depicted in Figures 6.44a, b; they used air as the working fluid in

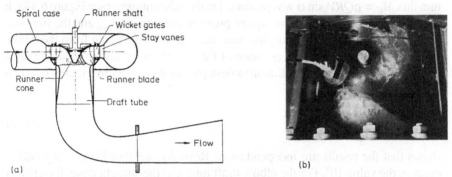

Figure 6.43. (a) Sketch of Francis turbine and draft tube (from Escudier, 1987). (b) Cavitating vortex flow in a draft tube (9% cone; from Kubota and Aoki, 1980).

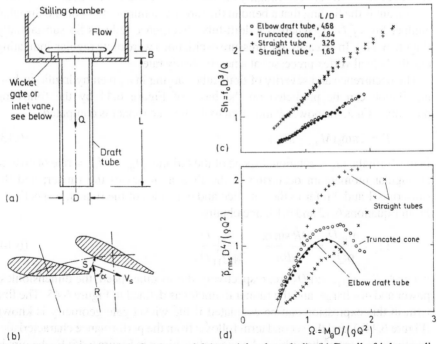

Figure 6.44. (a) Idealized draft-tube model (type 'circular tube'); (b) Detail of inlet conditions. (c, d) Strouhal number and dimensionless pressure amplitude as a function of the swirl parameter Ω; $Re = 4Q/(\pi D v) \geq 10^5$ (after Falvey & Cassidy, 1970).

order to eliminate two-phase effects. Three types of tubes were studied: (a) a model of an actual elbow draft tube (Figure 6.43a); (b) a straight, truncated cone with the same L/D ratio and the same divergence angle (12.25°, total, in draft-tube throat); and (c) straight circular tubes of uniform diameter. All draft tubes had sharp-edged entrances and diameters $D = 15.6$ cm. The discharge $Q = V\pi D^2/4 = V_s sBn$ passed between $n = 24$ vanes such that an angular momentum flux $M_\theta = \rho QRV_s \sin \alpha$ was produced at the tube entrance (see Figure 6.44a, b for definitions). The root-mean-square pressure amplitudes $\tilde{p}_{rms}$ and the frequencies f_o occurring during the surging were measured (a) upstream from the elbow of the draft tube; (b) near the entrance of the straight cone; and (c) near the open end of the straight tubes. A dimensionless plot of these data in Figures 6.44c, d with

$$\Omega = \frac{M_\theta D}{\rho Q^2} = \frac{DR \sin \alpha}{sBn} \tag{6.12}$$

shows that the results are independent of Reynolds number if Re $\equiv 4Q/(\pi D\nu)$ exceeds the value 10^5. For the elbow draft tube and the straight cone, the critical value of Ω for the onset of surging was found to be 0.4, and maximum amplitudes were observed between $\Omega = 1.3$ and 1.5 (Figure 6.44d). For straight tubes, the critical Ω changed between roughly 0.3 for $L/D \simeq 1$ and 0.4 for $L/D \simeq 10$. Most significant is the finding that a bend in the tube has a minor effect on the Strouhal number Sh $\equiv f_o D^3/Q$ whereas a draft-tube divergence reduces Sh substantially (Figure 6.44c). In all cases, the pressure oscillations were quite regular, indicating that the helical vortex precesses at a nearly steady rate.

The occurrence and severity of draft-tube surging in a given hydraulic-turbine installation can be predicted on the basis of Figure 6.44 by the following procedure. First, the power P imparted to the turbine runner is computed,

$$P = 2\pi\omega_R(M_{\theta_1} - M_{\theta_2}) \tag{6.13}$$

In this formula, ω_R = rotational speed of the turbine; $M_{\theta_1} - M_{\theta_2}$ = rate of change of angular momentum occurring in the flow as it passes the runner; and the subscripts 1 and 2 refer to the entrance and exit sides of the runner, respectively. From Equations 6.12 and 6.13, one obtains

$$\Omega_2 = \frac{M_{\theta_2} D}{\rho Q^2} = \frac{DR \sin \alpha}{sBn} - \frac{P_{11} D}{2\phi Q_{11}^2 D_2} \tag{6.14}$$

where ϕ, P_{11}, Q_{11}, and D_2 are, respectively, the specific speed, the dimensionless power and discharge, and the runner diameter as defined in Figure 6.45. The first term in this expression can be evaluated if the wicket-gate geometry is known (Figure 6.44b), and the second term follows from the performance characteristics and the geometry of the turbine. An example is given in Figure 6.45: Each point in the efficiency diagram of Figure 6.45a yields a particular value of Ω_2 that helps to determine areas of potential surging (i.e., $\Omega \geq 0.4$) and the locus of maximum

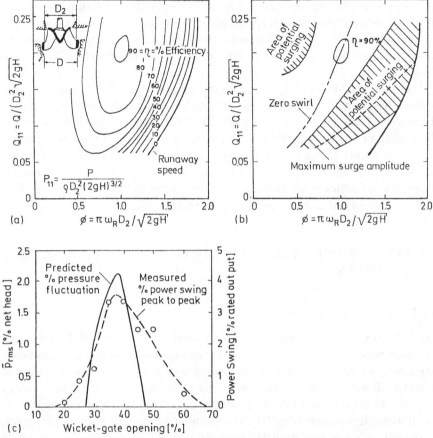

Figure 6.45. (a) Efficiency diagram, (b) predicted regions of surging, and (c) comparison of predicted and measured surging effects (net head on the unit H = 30.5 m; rated output = 11.9 MW) for a particular turbine installation (after Falvey & Cassidy, 1970).

surge amplitude (i.e., $\Omega \simeq 1.3$, Figure 6.44d) as shown in Figure 6.45b.

To predict the pressure fluctuations $\tilde{p}_{rms}$ for a particular hydropower installation, Falvey & Cassidy (1970) used Equation 6.14 in conjunction with a parabolic expression

$$\frac{\tilde{p}_{rms}D^4}{\rho Q^2} = 1.1 \left[1 - 1.563 \, (\Omega - 1.3)^2 \right]$$

fitted to the data for the elbow draft tube in Figure 6.44d. The results of this computation are shown in Figure 6.45c along with relative power swings actually measured at the plant. The agreement for the wicket-gate opening of maximum surging is excellent, even though the coupling between the dynamics of the mechanical and electrical systems had not been explicitly accounted for. Hydraulic turbines can thus be analyzed for possible surging by the method presented. Strictly, however, this method is applicable only for 'hard-core' vortex conditions. In cases of low tailwater levels in which the vortex core becomes 'unwatered', pressure surges are usually somewhat milder due to the effects of two-phase flow.

The influence of draft-tube shape on the surging characteristics of hydraulic turbines has been addressed by Palde (1972). Important for the understanding of the transient behavior of surges associated with the startup or shutdown of units is the finding of Sarpkaya (1971): if either the flow rate Q or the vane angle α changes suddenly, the swirl-flow pattern overshoots past and then relaxes toward the equilibrium pattern for the new flow rate or valve setting. Information on the suppression of vortex breakdown or draft-tube surging as well as on swirling-flow problems associated with pumps is given in Section 6.8.5.

6.8.   CONTROL OF INSTABILITY-INDUCED EXCITATION / PRACTICAL EXAMPLES

6.8.1 *Basic design criteria*

The best design strategy regarding instability-induced excitation (IIE) is to avoid excitation altogether by modifying the system's geometry, its dynamic character-istics, or its operating conditions such that neither a structure or structural part nor a fluid oscillator is excited to undergo resonant vibrations. Priority in this strategy is given to an optimization of geometry or to what is also referred to as 'fluid-dynamic attenuation' of IIE. Since definite limitations exist in this regard, an obvious solution consists in designing the resonant frequencies of the body or fluid oscillators, f_N or f_R, to lie well outside the range of excitation frequencies. Whenever possible, in other words, f_N and f_R should be so large that f_o/f_N and f_o/f_R (or the equivalent product of Strouhal number and reduced velocity, $\mathrm{Sh}V_r$) lie below the lower limits of the lock-in or resonance ranges (e.g., Figure 3.23 or Figure 8.13). In cases in which these lower limits are not known, a conservative design criterium is to keep f_N or f_R 40% away from the instability-induced flow frequency f_o unaffected by vibrations of the body or fluid oscillators (e.g., Paidoussis, 1983).

However, effective attenuation at the source of excitation, or avoidance of resonance, is not always achievable. Then, one can only increase the damping of the system so as to reduce the vibration amplitude (e.g., Figure 6.16a or Figure 8.1), or employ one of a variety of added damping devices (e.g., Sections 5.3 and 7.7.3). If none of these solutions can be achieved, one may apply, as a last resort, a tuned mass damper to avoid excessive vibrations of a structure (Section 7.7.3) or a filter to prevent the excitation of a fluid oscillator (Section 8.5.2).

Several basic concepts of fluid-dynamic attenuation are illustrated in the following section. Even though vortex shedding is used for this illustration, many of the attenuation concepts may be employed in the suppression of other flow instabilities. Recent review articles on fluid-dynamic attenuators are those by Zdravkovich (1981, 1984b) and Every et al. (1982b). Basic techniques for attenuating the instabilities giving rise to vortex shedding have been described by Huerre & Monkewitz (1990).

6.8.2 *Concepts of fluid-dynamic attenuation/control of vortex shedding*

(a) *Vortex trapping.* For elongated bodies such as vanes, blades, and plates, a *concave trailing edge* was shown in Figure 6.15a to reduce vortex-induced excitation considerably. A possible explanation of this attenuating effect is the trapping of vortices in the concave space that is brought about by the deflections of the trailing edge during vibration. According to Reed (1980), concave trailing edges have been successfully used to attenuate the vibration ('singing') of ship propellers.

(b) *Dephased separation (transverse).* An effective means of attenuation in the case of vanes, blades, and plates is a *beveled trailing* edge with angle $\alpha \leq 30°$ (Figure 6.46). In tests with flat plates rigidly mounted in an air tunnel, Greenway & Wood (1973) measured the pressure fluctuations $\tilde{p}$ at an antinode of the standing pressure wave that was excited between plate and tunnel walls because of vortex shedding. The Strouhal number was $f_o d/V \simeq 0.24$ and the Reynolds number $VL/v \simeq 1.3 \times 10^6$, so that probably the boundary-layer separation was turbulent. Beveling the trailing edge destroyed the symmetry of vortex shedding and thereby reduced $\tilde{p}$ as shown in Figure 6.46b.

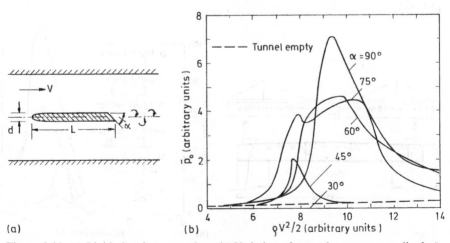

Figure 6.46. (a) Rigid plate in test section. (b) Variation of acoustic pressure amplitude $\tilde{p}_o$ with tunnel dynamic pressure for different trailing-edge angles (after Greenway & Wood, 1973).

The effects of the trailing-edge shape on the vortex-induced vibration of plates have also been investigated by Blake (1986, Vol. II, p. 761), Heskestad & Olberts (1960), and Donaldson (1956). Particularly effective attenuations were obtained with trailing edges shaped according to Figures 6.47 H, I, and J. Donaldson (1956) found only irregular vibrations with these plates, and the trailing-edge amplitudes were 1% of those obtained with plate C. The location of separation along the curved surfaces seems to be not well defined so that the vortex shedding is 'dephased' even in the case of trailing-edge shapes H and I. Applications of some of these configurations to hydraulic machinery have been reported by Chen, Kirchner, Neilson, and Ulith (summarized in Naudascher & Rockwell, 1980).

An unusual utilization of the principle of dephased separation is shown in Figure 6.48. Although bridge decks may experience excitation due to various mechanisms including movement-induced excitation, one is associated with vortex shedding. As shown in Figure 6.48b, the decrease in vibration amplitude with decreasing angle of fairing α is astonishingly similar to the decrease in pressure amplitude $\tilde{p}_o$ with decreasing α, depicted in Figure 6.46b. In fact, nearly

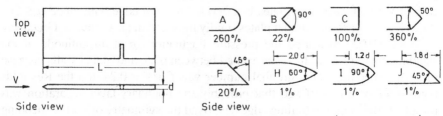

Figure 6.47. Trailing-edge shapes investigated in the range $2{\times}10^4 < Vd/\nu < 6{\times}10^4$ by Donaldson (1956). (Numbers give trailing-edge amplitude in percent relative to that of plate C.)

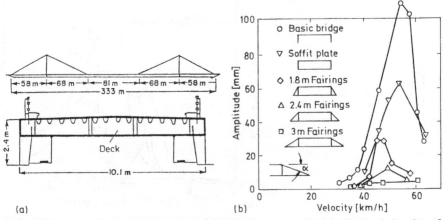

Figure 6.48. (a) Side and end view of bridge deck with support. (b) Amplitude of vibration of various bridge-deck cross-sections (after Wardlaw, 1980).

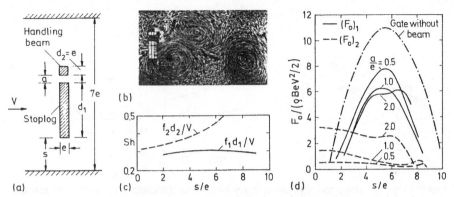

Figure 6.49. (a) Definition sketch of stoplog (height d_1) and handling beam (height $d_2 = e$). (b) Visualization of shed vortices (camera moving). (c) Strouhal numbers $Sh_{1,2}$ and (d) maximum amplitudes $(F_o)_{1,2}$ of major components of fluctuating vertical load as a function of position (after Naudascher & Farell, 1965).

complete attenuation is reached for $\alpha \simeq 30°$ in both cases. However, this fairing does not protect the bridge deck from torsional stall flutter at larger velocities (Figure 7.17, profile 3).

During closure of stoplogs or multiple-leaf gates, the individual elements are often lowered by means of a handling beam (cf. Figure 9.50). An idealized configuration of stoplog and handling beam is shown in Figure 6.49a. The interaction between and mutual disturbance of the vortex shedding past these two cylindrical bodies (Figure 6.49b) depend on, among other factors, the spacing '*a*' between them. As one would expect, the total hydrodynamic load F acting vertically on the system of stoplog and beam contains two fluctuating components with dominant frequencies f_1 (stoplog) and f_2 (beam) as shown in Figure 6.49c, and with maximum amplitudes $(F_o)_1$ and $(F_o)_2$ depending on the wake interaction (Figure 6.49d). The 'price' one has to pay for attenuating $(F_o)_1$, evidently, is an increase in the high-frequency component $(F_o)_2$. For a spacing ratio of roughly $a/e = 0.5$, the tests revealed a fluctuating force component of even higher frequency than f_2 that could be traced to shear-layer impingement of the jet flowing through the gap between stoplog and handling beam.

(c) *Dephased separation (spanwise) and 3-D disturbance.* Spanwise dephasing along a stationary circular cylinder in the critical and, possibly, the postcritical regime, $1.3 \times 10^5 < \mathrm{Re} < 3.5 \times 10^6$, can be achieved by *staggered separation wires* (Figure 6.50a), provided that a critical angle difference $\Delta\alpha$ is exceeded. The magnitude of $\Delta\alpha_{\mathrm{crit}}$, for a three-wire arrangement, is given in Figure 6.50b. If separation is controlled at its origin in this way, the separation wires can be extremely thin (e.g., $0.013d$) and still disrupt regular vortex shedding effectively if staggered in this way. If fixed parallel to the cylinder axis along the entire span,

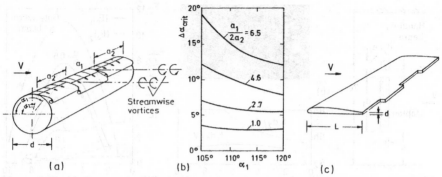

Figure 6.50. (a) Staggered separation wires of 0.013d thickness on a stationary circular cylinder between endplates create strong three-dimensional disturbances; (b) Critical angle difference $\Delta\alpha = \alpha_1 - \alpha_2$ for suppressing regular vortex shedding (after Nauman & Quadflieg, 1974). (c) Stepped trailing edge of a vane generates similar dephased separation with streamwise vorticity (Tanner, 1972).

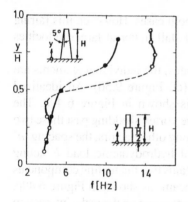

Figure 6.51. Shedding frequency along the axis of a stationary cone and a stepped cylinder (after Naumann et al., 1974).

the separation wires can enhance the regularity of vortex shedding throughout the supercritical regime. This finding of Naumann & Quadflieg (1974) highlights the danger of *longitudinal fins* which are sometimes used to stiffen cylindrical shells. According to Mahrenholtz & Bardowicks (1980), five longitudinal fins without stagger may render a circular cylinder susceptible even to dangerous galloping vibrations.

A possibility of effecting dephased separation from a blunt-trailing-edge wing (or a plate) is depicted in Figure 6.50c. A side benefit of this measure is a reduction in base drag according to Tanner (1972).

Figure 6.51 shows the similarity in the dephasing achieved with a *slender cone* (d_{max} = 16.5 cm, d_{min} = 6 cm, H = 60 cm, Re = 3×10^4) and a *stepped cylinder* (d_{max} = 13.8 cm, d_{min} = 5.8 cm, H = 60 cm, Re = 3.7×10^4). In both cases, the wake contained branched vortex filaments, and the maximum periodic load on the stationary cylinder was remarkably reduced as compared to the case of a comparable straight circular cylinder. Still, structural vibrations have a potential to reorganize the vortex shedding (cf. discussion of Figure 6.22). In fact, collars and rings which have been tried to suppress vortex excitation of marine cables through dephasing (Figures 6.52b, c) have generally proven to be ineffective.

Typical applications of marine cables are illustrated in Figure 6.52a. With towing speeds as high as 10 m/s, the towed-cable Reynolds number may rise to 10^5 and more. Vortex-induced vibrations of marine cables and long pipes, which occur at ever higher modes with increasing velocity, have been described by Blevins (1977), Whitney et al. (1981) and Every et al. (1982). The excitation of cables can be attenuated, and cable life extended, by a number of *flexible devices* interwoven into the cable (Figures 6.52d-f). To be effective, ribbons should be 4-6d long, 1-2d wide, and spaced 1-2d apart. Similarly, tufts of fibers trailing from a cable (like a brush or fringes) should be 4-6d long and spaced at 1d intervals. The disadvantage of ribbons and fringes is the increased mean drag (C_D ranging from 2 to 6). With 6d long fibers attached individually to the cable at 1/2d intervals (Figure 6.52f), vortex shedding was found to be well attenuated without an increase in drag; in some cases C_D was even lower than for the bare cable.

Occasionally, vibration problems exist only during the construction stage so that measures to control vibration need not be permanent. For illustration, Figure 6.53 depicts some stages of the underwater installation of foundation piles for a Cognac platform. According to model tests by Fischer et al. (1980), transverse pile vibrations can be attenuated effectively by attaching *flags* made of flexible plastic (0.02 to 0.1 mm thick in the model) to the free end of the pile as shown in Figures 6.53a, b. The most effective flag was 10d high and 5d wide with the sheet being cut into ribbons of unspecified size. During the stage of 'stabbing' operation (Figure 6.53b) where the principle mode of excitation is the transverse pendulum

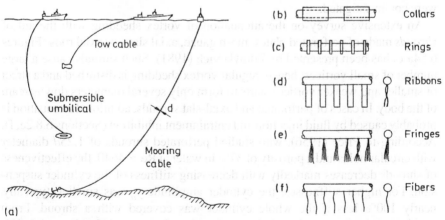

Figure 6.52. (a) Marine-cable applications. (b-f) Various devices tested for the suppression of vortex excitation of cables (after Every et al., 1982).

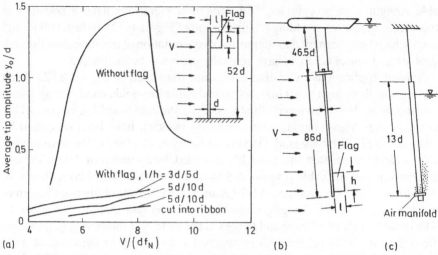

Figure 6.53. (a) Transverse tip amplitudes of Cognac-pile vibration during 'driving' operation without and with plastic flag (Re ≃ 6×10³) (after Fischer et al., 1980). (b, c) Vibration attenuating devices for Cognac piles during 'stabbing' operation; (b) Plastic flag (after Fischer et al., 1980); (c) Air-manifold device (after Every et al., 1982b).

mode with collinear pile and cables, a $10d/5d$ flag was effective in reducing the tip amplitudes to around 15% of the previously observed maxima [near $V/(df_N) \simeq 5.6$]. A further attenuating device for piles suspended from a barge is an *air manifold* at the lower end through which air is pumped to produce a stream of bubbles in the near wake. The vibration was reduced substantially for a relatively short pile of $13d = 10$ m at Re $\simeq 3×10^5$ with an air-flow rate of 0.013 m³/s (free air). The effects of flags and air manifold on streamwise vibrations were not investigated.

An extensive survey on the attenuation of vortex shedding with the aid of *shrouds* made of perforated plates, mesh gauze, axial slats, or axial rods (Figures 6.54a-c) has been presented by Zdravkovich (1981). Such shrouds cause a large number of small vortices; hence, regular vortex shedding is disturbed and a street of smaller and weaker vortices starts to form only several diameters downstream of the body. In cases of perforated and axial-slat shrouds, additional attenuation is probably caused by fluid injection and entrainment inhibition (Sections 6.8.2e, f). According to Price (1956), who studied perforated shrouds of $1.25d$ diameter with circular holes and a porosity of 37% in water at Re $\simeq 4640$, the effectiveness of shrouds decreases markedly with decreasing stiffness of the cylinder suspension. For higher stiffnesses, the cylinder amplitude y_o was smaller, typically, nearly 100% less if the whole cylinder was covered with a shroud. From supplementary tests in the wind tunnel within the range $10^5 <$ Re $< 4.5×10^5$, Price (1956) found that the drag coefficient C_D of the shrouded cylinder, based on the shroud diameter, remained approximately constant at a value 0.6, which is intermediate between subcritical and supercritical values for the cylinder alone. Even though this lack of C_D variation implies a weak Reynolds-number dependence, Price's findings may not apply to other ranges of Re. Similarly, there are a number of comparisons in the literature, such as the one in Figure 6.54d, that must not be generalized because they reflect different influences of Reynolds number, background turbulence, cylinder end conditions, etc. Significant in Figure 6.54d is the increased steepness of curves B to E as compared to curve A; it shows that shrouds and strakes work reasonably well, in general, as long as the mass-damping parameter does not fall below a critical value (δ_s = logarithmic decrement of damping). Experiments by Ruscheweyh (1981) show clearly that without sufficient damping, the body movement becomes large enough to enhance the flow coherence in the near wake despite the attenuating devices. For similar reasons, shrouds and strakes are usually ineffective with respect to such movement-induced excitations as wake flutter (Cooper, 1974), interference galloping (Ruscheweyh, 1981), or in-line vibration due to wake breathing (Wootton et al., 1974).

A possible exception of this rule appears to involve axial-rod shrouds (Figure 6.54c), which have the additional advantage of lowering the drag coefficient ($C_D \simeq 0.9$) for cases of high porosity. The most effective axial-rod shroud, according to Zdravkovich (1981), is that with $1.25d$ diameter and 63% porosity, or with $1.16d$ diameter and 81% porosity. An arrangement of four rods proved to be highly effective in suppressing fluid-elastic vibrations in tube banks when placed around the tubes of the first two rows one rod diameter away from the tube surface

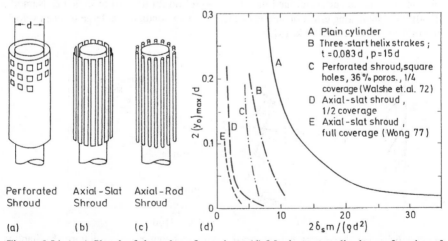

Figure 6.54. (a-c) Sketch of shroud configurations. (d) Maximum amplitude as a function of mass-damping coefficient for various fluid-dynamic attenuators after Wong & Cox (1978). (Strakes: see Figure 6.56a.)

as shown in Figure 6.55a $(8.6 \times 10^4 < \mathrm{Re}_{\mathrm{gap}} < 2.1 \times 10^5)$; even the interference galloping was suppressed by a similar rod arrangement fitted on the two cylinders (Section 7.7.2; Zdravkovich, 1974b). In cases of vibration due to severe buffeting conditions, Zdravkovich & Volk (1972) claim that a shroud made of a fine-mesh gauze is more effective in reducing the amplitude at high reduced velocities than shrouds with circular or square holes of the same porosity (Figure 6.55b); however, the mean drag reaches a maximum in that case with $C_D \simeq 1.3$.

Perhaps the best-known method of attenuating vortex shedding is the use of *helical strakes* (Figure 6.56). Since the early investigations of Scruton & Walshe (1956), the optimal pitch has been determined as $5d$ by Woodgate & Mabey (1959) and between $4d$ and $5d$ by Hirsch et al. (1975), whereas the minimum thickness of the strakes, t, for effective vibration suppression (with $4.8d$ pitch)

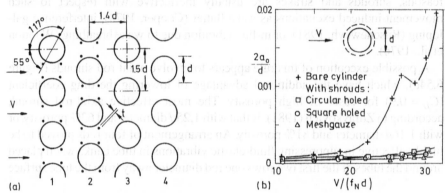

Figure 6.55. (a) Tube vibration attenuated by mounting four rods of $t = 0.026d$ thickness at a distance t from the surface around the tube rows 1 and 2 (after Zdravkovich & Namork, 1980). (b) Attenuating effect of shrouds under buffeting conditions at large reduced velocities (after Zdravkovich & Volk, 1972).

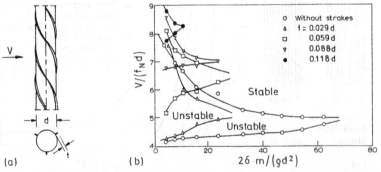

Figure 6.56. (a) Three-start helical strakes with rectangular cross-section. (b) Stability diagram for a two-dimensional circular cylinder fitted with three-start helical strakes of $15d$ pitch (after Scruton & Walshe, 1957).

was obtained as $t \simeq 0.08d$ by Ruscheweyh (1978b). For marine applications, Every, King & Weaver (1982) recommend at least $t = 0.1d$. Among the disadvantages of helical strakes are, first, the substantial increase in mean drag ($C_D = 1.35$ and 1.45 for $t = 0.06d$ and $0.12d$, respectively, throughout the range $10^5 < \text{Re} < 4 \times 10^6$, according to Cowdrey & Lawes, 1959); second, the significant reduction in effectiveness in turbulent approach flow (Gartshore et al., 1979) as well as downstream from neighboring cylinders (Wardlaw, 1980; Vickery & Watkins, 1964); and, third, the inefficiency below a critical mass-damping parameter (Figure 6.54d; Ruscheweyh, 1981).

In view of the findings on helical strakes, the observations on thin *helical wires* are surprising (Figure 6.57). In tests of wires of only $0.04d$ thickness on flexibly mounted cylinders at supercritical Reynolds numbers, Nakagawa et al. (1959, 1963) found that a winding of eight wires at $64d$ pitch actually increased the maximum vibration amplitude by a factor of two whereas a single wire at $0.5d$ pitch produced the same amplitudes as a bare cylinder. With four wires at $8d$ or $16d$ pitch, vibrations were almost completely suppressed throughout the range tested, and C_D was found to be 0.6 and smaller. Weaver's (1961) investigation on the optimum number, pitch, and size of helical wires confirms that three or four helical wires are most effective at a pitch between $8d$ and $16d$, i.e., far away from the $4d$ to $5d$ obtained for three helical strakes of rectangular cross-section. The attenuating influence of the relative wire thickness t/d is shown in Figure 6.56b.

Sallet (1980) used an unusual arrangement of helical wires to cure excessive vibrations of a cylindrical buoy of aspect ratio 5.15, floating freely at the end of a long moving cable. Using three equally long patches of a 45° helix made of twelve nylon ropes ($t/d = 0.03$), and reversing the sense of rotation in the middle patch, he was able to reduce the vibration amplitudes from roughly $3d$ and $2.8d$ for the first and second modes of oscillation (near Re = 1.7×10^5 and 4.8×10^5) to $1.4d$ and $0.5d$, respectively.

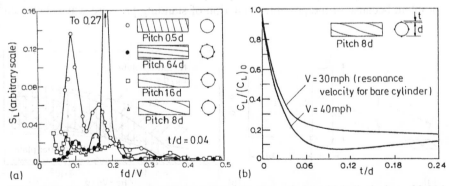

Figure 6.57. (a) Power spectra of lift forces for a circular cylinder with helical wires of $0.04d$ thickness in supercritical Reynolds-number flow (after Nakagawa et al., 1959, 1963). (b) Effect of relative wire thickness, t/d, on lift coefficient C_L normalized by $(C_L)_o$ for a bare cylinder (after Weaver, 1961).

(d) *Instability inhibition via shear-layer reattachment.* One of the earliest techniques for attenuating vortex excitation involves the use of a *splitter plate* either touching the cylindrical body or immediately downstream of it. As shown in the IAHR Monograph on Hydrodynamic Forces (Naudascher, 1991, Figure 2.80b), a splitter plate can significantly affect the drag and Strouhal number of vortex shedding (Roshko, 1954; Gerrard, 1966). Appelt et al. (1973, 1975) found for subcritical Reynolds numbers that vortex formation was postponed to a region downstream of the trailing edge of the splitter plate, and that the wake width immediately behind the cylinder was decreased as the length of the splitter plalte, L_s, was increased. These observations were most evident in the range $L_s/d < 2$. For $L_s/d > 5$, the flow was observed to reattach to the splitter plate at a distance of about $5d$, irrespective of the plate length. Although the vortex shedding was eliminated and C_D tended to remain constant at a value of about 0.8, a vortex street did form at roughly $17d$ downstream of the cylinder.

Vortex-excited vibration of a cylindrically shaped buoy used for measuring ocean currents can lead to incorrect measurements. For the particular case depicted in Figure 6.58a ($d = 0.5$ m, $L/d = 5$), Sallet (1980) was able to reduce vibration amplitudes of as much as $1.75d$ to almost zero by a splitter plate with a length of $3d$ (Figure 6.58b). The splitter plate should be sufficiently rigid not to be induced to flutter on its own. For the necessary plate width W_s, Sallet (1970) suggests the formula

$$\frac{W_s}{L} = \frac{C_D \text{ (for finite } L/d)}{C_{D_\infty} \text{ (for } L/d = \infty)}$$

Since turbulent separation for flow around a circular cylinder takes place further downstream than does laminar separation, the former requires a smaller L_s than the latter. For extremely short cylinders, end effects become important (Welsh, 1966). When vortex excitation is to be attenuated in a ducted flow, one must keep in mind that splitter plates can easily aggravate rather than cure a vibration problem because of the possible excitation of standing pressure waves between

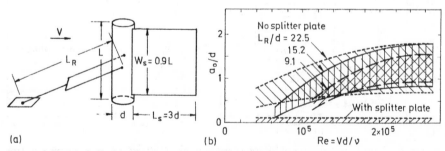

Figure 6.58. (a) Cylindrically shaped buoy with stabilizing splitter plate and center-of-drag mooring. (b) Amplitude a_o of pendulum-like vibration as a function of Reynolds number (after Sallet, 1980).

the splitter plate and the duct walls (Figure 8.13; Parker, 1966, 1967). It is dangerous, also, to attach a splitter plate to a flexible body; for short plates the amplitude of vibration may increase, probably because a weaker fluctuating pressure has a larger area over which to act.

A variation of the splitter-plate concept has been applied by Baird (1955) to prevent vortex-excited vibrations of a pipeline suspension bridge; he attached triangular plates, with $L_s = 1.5d$ and $W_s = 12d$, on both the upstream and downstream sides at intervals of $48d$ in staggered arrangements. With this double saw-tooth pattern, the wind could occur from either side along the river without leading to detectable vibrations.

(e) *Instability inhibition via fluid injection.* Another means of affecting the near wake region is to *bleed fluid* from the interior of the body into the wake (Wood, 1964, 1967; Bearman, 1967b). At a Reynolds number of $Ve/v \simeq 1.25 \times 10^4$, Wood found bleeding according to Figure 6.59a to increase the vortex-shedding frequency f_o only slightly, but the vortices diminished in strength and formed further away from the body as the bleed coefficient $C_q = V_j/V$ was increased (V_j = mean bleed velocity over the base height d). Since the location of vortex formation is difficult to quantify, Wood evaluated the 'entrainment length' L_E (Figure 6.59) which is closely related to the 'formation length'. At higher bleed coefficients ($C_q \geq 0.146$), the wake was highly intermittent, and the Strouhal number $\mathrm{Sh} \equiv f_o d/V$ was ill defined.

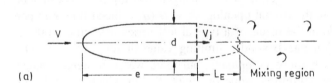

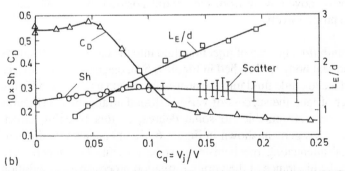

Figure 6.59. (a) Cylindrical body bleeding fluid from the interior; (b) Corresponding effects of base bleed on Strouhal number, mean-drag coefficient, and relative entrainment length (after Wood, 1964).

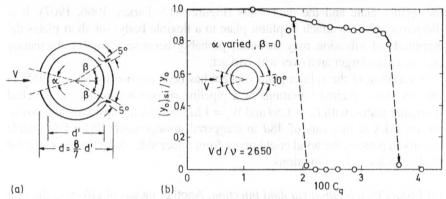

(a)　　　　　　　　　　(b)

Figure 6.60. (a) Model for self-sustained fluid injection; (b) Tip amplitude with self-injection normalized by tip amplitude without self-injection as a function of injection coefficient (after Wong, 1985).

Igarashi (1978) investigated the characteristics of flow past a circular *cylinder with a slit* cut along the whole span for two slit widths, 0.08d and 0.18d, at Re = 4.5×10^4. The corresponding mean drag coefficients were 0.9 and 0.75 and the Strouhal numbers 0.26 and 0.32, respectively, and the vortex-formation region was found to be displaced in a way similar to that for cases of artificial base bleed. The effect of *self-sustained injection* on the attenuation of vortex-induced vibration has been studied by Wong (1985). Using a model as shown in Figure 6.60a, mounted with gimbals at one end, Wong changed the injection coefficient $C_q = q_b/(dV)$ by varying the frontal opening angle α (q_b = bleed flow rate per unit width). As seen from the result of a typical test series in Figure 6.60b, a relatively small injection rate beyond a limiting value of roughly 3.5% was sufficient to suppress vibrations completely, but there was a hysteresis range between this and a lower limiting value of about 2%, in which the cylinder remained stationary unless excited by an appropriate disturbance. As the location of self-injection was moved away from the central position, $\beta \neq 0$, vibrations tended to increase progressively.

(f) *Entrainment inhibition.* Instead of injecting fluid into the base region from the interior of a cylindrical body as sketched in Figure 6.59a, one can mount devices near the body surface to guide fluid into the near-wake region as shown in Figure 6.61b. As a matter of fact, the success of various shroud arrangements (Figures 6.54a, b) would seem attributable, to some degree, to this bleeding effect. However, since the arrangement shown in Figure 6.61a attenuates vibrations as well, an additional attenuating mechanism must be present. This mechanism probably involves the hindrance of the vortex formation process by an inhibition of the entrainment of external, irrotational fuid (Figure 6.11a).

Grimminger (1945) tested *guide vanes* roughly one diameter long, fixed rigidly

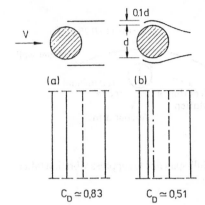

Figure 6.61. Sketch of rigid guide vanes and corresponding drag coefficients (after Grimminger, 1945).

at a distance of $0.1d$ from the cylinder surface, in flowing water at Re = 1.7×10^4 and found that both shapes depicted in Figure 6.61 attenuated vibrations markedly. The achieved mean-drag coefficients shown are considerably smaller than the value of $C_D = 1.06$ for the bare cylinder. Further improvement was achieved by extending the leading edges of the guide vanes around the cylinder upstream to points roughly ±45° from the center plane and by reducing the width between the vanes downstream as shown in Figure 6.61b. Of course, guide vanes like these can be used only in unidirectional flows. As reported by Grimminger (1945), they were successfully employed to suppress vibrations of submarine periscopes.

(g) *Miscellaneous devices.* Effective vibration reduction using two full-span *longitudinal fins* at ±45° from the center plane of a circular cylinder close to a wall has been achieved by King & Jones (1980). With reference to Figure 6.62, rotation of an anchor agitator produced vibration of the vertical cylindrical arms in a predominantly radial mode about the agitator axis. The unusual vibration mode and the proximity of the wall appear to be significant for the successful modification. In several other cases, longitudinal fins enhanced rather than suppressed vibration (cf. discussion of Figure 6.50).

A somewhat related problem concerns the lowering of cylindrical stoplogs into a sluiceway in free-surface flow. As reported by Kolkman (1980), plain circular cylinders cannot be used in such a case since they dance and sometimes even jump out of the water. Because of the economic advantage of using commercial pipes as stoplogs, model tests were performed; it was found that by attaching longitudinal fins to the pipes as shown in Figure 6.63, a steady downpull was produced and the stoplogs could be lowered. To stabilize the flow separation from the upper side ahead of the 90° point, even at postcritical Reynolds numbers, a wire of roughly $0.04d$ thickness was welded on that side at an angle of 65° from the stagnation point.

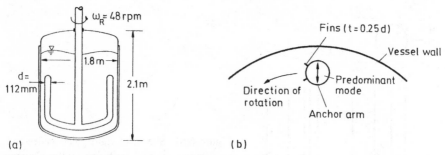

Figure 6.62. Rotating anchor agitator and location of full-span fins to suppress vibration (after King & Jones, 1980).

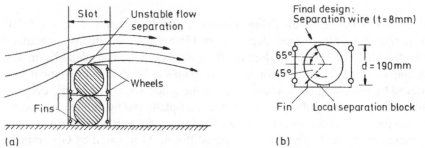

Figure 6.63. Circular stoplogs and location of fins and separation wires used to produce steady downpull and stable flow separation (after Kolkman, 1980).

A great variety of further practical examples and cures of vortex-induced vibrations are contained in the book by Naudascher & Rockwell (1980).

6.8.3 *Control of impinging shear layers*

Some of the attenuation concepts presented in the foregoing section are also applicable to problems involving shear-layer instability. The concept of a *splitter plate*, for example, has been employed successfully by Dougherty & Anderson (1976) to reduce pressure oscillations in a circular cavity. The suppression of flow oscillation by *injection of fluid* from the leading edge of a cavity has been studied by Willmarth et al. (1978). In a cavity on an axisymmetric body as shown in Figure 6.64a, pressure oscillations were essentially eliminated by combining fluid injection with a *leading-edge deflection* comparable to sketch H in Figure 6.65. For rectangular cavities, Heller & Bliss (1975) found that cavity-pressure oscillations can be effectively suppressed by means of a combination of leading-edge *spoilers* and a trailing-edge *ramp* (Figure 6.65, sketch J). Additional information on this geometry has been presented by Maurer (1976), while Rossiter (1964) reported on a means of attenuating cavity-pressure oscillations with spoilers only.

In the case of a planar jet impinging upon an edge (Figure 6.5a), the 'dephasing' of the upper and lower vortex sheets by a slanted jet nozzle (Figure 6.64c) does *not* attenuate the flow oscillations as one could have expected from the discussion of the beveled trailing edge (Figure 6.46). Instead, a substantial

CONCEPT	EXAMPLE	REFERENCE
Instability inhibition via mass injection	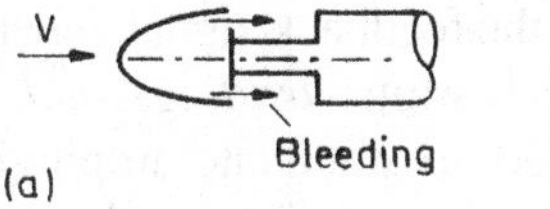 (a) Bleeding	Willmarth et al. 1978 Sarohia & Massier 1976
Dephased separation (spanwise) and 3D disturbance	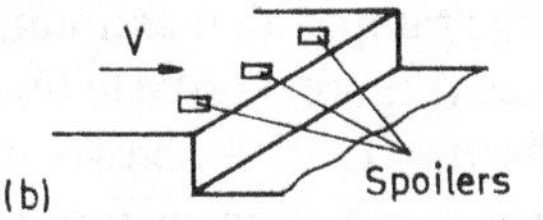 (b) Spoilers	Heller & Bliss 1975; Maurer 1975; Rossiter 1964
Dephased separation (transverse)	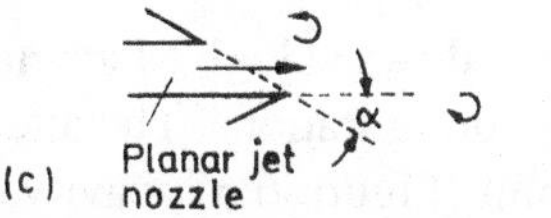 (c) Planar jet nozzle	Lucas & Rockwell 1987

Figure 6.64. Application of fluid-dynamic-attenuation concepts (Section 6.8.2) to problems involving shear-layer instability.

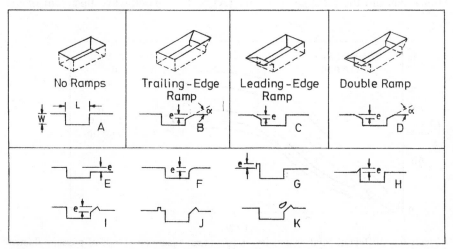

Figure 6.65. Geometrical variation of rectangular cavities employed to attenuate oscillation amplitudes (after Rockwell & Naudascher, 1978).

increase in oscillatory energy has been observed for nozzle-lip offsets smaller than one wavelength of the instability wave (Lucas & Rockwell, 1987). The example illustrates that one should meticulously examine attenuating devices even in cases based on successful attenuation concepts.

Feedback inhibition. An additional attenuation concept regarding an instability-induced excitation that depends so strongly on feedback as do impinging shear layers, obviously, is inhibition or disturbance of feedback. One of the most crucial events in the feedback process, depicted in Figure 6.5, is the interaction between the perturbed velocity field and the impingement edge (point II). By modifying this edge appropriately, therefore, the strength and coherence of the feedback signal can be weakened effectively. Another sensitive region is the shear layer origin (point I) where the feedback signal generates new perturbations.

Figure 6.65 summarizes various *trailing- and leading-edge modifications* which have been employed to attenuate impinging-shear-layer oscillations. Ethembabaoglu (1973, 1978) examined shapes E and F at low Mach numbers and found the gradual trailing-edge ramp with maximum recess (cavity F, $e/W = 0.2$) to be the most effective device (Figure 6.66). He observed that the spectral peaks of each of the three modes $n = 1, 2, 3$ shown in Figure 6.25 are generally attenuated in proportion to the reduction in root-mean-square pressure amplitude.

In an extensive study using flows at high Mach number, Franke & Carr (1975) tested sixty different cavity configurations. The attenuation of the spectral peaks achieved with a double ramp (Figure 6.65, cavity D) amounted to as much as 20 dB for the first and second modes, although the overall level of pressure oscillations was only slightly affected. When an expansion-wave formed upstream of the cavity, the double-ramp cavity produced greater pressure oscillations than did the rectangular cavity (A); with a shock-wave upstream of the

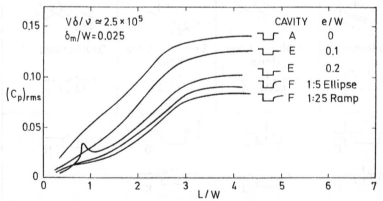

Figure 6.66. Effect of cavity geometry on root-mean-square amplitude of pressure fluctuations at downstream cavity wall for different length-to-depth ratios (after Ethembabaoglu, 1973).

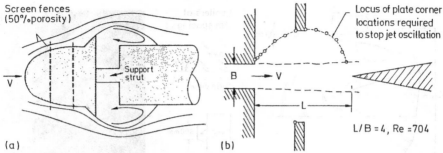

Figure 6.67. (a) Application of fences and ramp to control flow over a cavity in an axisymmetric body (after Willmarth et al., 1978). (b) Suppression of jet-edge tone oscillations by feedback inhibition through two symmetrically placed normal plates (after Karamcheti et al., 1969).

cavity, on the other hand, the double ramp was again effective in suppressing oscillations. The example illustrates the extent to which the character of an upstream pressure wave can influence the attenuation characteristics.

The effectiveness of leading-edge deflectors (Figure 6.65, cavity H) has been investigated by Willmarth et al. (1978), Falvey (1980), and, so far as cavitation aspects are concerned, by Ball (1959). As seen in Figure 6.67a, the deflector changes the streamline pattern so that shear-layer impingement on the downstream edge is avoided. The screen fences were added to suppress sound radiation completely. An application of a leading-edge deflector in the case of free-surface flow is presented in Figure 8.28. Sarohia & Massier (1976) used *fluid injection* from the base of the cavity to deflect the shear layer. Although the required fluid injection varied with flow conditions and geometry, effective suppression was achieved with very small injection rates. An interesting possibility of feedback inhibition has been demonstrated by Karamcheti et al. (1969), who succeeded in eliminating jet oscillations and edge tones by interrupting the feedback path through two *normal plates* as shown in Figure 6.67. Edge tones could be suppressed even with a single plate.

6.8.4 *Control of interface instabilities*

The common method of attenuating nappe oscillations (Figure 6.34) fluid-dynamically is based on the concept of spanwise dephasing and three-dimensional disturbances. Several devices have been tested and used including *dents* in the flap skinplate, a flap with *stepped trailing edge*, and a variety of *spoilers* (e.g., Seifert, 1940, 1941). Two examples for a successful suppression of flap vibrations excited by an oscillating nappe are shown in Figures 6.68, 6.69. In both cases, the vibration frequency was 4 Hz and maximum amplitudes reached 2 and 1.8 mm at the tip of the flap. If the spoilers are too small to rip open the nappe,

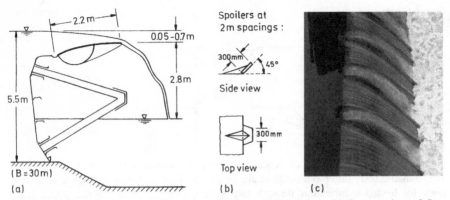

Figure 6.68. (a) Flap on gate vibrated at small overfall heights. (b) Suppression of flap vibration by nappe spoilers spaced at 2 m along the crest of the flap. (c) Photograph of prototype with nappe spoilers (after Petrikat, 1955)

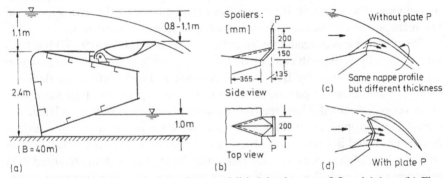

Figure 6.69. (a) Flap on gate vibrated at overfall heights between 0.8 and 1.1 m. (b) Flap vibration suppressed by nappe spoilers when arranged alternately with and without a plate P in 1.5 m spacings along the flap crest. (c, d) Flow patterns near spoiler: (c) without and (d) with plate P (after Ogihara & Ueda, 1980).

according to Petrikat (1955, 1964), they should produce nappe portions with phase shifts in nappe oscillation of roughly 180° (Figure 6.68c). From natural-frequency computations, Ogihara & Ueda (1980) concluded that the flap in Figure 6.69a vibrated in a flexural mode like a beam between two supports, rather than in a rotational mode as a rigid body. Important for the success of the spoilers shown in Figure 6.69b is the additional aeration of the nappe underside that they provide (Figure 6.69d).

6.8.5 *Control of swirling-flow instabilities*

The most elegant way of controlling the oscillatory behavior of any flow involving instability-induced excitation is to shape the crucial flow passages so that the potential sources of excitation are eliminated. In the case of a swirling flow in an exit chamber of a turbomachine, for example, Merkli (1978) achieved not only complete suppression of any periodic pressure fluctuations but also a 20% reduction in energy losses by modifying the geometry of the exhaust chamber appropriately. Suppression by means of a splitter plate in the exit chamber, in comparison, increased the pressure-loss coefficient by 25% (Escudier, 1987).

The most significant parameters affecting the swirl, and hence the surging, in draft tubes are the draft-tube geometry (Palde, 1972; Kubota & Aoki, 1980), the wicket-gate geometry, and the turbine characteristics (Grein, 1980; Falvey, 1980). Corrective measures to attenuate the surging accoring to Falvey (1971, 1980) have included modifying the runner characteristics; eliminating the feedback path by filling the reverse-flow region with air or vapor bubbles (Ulith, 1968; Kenn et al., 1980); reducing flow asymmetry by extension of the runner cone into the draft tube or by a cone rising from the draft-tube floor (Chen, 1980, p. 269); damping the angular momentum with fins on the draft-tube walls or fins on a shaft; and introducing counter-swirl through flush-mounted nozzles located in the draft tube immediately downstream of the turbine rotor (Seybert et al., 1978). As the amount of vapor or entrained air increases, the surge intensity first increases to a maximum and then decreases again to a negligible level. According to Ulith (1968), the necessary amount of injected air for optimal surge attenuation is between 0.1% and 2% by volume, and the amount depends on whether the air is injected upstream or downstream of the runner. Seybert et al. (1978) claim that

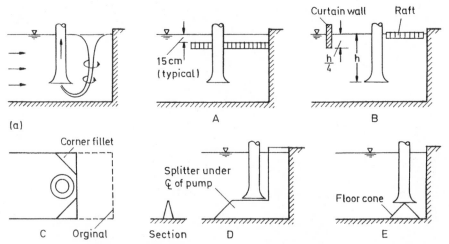

Figure 6.70. (a) Pump intake with surface vortex. (A-E) Schematics of vortex suppressing measures: (A) Horizontal floor grating over entire width; (B) Floating raft and curtain wall; (C) Reduction of wall clearance; (D) Floor and wall splitter; (E) Floor cone (after Padmanabhan, 1987).

their method using counter swirl is one of the few that does not reduce the efficiency of the machine.

Swirling flow is generated also at intakes of pumps, fans, or compressors. If a pump intake is located near a free surface (Figure 6.70a), a 'surface vortex' may form that can suck air and debris intermittently into the intake pipe. Whether air (and debris) is sucked into the surface vortex depends on the depth of submergence of the pump entrance, the pump discharge, and the boundary geometry of the pump intake (Knauss, 1987). Since surface vortices can produce low-frequency pressure fluctuations (Chen, 1980; Chow & Rudavsky, 1980), increased energy losses, and damage by entrained debris, they must be avoided by appropriate design. Surface vortices can be controlled by (a) increasing the submergence of the pump entrance or eliminating the free surface by means of an adequately inclined roof; (b) eliminating approach-flow nonuniformities through modifications of the pump geometry or the use of screens, gratings, or perforated walls; and (c) providing vortex suppression devices (Hydraulic Institute Standards, 1975; Prosser, 1977; Padmanabhan, 1987). Some of the measures for suppressing surface vortices are illustrated in Figure 6.70. More information is contained in the IAHR Monograph on Swirling Flow Problems at Intakes (Knauss, 1987).

CHAPTER 7

Vibrations due to movement-induced excitation

7.1 SOURCES OF MOVEMENT-INDUCED EXCITATION (MIE)

The cases of extraneously induced (EIE) and instability-induced excitation (IIE) described in Chapters 5 and 6 can be identified by examination of the flow without considering body motion. For example, turbulence incident upon a body (EIE) or vortex shedding from it (IIE) exist even if the body is stationary. To be sure, the nature of the vortex shedding due to IIE is altered if the body undergoes motion, but its source is still the same: it arises from a flow instablity in the near wake. Likewise, the incident turbulence is an inherent feature of the flow, irrespective of body motion.

Movement-induced excitation (MIE), in contrast, is inherently linked to body movements and disappears if the body comes to rest. Whenever a body is accelerated in a fluid, an unsteady flow is induced that alters the fluid forces on the body (Figure 3.1). If this alteration leads to negative damping or to a transfer of energy to the moving body, a self-excited body vibration is possible. This self-excitation is referred to as MIE in this Monograph.

The special danger inherent in MIE is that it cannot be detected from model tests unless one anticipates its occurrence and plans the tests accordingly. Obviously, movement-induced exciting forces can be simulated only if the significant parts of the structure are reproduced with similar mass, damping, and elastic characteristics.

In the following brief summary, some MIE mechanisms are presented under four main headings (Figure 7.1):
1. MIE not dependent on coupling;
2. MIE involving coupling with fluid-flow pulsations;
3. MIE involving mode coupling;
4. MIE involving multiple-body coupling.

The primary feature of the *coupling-independent* type of MIE is that motion of a single body in a single mode suffices to change the fluid forces in a way that energy is transferred from the flow to the moving body. For some cases (e.g., galloping), the exciting force can be described in a more or less quasi-steady

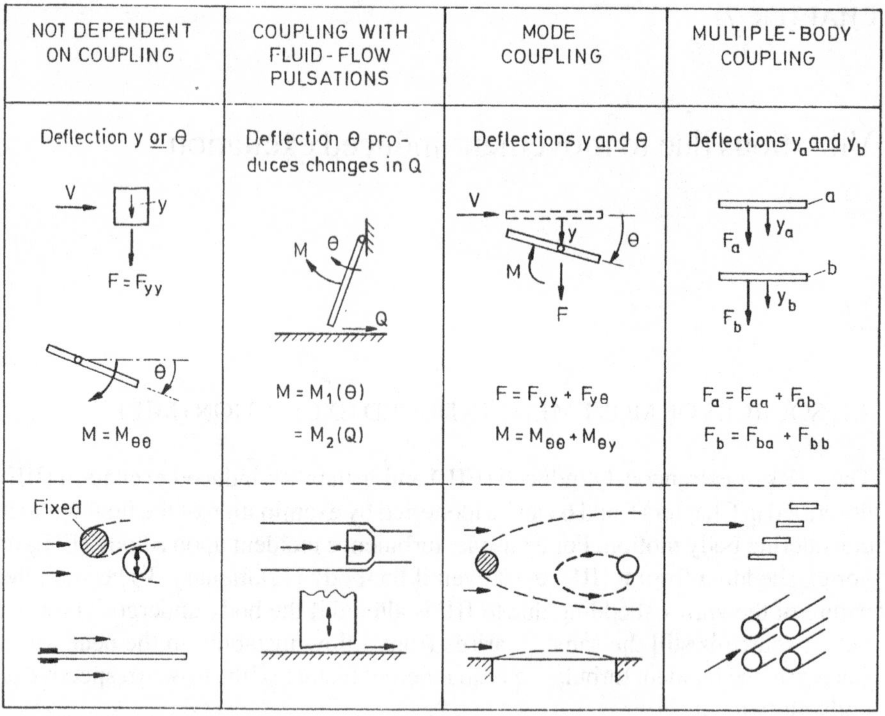

NOT DEPENDENT ON COUPLING	COUPLING WITH FLUID-FLOW PULSATIONS	MODE COUPLING	MULTIPLE-BODY COUPLING
Deflection y or θ	Deflection θ produces changes in Q	Deflections y and θ	Deflections y_a and y_b
$F = F_{yy}$ $M = M_{\theta\theta}$	$M = M_1(\theta)$ $= M_2(Q)$	$F = F_{yy} + F_{y\theta}$ $M = M_{\theta\theta} + M_{\theta y}$	$F_a = F_{aa} + F_{ab}$ $F_b = F_{ba} + F_{bb}$
Fixed			

Figure 7.1. Classification of movement-induced excitation (Subscript $y\theta$ denotes contribution of θ motion to y force; subscript ba denotes contribution of body a on body b).

manner. In other cases (e.g., stall flutter), the energy transfer and excitation is brought about by the deviation from quasi-steadiness.

In the category of *coupling with fluid-flow pulsations*, the major feature of excitation is a pulsation in discharge and/or head that arises in conjunction with the body oscillation. One may distinguish cases of underflow, overflow, and flow through gaps formed by seals or valves. In general, any flow system with structural movements giving rise to appreciable changes in discharge or head are susceptible to this type of MIE. For example, if the fluid-inertia forces accompanying the discharge variations are likely to act in the direction of body velocity, energy will be transferred to the body and the necessary condition for MIE will be satisfied.

The *mode-coupling* type of MIE involves simultaneous movements of a body in two or more modes. Even though each mode by itself may be free of MIE, vibration in the one mode can produce fluid forces which transfer flow energy to the body in the other mode(s). With this coupling, what is called frequency merging must occur; that is, the frequency of each mode must be a function of flow velocity such that two of them merge to a common value at a certain velocity.

In cases of MIE with *multiple-body coupling*, finally, the motion of neighboring bodies, influencing one another through fluid-dynamic coupling, plays an essential role in the excitation.

The simplified classification scheme used herein involves sharp lines of demarcation between the proposed types of MIE. Such distinctions may not always be appropriate. Moreover, some risk is involved in codifying existing knowledge at a time when the state of understanding is as incomplete as it is on this subject. For this reason, the classification is intended as a preliminary tool for organizing the assessment of flow-induced vibrations in engineering systems rather than a code; it is a tool that will undoubtedly be improved in the future. The objective of the following sections is to provide a framework in the search for possible sources of MIE, rather than to give a comprehensive overview of such sources or an in-depth review of information available on specific MIE cases. Emphasis is placed on the MIE of body oscillators. Examples of MIE of fluid oscillators are presented in Section 8.4.

To identify sources of MIE for a particular flow-structure configuration, one must determine how structural movements affect the fluid forces in such a way that they do positive work on the structure. This task is impossible without some analysis. Hence, some rudimentary analytical tools are introduced in the following section. They are also the basis for the notation in Figure 7.1.

7.2 METHODS OF ANALYZING MOVEMENT-INDUCED EXCITATION

7.2.1 *Use of added coefficients*

As defined in Section 2, the equation of motion for a linear body oscillator with a translatory or a torsional degree of freedom can be written, respectively, as

$$m\ddot{y} + B\dot{y} + Cy = F_y(t)$$
$$\ddot{y} + 2\zeta\omega_n\dot{y} + \omega_n^2 y = F_y(t)/m, \qquad \omega_n^2 = C/m \tag{7.1}$$

or

$$I_\theta\ddot{\theta} + B_\theta\dot{\theta} + C_\theta\theta = M(t)$$
$$\ddot{\theta} + 2\zeta_\theta\omega_{\theta n}\dot{\theta} + \omega_{\theta n}^2\theta = M(t)/I_\theta, \qquad \omega_{\theta n}^2 = C_\theta/I_\theta \tag{7.2}$$

where m = mass and I_θ = mass moment of inertia, both per unit width, and ω_n and $\omega_{\theta n}$ are the (uncoupled) circular natural frequencies. In cases where the force $F_y(t)$ or the moment $M(t)$ exerted by the fluid on the body is solely due to body movements, it can be described in terms of components containing the corresponding accelerations $\ddot{y}$, $\ddot{\theta}$, velocities $\dot{y}$, $\dot{\theta}$, and displacements y, θ. Simplified by linear expressions, this description yields, by analogy with Equation 3.1 or 3.4,

$$F_y(t) = -A'\ddot{y} - B'\dot{y} - C'y \quad \text{or} \quad F_y(t) = -A'\ddot{y} - B'\dot{y} \qquad (7.3\text{a, b})$$

$$M(t) = A'_\theta\ddot{\theta} - B'_\theta\dot{\theta} - C'_\theta\theta \quad \text{or} \quad M(t) = -A'_\theta\ddot{\theta} - B'_\theta\dot{\theta} \qquad (7.4\text{a, b})$$

In Equations 7.3b and 7.4b, the terms $C'y$ and $C'_\theta\theta$ have been included in the first terms on the right-hand sides on the basis of the relationships $y = -\omega_n^2\ddot{y}$ and $\theta = -\omega_{\theta n}^2\ddot{\theta}$, which are valid for cases of harmonic motion. (The coefficients A' and A'_θ in Equations 7.3b and 7.4b are different from those in Equations 7.3a and 7.4a.)

If the body has two degrees of freedom as the one shown in Figure 7.2, one needs two equations to describe its motion:

$$m\ddot{y} + S\ddot{\theta} + B\dot{y} + Cy = F_y(t) \qquad (7.5)$$

and

$$I_\theta\ddot{\theta} + S\ddot{y} + B_\theta\dot{\theta} + C_\theta\theta = M(t) \qquad (7.6)$$

(e.g., Simiu & Scanlan, 1978, p. 222, or Dowell et al., 1989, p. 58). Herein, y and θ are the translatory and torsional displacements, respectively; and S is the static imbalance per unit width defined as

$$S \equiv ma = I_\theta a/r_g^2 \qquad (7.7)$$

where a = distance separating the center of gravity from the elastic center (Figure 7.1); and r_g = radius of gyration of the body about the elastic center (the sign of S depends upon whether the center of gravity is upstream or downstream of the elastic center). The second terms in Equations 7.5 and 7.6 describe the *mode coupling* of the body itself as it would occur in a perfect vacuum. Clearly, angular accelerations $\ddot{\theta}$ contribute an inertia term $S\ddot{\theta}$ in the y direction, much as translatory accelerations $\ddot{y}$ contribute a term $S\ddot{y}$ in the θ direction.

In addition, mode coupling may also be associated with the fluid loading. This part of mode coupling is strongly affected by the offset of the elastic axis from the *fluid-dynamic center*, defined as the location about which the moment is indepen-

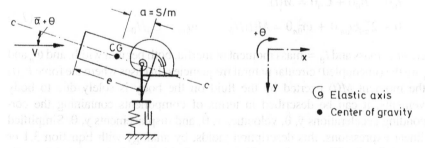

Figure 7.2. Schematic for cylindrical structure with two degrees of freedom in a cross-flow ($\tilde{\alpha}$ = mean angle of incidence).

dent of the angle of incidence. Even if this offset is zero, however, coupling terms are still caused by the lack of symmetry of the flow pattern and the corresponding fluid loading. In general, the motion in the y direction produces inertia, damping, and restoring moments in the θ direction designated as $A'_{\theta y}\ddot{y}$, $B'_{\theta y}\dot{y}$, $C'_{\theta y}y$, where the subscript θy denotes the contribution of y direction motion to θ direction moment. Likewise, motion of the body in the θ direction produces corresponding forces $A'_{y\theta}\ddot{\theta}$, $B'_{y\theta}\dot{\theta}$, and $C'_{y\theta}\theta$ in the y direction. Of course, as indicated in Equations 7.3 and 7.4, the motions in the y and θ directions also produce their own inertia, damping, and restoring forces and moments in the respective directions. Hence, $F(t)$ and $M(t)$ in Equations 7.5 and 7.6 assume the form

$$F_y(t) = -A'_y\ddot{y} - B'_y\dot{y} - C'_y y - A'_{y\theta}\ddot{\theta} - B'_{y\theta}\dot{\theta} - C'_{y\theta}\theta \qquad (7.8)$$

$$M(t) = -A'_{\theta y}\ddot{y} - B'_{\theta y}\dot{y} - C'_{\theta y}y - A'_{\theta}\ddot{\theta} - B'_{\theta}\dot{\theta} - C'_{\theta}\theta \qquad (7.9)$$

The magnitudes of the 'added' coefficients in these equations depend on the MIE mechanism (e.g., unstalled or stalled flutter) and must be determined, by analogy with Equation 3.26, from (a) free- or (b) forced-oscillation experiments as a function of the geometry of the flow boundaries, the approach-flow conditions, and various flow parameters. The latter include the reduced velocity and (a) the relative damping or (b) the relative amplitude of vibration. Some examples are given in the following sections.

The drawback of the methods of analysis using added coefficients, or stability diagrams based on them, is their failure to predict nonlinear processes or behavior. They are appropriate if one wishes to predict processes involving small amplitudes such as the onset of vibration from rest ('soft' excitation). In cases involving large-amplitude fluid forces such as 'hard' excitation, for which a finite disturbance is required to trigger the body vibration, the expression for $F_y(t)$ and $M(t)$ would actually have to contain higher-order contributions (e.g., $B'_2\dot{y}^2$, $B'_3\dot{y}^3$, ...), in addition to $B'\dot{y}$. An example of this sort is described in conjunction with Figure 7.21d.

7.2.2 Use of stability criteria

Section 3.3.3 illustrates how stability diagrams can be deduced from added coefficients for the prediction of ranges of vibration. For example, Figure 3.27 depicts the range of galloping vibrations for the case of a square prism in cross-flow, obtained from an equation analogous to Equation 7.3b. (A comparison of Equations 7.3b and 3.34 shows that $A' = +\rho d^2 (\omega_n/\omega)^2 h_a$ and $B' = -\rho d^2\omega_n k_a$.) The two most important parameters characterizing dynamic stability according to this figure are the reduced velocity V_r and the mass-damping parameter, or Scruton number, Sc:

$$V_r \equiv \frac{V}{f_n d} \quad \text{and} \quad Sc \equiv 2m_r\zeta, \quad m_r \equiv \frac{m}{\rho d^2} \qquad (7.10)$$

where m_r = reduced mass. Parameters of this kind play a dominant role in the formulation of dynamic stability of *any* flow/structure system, as well as in the description of similitude in physical-model investigations of movement-induced excitation.

If one wishes to find whether the system described by Equation 7.1 is susceptible to movement-induced excitation, one can combine it with Equation 7.3b in the form

$$\ddot{y} + (2\omega_n \zeta + B'/m)\dot{y} + \omega_n^2 y = 0 \qquad (7.11)$$

where m includes the added mass A'. The criterion for dynamic stability with respect to infinitesimal disturbances (i.e., with respect to 'soft' excitation) follows from the expression in parentheses as

$$B' + 2m\zeta\omega_n \geq 0 \qquad (7.12)$$

Since $F(t)$ in Equation 7.3b is a fluid force per unit width, one may assume its amplitude to be proportional to $\rho dV^2/2$, where ρ is the fluid density, and d and V are the characteristic length and velocity of the flow system. Hence, the component $-B'\dot{y}$ of that force will also scale with $\rho dV^2/2$, i.e.,

$$\frac{-B'\dot{y}}{\rho dV^2/2} = \frac{-B'}{\rho dV/2}\frac{\dot{y}}{V} = a_1\frac{\dot{y}}{V} = \text{Fct}\begin{bmatrix}\text{Parameters describing}\\\text{geometry, flow, and}\\\text{structural dynamics}\end{bmatrix} \qquad (7.13)$$

where a_1 is a dimensionless parameter similar to k_a (Equation 3.37). Incorporating this expression in Equation 7.12, one obtains

$$\frac{V}{f_n d} \leq \frac{4\pi}{a_1}\left(\frac{2m\zeta}{\rho d^2}\right) \qquad \text{or} \qquad (V_r)_{cr} = \frac{4\pi}{a_1}\,\text{Sc} \qquad (7.14)$$

where $(V_r)_{cr}$ is the critical reduced velocity $V/(f_n d)$ at the onset of movement-induced excitation. As shown in the following sections, the stability criteria for soft excitation can be expressed in the form of $(V_r)_{cr}$ as a function of Sc for a variety of MIE mechanisms (e.g., Equations 7.22, 7.23, 7.41).

7.2.3   Use of Argand diagrams

One can view the onset of movement-induced excitation (MIE) as a 'free' vibration (Section 2.2) if one combines the added coefficients (Equation 7.3) with their structural counterparts (Equation 7.1) in the form

$$m^*\ddot{y} + B^*\dot{y} + C^*y = 0 \qquad (7.15)$$

where $m^* = m + A'$, $B^* = B + B'$, and $C^* = C + C'$. The solution of this equation is

$$y(t) = y_0 e^{-(B^*/2m^*)t} \sin\left[\left(\sqrt{1 - \frac{B^{*2}}{4C^*m^*}}\sqrt{\frac{C^*}{m^*}}\right)t + \phi\right] \qquad (7.16)$$

In many analyses of MIE (flutter), the mechanical damping B is omitted in determining the potential for amplified vibrations. Since the criterion for amplification of the vibration is $B^* = B + B' < 0$, omission of B generally yields a conservative estimate for the onset of amplified body motion. (Exceptions like certain cases of coupled-mode flutter do occur, in which structural damping can have a destabilizing effect and its omission can lead to severe mispredictions, e.g., Benjamin, 1963, and Gregory & Paidoussis, 1966.)

If mechanical damping is omitted, the exponent $(B^*/2m^*)t$ may be expressed as $B'_r t_r$,

$$y(t) \propto y_0 e^{-B'_r t_r} \tag{7.17}$$

where B'_r and t_r are the reduced damping and time, both made dimensionless using mass, elastic, and geometric parameters (Bisplinghoff et al., 1955; Paidoussis, 1966). The corresponding reduced natural frequency is $(\omega_n)_r$. The term B'_r represents an amplification factor similar to k_a that is in general strongly dependent on the reduced velocity $V_r \equiv V/(f_n d)$, in close correspondence to Equation 3.37. In cases of distributed-mass body oscillators such as rods, pipes, and shells, B'_r symbolizes the integrated contribution to the reduced damping of all parts of the structure.

The critical reduced velocity, $(V_r)_{cr}$, at which vibration will start to amplify $(B'_r \leq 0)$ and the corresponding reduced frequency, ω_{cr}, can be graphically represented with the aid of an Argand diagram depicting B'_r versus $(\omega_n)_r$ with V_r as the curve parameter (Figure 7.3). The actual shapes of the curves, the expressions for B'_r and ω_r, and the magnitudes of $(V_r)_{cr}$ and ω_{cr} depend on the specific flow and body-oscillator characteristics. Suffice it here to illustrate the use of Argand diagrams for various types of instabilities and the central role played by the fluid damping (or amplification) B'_r.

Figure 7.3a shows how the Argand diagram is constructed by combining the B'_r versus $(\omega_n)_r$ and the B'_r versus V_r characteristics; the arrow in each diagram indicates the direction of increasing V_r. An advantage of this representation is that it discloses dynamic as well as static instabilities. For the first example depicted in Figure 7.3a, the vibration remains damped $(B'_r > 0)$ as V_r increases until a critical velocity $V = V_D$ is reached at $(\omega_n)_r = 0$; at this point, B'_r 'diverges' in the positive and negative directions. Since this instability occurs at zero frequency, it is termed static instability (or divergence), and V_D is called divergence velocity (Section 3.4).

Another advantage of the Argand diagram is its applicability to movement-induced excitations of both the single-mode and coupled-mode variety. Figures 7.3b, c show applications of the first variety. For axial flow past a cantilevered rod, an increase in V_r results in an initial increase and then a decrease in B'_r until the condition for onset of flutter vibration is reached $(B'_r = 0)$, followed by increased amplification $(B'_r$ increasingly negative). In the event that the natural frequency of the structure is independent of flow velocity, as in the case of bridge decks,

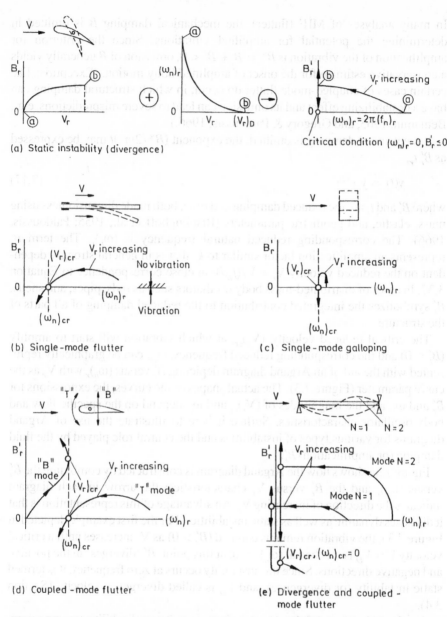

Figure 7.3. Argand diagrams showing variation of reduced fluid damping B'_r with reduced natural frequency $(\omega_n)_r$ as the reduced velocity V_r is varied for several flow-structure systems (Examples in b, d, and e are from Paidoussis, 1973, Carta, 1978, and Paidoussis & Issid, 1974, respectively).

hydraulic gates, etc., the Argand diagram is considerably simpler. For a square prism in cross-flow with a transverse degree of freedom, for example, the locus of increasing V_r is a vertical line (Figure 7.3c). If $V_r = (V_r)_{cr}$, B'_r becomes zero and the vibration is amplified.

If a structure possesses more than one possible mode of vibration, as for an air- or hydrofoil having torsional ('T') and bending ('B') modes, the possibility for coupled-mode flutter exists. As shown in Figure 7.3d, the variation of B'_r with increasing V_r is, in general, quite different for the two independent modes. Although both show an initial increase in damping with incrasing velocity, it is the torsional mode for which the increasing V_r eventually leads to $B'_r \to 0$. The remarkable feature of coupled-mode flutter is that this negative damping of only one of the two modes suffices to excite vibration if the natural frequencies of the two modes become equal so that energy can be exchanged between the modes. There are, however, cases of coupled-mode flutter for which not only the values of natural frequency but also the values of negative damping are equal at the onset of vibration. Examples include flutter of pipes with internal flow (Paidoussis, 1975) and coaxial cylindrical shells conveying viscous fluid (Paidoussis et al., 1985).

For certain configurations, such as pipes pinned at both ends or rods with axial flow, for which divergence is a prelude to flutter, the Argand diagram takes on a markedly different shape (Figure 7.3e). For increasing V_r, there is initially an increase in damping B'_r for both modes $N = 1, 2$ followed by divergence in the form of buckling in each mode $[(\omega_n)_r = 0]$. If the velocity increases further, the damping parameters eventually merge at $V_r = (V_r)_{cr}$, and the pipe or rod goes into coupled-mode flutter at a frequency just above $(\omega_n)_r = 0$. A further increase in V_r results in larger magnitudes $|B'_r|$ of the negative damping and therefore more rapidly amplified vibrations. In contrast to the case depicted in Figure 7.3d, the values of B'_r of both modes are the same at the onset of flutter, and divergence sets in for both modes prior to this onset.

7.2.4   *Use of quasi-steady approach*

For a certain type of movement-induced excitation known as *galloping*, the various phases of vibration are quasi-steady in the sense that there is no phase shift between fluid force and body velocity. The instantaneous exciting force acting on the moving structure in this case is nearly equal to the 'static' force evaluated at the instantaneous angle of flow incidence, provided that the body velocity (of order $f_n d$) is small compared to the flow velocity (V), or $V_r \equiv V/f_n d$ is sufficiently large (say $V_r >$ about 20). For the cylindrical structure in Figure 7.4 at an instant when it has a velocity $\dot{y}$ in the transverse direction, the approach velocity, relative to the body, has a magnitude and an angle with the vector V

$$V_{rel} = \sqrt{V^2 + \dot{y}^2}, \qquad \theta = \tan^{-1}(\dot{y}/V) \qquad (7.18)$$

If F_D and F_L denote the components of drag and lift at that instant, then the instantaneous transverse force F_y in the positive y direction is

$$F_y = -F_L \cos \theta - F_D \sin \theta \qquad (7.19)$$

with

$$F_L = C_L \, d\rho V_{rel}^2/2, \qquad F_D = C_D d\rho V_{rel}^2/2$$

Here, C_L and C_D are the lift and drag coefficients obtained from 'static' tests as a function of $\alpha = \bar{\alpha} + \theta$ and a variety of other independent parameters such as cross-sectional form, Reynolds number, turbulence intensity, aspect ratio, etc. (see Naudascher, 1991, Section 2.3.4, for information on C_L and C_D). In terms of coefficient C_y, hence,

$$F_y = C_y d\rho V^2/2, \qquad C_y = -\frac{1}{\cos \theta}(C_L + C_D \tan \theta) \qquad (7.20)$$

The criterion of stability with respect to *soft excitation* (small-amplitude excitation from rest) can now be obtained by using Equation 7.3b, which expresses F_y in terms of added coefficients. Since $A'\ddot{y}$ can be assumed to be invariant with V_r for the velocity range considered ($V_r' \gg 1/\mathrm{Sh}$), this part can be lumped together with the inertia term $m\ddot{y}$, so that the equation of motion (Equation 7.1) becomes, after substitution of the expression from Equation 7.13 for $B'\dot{y}$,

$$m\ddot{y} + m2\zeta\omega_n\dot{y} + m\omega_n^2 y = C_y d\frac{\rho V^2}{2} = \left(d\frac{\rho V^2}{2}\right)a_1\frac{\dot{y}}{V}$$

$$\ddot{y} + \left(2\zeta\omega_n - \frac{\rho V d}{2m}a_1\right)\dot{y} + \omega_n^2 y = 0 \qquad (7.21)$$

The structure in question is stable with respect to small-amplitude disturbances if either the combined damping term is larger than zero or if

$$V_r \equiv \frac{V}{f_n d} \le \frac{4\pi}{[dC_y/d\alpha]_{\alpha = \bar{\alpha}}} \mathrm{Sc}, \qquad \mathrm{Sc} \equiv \frac{2m\zeta}{\rho d^2} \qquad (7.22)$$

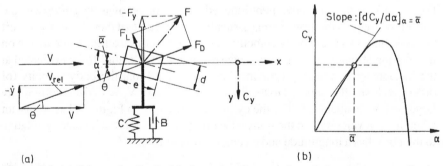

(a)   (b)

Figure 7.4. (a) Schematic of cylindrical structure in cross-flow during transverse galloping showing quasi-steady fluid forces. (b) Corresponding force coefficient $C_y \equiv F_y/(d\rho V^2/2)$ ($\bar{\alpha}$ = angle of flow incidence for body at rest).

Herein, a_1 has been replaced by

$$a_1 = \left[\frac{dC_y}{d\theta}\right]_{\theta = 0} = \left[\frac{dC_y}{d\alpha}\right]_{\alpha = \bar{\alpha}} \tag{7.23}$$

in accordance with the expression $C_y = a_1 \dot{y}/V$. Transverse galloping from rest develops, according to Equation 7.22, if the slope of the curve C_y versus α at the equilibrium position $\alpha = \bar{\alpha}$ (Figure 7.4b) is positive (Den Hartog criterion) *and* if the reduced velocity V_r exceeds its critical value, which is

$$(V_r)_{cr} = \frac{4\pi \, Sc}{[dC_y/d\,\alpha]_{\alpha = \bar{\alpha}}} \tag{7.24}$$

The analysis of *torsional galloping* is more complicated than transverse galloping because the instantaneous angle of flow incidence and the relative velocity V_{rel} are functions of two factors: the instantaneous angular displacement θ and the instantaneous linear velocity in a direction normal to the chord c-c (Figure 7.2); the latter depends in turn on the location of the elastic axis and the position along the chord. The elastic axis is the line about which a vertical static force produces displacement but no torsion; it can thus be interpreted as the 'center of twist'.

In summary, the stability criterion with respect to soft excitation, deduced from Equation 7.2 with $M = C_M d^2 \rho \, V_{rel}^2/2$, is

$$V_r \equiv \frac{V}{f_n d} \le \frac{4\pi}{[dC_M/d\alpha]_{\alpha = \bar{\alpha}}} Sc_\theta, \qquad Sc_\theta = \frac{2I_\theta \zeta_\theta}{\rho R d^3} \tag{7.25}$$

(e.g., Blevins, 1977, p. 62, or 1990, p. 110), wherein C_M = moment coefficient obtained from static measurements about the axis of rotation (or elastic axis), I_θ = mass moment of inertia of the body including added-mass effects, and R = 'characteristic radius' used to evaluate the angle of attack induced by the body rotation. (If the elastic axis is downstream of the fluid-dynamic axis, defined as the axis around which $M = 0$, R is negative.)

In the mass-damping ratio Sc_θ of Equation 7.25, a typical value for R in the denominator is $e/4$ for a rectangular prism of chord length e galloping around its geometric center (Nakamura & Mizota, 1975). For long bluff bodies ($e/d \gg 1$), however, the quasi-steady assumption does not apply, and Equation 7.25 is, hence, invalid (Section 7.3.2).

In cases of transverse galloping, the quasi-steady approach can be used to predict not only the structural response for $V_r > (V_r)_{cr}$ (Parkinson, 1974) but also the onset of *hard excitation*, i.e., the onset of vibration triggered by a finite-amplitude disturbance (Novak, 1969, 1970, 1972). In this case, however, one must account for the higher-order terms in the equation of motion, which, in the above quasi-steady analysis, follow from a power-series description of the C_y versus α curve in Figure 7.4b. On the basis of $\alpha = \bar{\alpha} + \theta$ and $\theta = \tan^{-1}\dot{y}/V$ (Figure 7.4a), Parkinson & Smith (1964), and subsequently Novak (1969, 1972), used the following expression

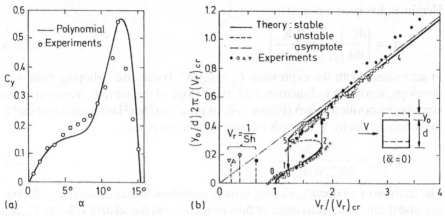

Figure 7.5. (a) Steady-state transverse-force coefficient for a square prism in low-turbulence cross-flow. (b) Galloping response of square prisms with different mass-damping ratios Sc in low-turbulence cross-flow (after Parkinson & Smith, 1964).

$$C_y = \sum_{j=1}^{n} a_j \left(\frac{\dot{y}}{V}\right)^j = a_1 \frac{\dot{y}}{V} + a_2 \left(\frac{\dot{y}}{V}\right)^2 + a_3 \left(\frac{\dot{y}}{V}\right)^3 + \dots \qquad (7.26)$$

to describe the curve for a variety of cross-sectional shapes and flow conditions. A solution for a square prism, obtained numerically by employing the method of Krylov and Bogoliubov (Minorsky, 1962) on the basis of a four-term polynomial with positive constants a_1, a_3, a_5, a_7 (Equation 7.26), is shown in Figure 7.5. Solutions for other profile shapes and different levels of free-stream turbulence are presented in Figures 7.22 and 7.24. According to more recent findings (Parkinson, 1980; Parkinson & Wawzonek, 1981; Bokaian & Geoola, 1984), the theoretical asymptote in Figure 7.5b is also approached for cases in which $(V_r)_{cr}$ is very close to, or even below, the value $V_r = 1/Sh$ that corresponds to the resonance point for vortex shedding (Figure 7.25). A mathematical model describing the interaction of vortex shedding and galloping by a combination of the oscillator approach (Section 3.3.2) and the quasi-steady approach has recently been presented by Corless & Parkinson (1988).

7.3 EXCITATION NOT DEPENDENT ON COUPLING

7.3.1 *Nonstall flutter in cross-flow*

If a slender body (blade or airfoil) in a parallel stream at small or zero incidence angle performs heaving or pitch vibrations, vortices are shed from the trailing edge in harmony with the body motion even for extremely small vibration amplitudes. These movement-induced vortices (which are not to be confused with vortices shed from a stationary body, Section 6.3) are due to changes of the bound circulation about the body. Such vortices are responsible for the crucial phase shift of the fluid loading relative to the body motion, without which the flutter type of movement-induced excitation would be impossible. Physically, this phase shift is explained by the fact that a vortex, produced near the trailing edge at time t, has its major effect on the body at time $t + \Delta t$.

Figure 7.6 shows schematically the basic components of flow for a stationary, inclined plate in a parallel stream (Figures 7.6a-c) and a plate performing heaving motions in a stagnant fluid (Figure 7.6d). The condition that the flow leaves the plate tangentially (Kutta condition) becomes satisfied by the superposition of appropriate circulatory flows to the noncirculatory flows shown in Figures 7.6a and d. In case of a plate with periodic heaving (or bending) motion depicted in Figure 7.6f, the flow pattern can be viewed as the resultant of noncirculatory, circulatory, and parallel flow contributions leading to a chain of alternately rotating vortices. These vortices are the counterpart of the bound circulations about the body from successive cycles of the body motion. Movement-induced vortex shedding of this kind is a side effect of *any* movement-induced excitation involving cylindrical bodies and is not to be confused with the instability-induced type of vortex shedding presented in Section 6.3.

For small-amplitude vibrations of the plate in stagnant fluid (Figure 7.6d), the fluid force may be expressed as

$$F_y = -A'\ddot{y}, \quad A' = \pi\rho(e/2)^2 \tag{7.27}$$

where e = length of the plate. With the combined effects of the circulatory and

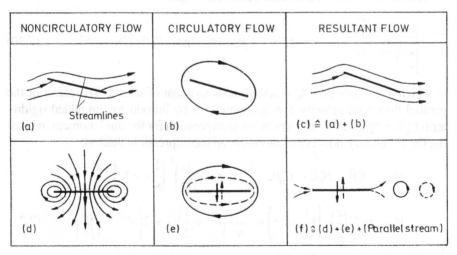

Figure 7.6. Schematic of basic flow elements: (a-c) Stationary inclined plate in parallel stream; (d) Vibrating plate in stagnant fluid; (e, f) Vibrating plate in parallel stream.

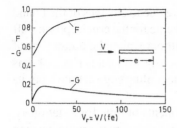

Figure 7.7. Theodorsen functions for an unstalled airfoil vibrating about zero angle of attack ($\bar{\alpha} = 0$).

parallel flow (Figure 7.6f), this force becomes

$$F_y = -A'\ddot{y} - B'\dot{y} - C'y$$

$$B' = \pi\rho VeF, \quad C' = -\pi\omega\rho VeG \tag{7.28}$$

in which F and G are functions of the reduced velocity as shown in Figure 7.7 (Bisplinghoff et al., 1955).

Due to the added damping B' and added rigidity C', a phase lag $\Phi_c = \tan^{-1}G/F$ is produced which remains significant up to very large reduced velocities V_r (G approaches the quasi-steady limit of zero at $V_r = \infty$ very slowly). The added damping B' is always positive. Consequently, one may state:

> Movement-induced excitation in the pure heaving mode of slender bodies (plates) placed parallel to a free stream is not possible. (7.29)

For a plate performing pitch (or torsional) vibration around an elastic axis a distance 'a' downstream from the midchord, the fluid moment due to the noncirculatory flow, induced in an otherwise stagnant fluid, is

$$M = -A'_\theta\ddot{\theta}, \quad A'_\theta = \pi\rho(e/2)^4 (1/8 + a^{*2}) \tag{7.30}$$

in which $a^* = 2a/e$ and A'_θ = added mass moment of inertia. The effect of the parallel flow with velocity V is accounted for by introducing an added-rigidity term $C_\theta\theta = \pi\rho(e/2)^2V^2\theta$. The complete expression for the fluid moment, including the circulatory-flow contributions in an incompressible fluid, is

$$M = -A'_\theta\ddot{\theta} - B'_\theta\dot{\theta} - C'_\theta\theta, \qquad A'_\theta = \pi\rho \left(\frac{e}{2}\right)^4 \left(\frac{1}{8} + a^{*2}\right)$$

$$B'_\theta = \pi\rho V \left(\frac{e}{2}\right)^3 \left[\left(\frac{1}{2} - a^*\right) - \frac{2V_r}{\pi}\left(a^* + \frac{1}{2}\right)G + 2\left(a^{*2} - \frac{1}{4}\right)F\right] \tag{7.31}$$

$$C'_\theta = -\pi\rho V^2 \left(\frac{e}{2}\right)^2 \left[2\left(a^* + \frac{1}{2}\right)F + \frac{2\pi}{V_r}\left(a^{*2} - \frac{1}{4}\right)G\right]$$

In contrast to the case of pure heaving, the added damping in torsion is dependent upon F and G and can undergo a change in sign at sufficiently high values of $V_r \equiv V/(fe)$, provided that the elastic axis is located forward of the quarter-chord location. With the elastic axis at the leading edge, for example, B'_θ becomes negative, indicating the onset of movement-induced excitation if $(V_r)_{cr} \geq 82.5$ (Bisplinghoff et al., 1955).

For light structures such as airfoils and blades, the frequency of vibration f will be nearly equal to the natural frequency $f_n = \omega_n/2\pi$. According to

$$f_n = \frac{1}{2\pi}\sqrt{\frac{C_\theta + C'_\theta}{I_\theta + A'_\theta}} \tag{7.32}$$

which is the counterpart to Equation 3.3a for a body with torsional degree of freedom (Equation 2.18), f_n is a function of the velocity V because the added rigidity C'_θ depends on V (Equation 7.31c). After some manipulation, one obtains

$$\frac{f_n}{f} = \left[1 + \frac{C_{M_R}}{I_r}\right]^{-\frac{1}{2}}, \qquad I_r = \frac{I_\theta}{\pi\rho(e/2)^4} \tag{7.33}$$

where f_θ = natural frequency at zero velocity and C_{M_R} = (real) moment coefficient of the fluid moment. Since C_{M_R} is always negative, there is a second important condition for MIE: the reduced moment of inertia I_r must be larger than $|C_{M_R}|$ if the frequency f_n is to remain real. In summary:

> Slender cylindrical bodies (plates) placed in a free stream with their long side parallel to the direction of flow may undergo movement-induced excitation in the pure torsional mode if the elastic axis is sufficiently far ahead of the quarter-chord location and if both the reduced moment of inertia I_r and the reduced velocity V_r exceed certain threshold values. (7.34)

An example for the magnitudes of these threshold values is given in Figure 7.8. The figure also illustrates why nonstall flutter is more likely to occur for supersonic than for subsonic flow velocities (Carta, 1978).

For incompressible flow (Ma $\simeq 0$), the threshold value $(I_r)_{cr} \simeq 600$ is usually larger than the magnitude of I_r for any practical installation. If Ma $= 0.5$, however, this value is lower by more than a factor of two, and the range of dangerous reduced velocities is dramatically larger. Physically, the reason for this increased susceptibility to flutter is that the fluid-dynamic center moves from the quarter-chord ($a^* = -\frac{1}{2}$) to the midchord ($a^* = 0$) as the Mach number Ma varies from 0 to 1.0. Thus:

> Increasing the Mach number enhances the possibility of movement-induced excitation (nonstall flutter) in the pure torsional mode due to a shift of the fluid-dynamic center towards midchord. (7.35)

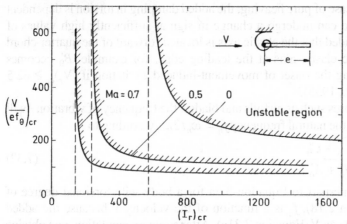

Figure 7.8. Regions of instability for nonstall flutter in the pure torsional mode about the leading edge as affected by Mach number (after Runyan, 1951).

7.3.2 *Stall flutter in cross-flow*

Conventionally, the term stall flutter is used to describe the type of movement-induced excitation associated with elongated structures from which the flow separates partially or completely during at least part of the cycle of structural vibrations. An essential ingredient of stall flutter is the existence of a phase shift between the fluid force and the body velocity (for galloping, in contrast, this phase shift is nearly zero). Stall flutter is possible in systems with both one and two degrees of freedom. The most frequent mode of vibration is torsional, but both transverse and longitudinal flutter are possible as well.

Many of the basic characteristics of nonstall flutter discussed in Section 7.3.1 are also found in cases of stall flutter. For example, the mentioned phase shift remains significant up to very large reduced velocities, and periodic movements induce vortex shedding (Figure 7.6f) for stalled flow as well. Here, however, the vortex shedding typically involves large-scale vortices initiating from the leading rather than the trailing edge of the body, and the effects of this large-scale shedding on the fluid forces dominate the excitation mechanism. These effects involve to a great extent feedback processes as discussed in conjunction with impinging shear layers (Section 9.1).

(a) *Slender bodies.* Stall flutter has been observed for a great variety of 'slender bodies', such as: Spoilers and flaps (Lang, 1976); cantilevered roofs (Quadflieg & Mankau, 1978; Mankau, 1980); bowl-shaped structures such as radio telescopes or solar collectors (Scruton, 1969); helicopter blades (McCroskey, 1977; Dowell et al., 1989); hydrofoils (Abramson, 1969); marine propellers (McCroskey, 1977); compressor blades (Fleeter, 1977; Mikojcsak et al., 1975); and cascades (Sisto, 1977; Perumal, 1976). Indeed, even the vibration of traffic signs about torsionally flexible posts belongs to this category.

A distinction between the two basic types of dynamic-stall patterns shown in Figure 7.9a, b is useful. Light stall is characterized by separation of the laminar boundary layer from the upper part of the leading edge followed by turbulent reattachment and then separation of the turbulent flow; thus, a separated-flow or stalled region is formed near the trailing edge of the air- or hydrofoil. Even in a qualitative sense, the behavior of this light-stall regime depends strongly upon geometry, maximum incidence angle, reduced velocity, and Mach number. Significant energy transfer to the vibrating foil can occur when this regime sets in. Deep stall (Figure 7.9b) occurs if both the maximum incidence angle $\alpha_{max} = \bar{\alpha} + |\theta_o|$ and the vibration amplitude θ_o are large. Under these conditions, the qualitative characteristics of the vortex-dominated stall region are nearly independent of the foil geometry, the Reynolds number, and even the Mach number, provided the velocity does not reach supersonic values near the leading edge. (With Ma = 0.4 in the approach flow, e.g., a local supersonic region develops as $\bar{\alpha}$ reaches 15°, McCroskey, 1982.)

The key features of the flutter mechanism for both light and deep stall are, first, the rather abrupt change in the flow pattern and the accompanying fluid force or moment brought about by structural movements, and, second, the delay time

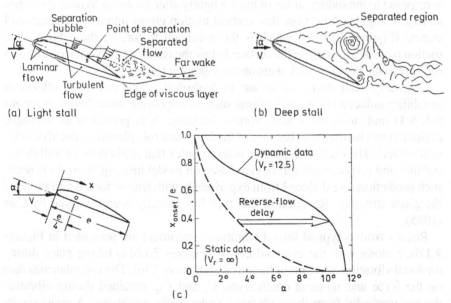

(a) Light stall

(b) Deep stall

(c)

Figure 7.9. (a, b) Basic stall patterns on an air- or hydrofoil vibrating in the torsional mode (after McCroskey, 1982). (c) Onset of reverse flow on a hydrofoil (modified NACA 0012) with incidence angle α increased according to $\alpha = 10° + 10°\sin(2\pi f_s t)$ at two different relative frequencies $f_s e/V = 1/V_r$ (after McAlister & Carr, 1979).

(phase shift) involved in the adjustment of local flow conditions to those associated with the instantaneous body location and velocity. Clearly, flow adjustment
from the attached to the separated state and back cannot occur instantaneously. It
takes place with a certain phase shift, accompanied by movement-induced vortex
generation. Excitation takes place if this phase shift and vortex generation change
the fluid force or moment in a way that the work done on the body is positive. (A
more detailed description of the excitation mechanism has been given by Ericsson
& Reding, 1988.)

Further insight into the dynamic-stall process was obtained by McAlister &
Carr (1979) in a water tunnel. At the relatively low Reynolds number of
$Ve/v = 21\,000$, they found the static-stall angle to be about 10°. With a forced
torsional (or pitch) vibration,

$$\alpha = \bar{\alpha} + \theta_0 \sin \omega_s t, \qquad \omega_s = 2\pi f_s \tag{7.36}$$

the onset of reverse flow was delayed during the quarter cycle during which α
increased from zero, as shown in Figure 7.9c. The process that sets in after flow
reversal has reached the leading edge is visualized in Figure 7.10. Numerous
shear-layer vortices with clockwise rotation form in the initially thin shear layer
along the separation streamline (Figure $7.10b_2$), and they gradually grow into a
dominant shear-layer vortex (Figure $7.10b_4$). As this vortex approaches the
trailing edge, a well-defined, unified vortical motion evolves at the leading edge
in response to the sudden influx of fluid. Slightly after the air- or hydrofoil reaches
its largest angular deflection, this vortical motion grows into the 'dynamic-stall
vortex' (Figures 7.11d-f), which is the dominant feature of these flows. The
suction peak developing on the surface below the center of the vortex contributes
strongly to the positive work done on the vibrating foil.

Obviously, stall flutter of an air- or hydrofoil is not free from effects of
instability-induced (IIE) mechanisms such as impinging shear layers (Sections
6.4, 9.1) and, to some extent, vortex shedding. It is presented as a unique
excitation mechanism here because of the important role played by the 'dynamic-
stall vortex'. The excitation process is so complex that predictions of stall-flutter
stability and response are still largely based on model investigations. In general,
such predictions are deduced from experiments with free or forced vibrations of
the given structure. Research of this type has recently been reviewed by Carr
(1985).

Results from a typical forced-vibration experiment are presented in Figures
7.12b, c along with the essential events (Figures 7.12d-i) taking place during
torsional vibration of an air- or hydrofoil (Equation 7.36). The instantaneous data
on the force and moment coefficients, C_y and C_M, obtained during vibration
deviate markedly from those obtained under static conditions. A quasi-steady
approach would hence be totally inadequate. The work done on the body per
vibration cycle is given by the hysteresis loops in Figure 7.12c according
to $W_e = \oint M d\alpha$. In other words, Figure 7.12c is the equivalent of a force-

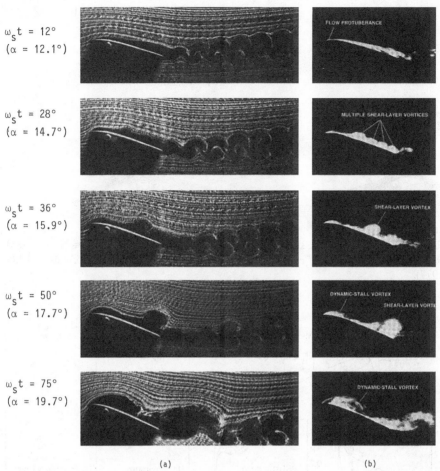

$\omega_s t = 12°$
$(\alpha = 12.1°)$

$\omega_s t = 28°$
$(\alpha = 14.7°)$

$\omega_s t = 36°$
$(\alpha = 15.9°)$

$\omega_s t = 50°$
$(\alpha = 17.7°)$

$\omega_s t = 75°$
$(\alpha = 19.7°)$

(a)   (b)

Figure 7.10. Critical stages of dynamic stall during torsional vibration of a hydrofoil about an axis at quarter chord according to $\alpha = 10° + 10°\sin(2\pi f_s t)$ at $f_s e / V = 1/12.5$ and $Re \equiv Ve/\nu = 21\,000$. (a) Hydrogen bubbles from free-stream electrode only; (b) from model electrodes only (after McAlister & Carr, 1979).

displacement diagram (Figure 2.6b) for a system with a torsional degree of freedom. The (cross-hatched) loop negotiated in the clockwise direction signifies positive work on the body, or the potential for self-excitation; the loop negotiated counter-clockwise signifies negative work on the body, or damping. Since the former contains a larger area than the latter, the test conditions are prone to stall flutter.

Figure 7.13 shows the effects of mean incident angle $\bar{\alpha}$ and reduced velocity V_r. Evidently, a contribution of positive work during part of the cycle comes about only if $\bar{\alpha}$ is close to the static-stall angle of roughly 14° (Figure 7.13b) and if V_r

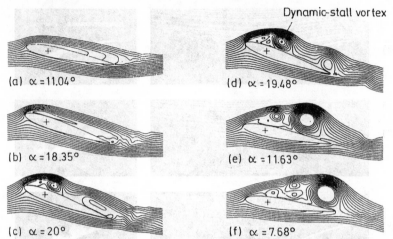

Figure 7.11. Instantaneous streamline patterns for vibrating airfoil (Re $\equiv Ve/\nu = 5000$, $\alpha = 10° + 10°\sin(2\pi f_s t)$ at $f_s e/V = 1/2\pi$) showing the 'dynamic-stall vortex' (after Mehta, 1977).

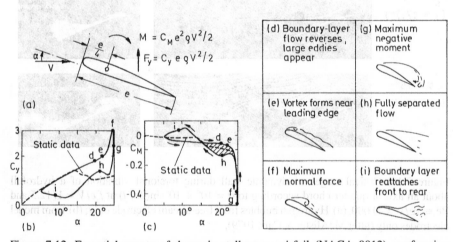

Figure 7.12. Essential events of dynamic stall on an airfoil (NACA 0012), performing torsional vibration ($\bar{\alpha} = 15°$, $\theta_o = 10°$) at $V_r \equiv V/(f_s e) = 42$, and associated effects on the normal-force and moment characteristics (after Carr et al., 1977).

exceeds a certain critical value (Figures 7.13e, f). At $\bar{\alpha} = 7.3°$, far below the static-stall angle, and at $\bar{\alpha} = 24.6°$ for which the airfoil remains fully stalled throughout the oscillation, the C_M loops are close to the static curve, indicating positive fluid damping (Figures 7.13a, c).

Figure 7.14a was obtained from measurements of flutter stress response by cross-plotting the data at constant stress. The figure illustrates the typical transi-

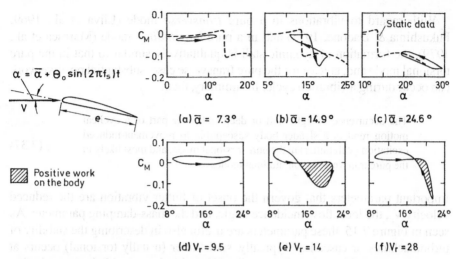

Figure 7.13. Moment coefficient of an airfoil as a function of angular displacement during forced torsional vibration. (a-c) Effect of mean incidence angle $\bar{\alpha}$ at fixed values of $V_r \equiv V/(f_s e) = 102$ and $\alpha_o = 4.85°$, for an axis at quarter chord (after Liiva et al., 1969); (d-f) Effect of reduced velocity V_r at fixed values of $\bar{\alpha} = 15°$ and $\theta_o = 6°$ (after Carta, 1978).

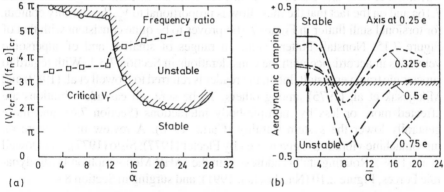

Figure 7.14. (a) Transition from classical flutter to stall flutter for a typical airfoil. (b) Effect of rearward movement of elastic axis on stall-flutter instability (from the review paper by Carta, 1978; f_n = natural frequency in pitching mode).

tion from classical flutter at small $\bar{\alpha}$ and high V_r to stall flutter at large $\bar{\alpha}$ and moderate V_r. According to such studies, the lower limit of V_r for torsional stall flutter seems to be $V/(f_n e) \simeq \pi$ for large values of incidence angle $\bar{\alpha}$ (f_n = natural pitch frequency). Figure 7.14b reveals the dramatic destabilizing effect obtained from moving the elastic axis downstream.

With regard to vibrations in a pure *transverse* mode (Liiva et al., 1969; Fukushima & Dadone, 1977) and in a pure *longitudinal* mode (Maresca et al., 1979), the formation of dynamic stall is qualitatively similar to that in the pure torsional mode and, at least for the pure transverse case, substantial positive work can occur during a vibration cycle. In summary, then:

> The occurrence of light stall or deep stall over part of a cyclical
> motion renders a slender body susceptible to movement-induced
> excitation (stall flutter) in the pure torsional mode, and most likely in   (7.37)
> the pure transverse or longitudinal modes as well.

Important parameters that govern the onset of flutter vibration are the reduced velocity V_r, the local flow-incidence angle, and the mass-damping parameter. As seen in Figure 7.15, these parameters are useful also in describing the stability of turbomachines or cascades. Typically, stall flutter (usually torsional) occurs at part-speed operation (i.e., at intermediate V_r) with the rotor blades set at higher than average incidence angle. Contours of constant flutter stress or tip amplitude have, in principle, the same appearance as the stability contour; their actual shape and location are functions of blade and casing geometry, stagger, Reynolds and Mach numbers, and, of course, mass-damping.

For a specific turbomachine, the flutter contours can be mapped in the machine-performance diagram. Based on the lines of constant incidence angle α in Figure 7.16a and on the fact that the mass flow is proportional to V_x, the stability contour for torsional stall flutter in Figure 7.16b proves to be quite consistent with that of Figure 7.15. Nonstall flutter occurs in ranges of small α and of supersonic velocities in accordance with the considerations in Section 7.3.1. With regard to supersonic transverse stall flutter, the reader is referred to Dowell et al. (1989) and Mikojcsak et al. (1975) among others. To be sure, all cascade vibrations are affected more or less by multiple-body interactions (Section 7.6), and these generally lower the system stability (Carta, 1978). A review of literature on turbomachine vibration is given, e.g., by Fleeter (1977); Sisto (1977); and Dowell et al. (1989). Rotating stall is addressed in the IAHR Monograph on Hydrodynamic Forces, Figure 2.10 (Naudascher, 1991), and surging in Section 8.4.

(b) *Long bluff bodies.* Whereas slender bodies such as airfoils or blades require a relatively large angle of incidence to develop light and deep stall, bodies having bluff leading edges, such as those shown in Figure 7.17, can experience dynamic stall at zero mean incidence ($\bar{\alpha} = 0$). Added coefficients (Section 7.2.1) for cross-sectional profiles that are used for bridge decks have been derived by Scanlan & Tomko (1971) and Nakamura (1978) with respect to vibrations in both the transverse (or bending) and the torsional mode. Since coupling can occur between the two modes, these coefficients are presented in Section 7.5. For certain profiles a possibility exists for uncoupled vibration in the pure torsional

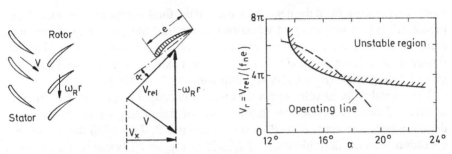

Figure 7.15. Typical stability contour for torsional stall flutter in a turbomachine or cascade (after Dowell et al., 1989).

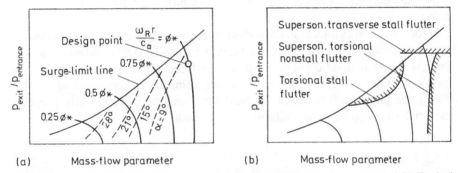

Figure 7.16. (a) Representative performance map of a multistage compressor and (b) Typical stability contours for that compressor (after Dowell et al., 1989). ($\omega_R r$ = speed of rotation, Figure 7.15a; c_a = speed of sound.)

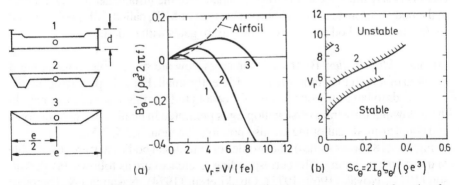

Figure 7.17. (a) Coefficient of fluid-damping B'_θ with respect to pure torsional motion for three long bluff bodies in low-turbulence cross-flow at zero incidence angle with elastic axis at midchord (after Scanlan & Tomko, 1971). (b) Corresponding stability diagram for torsional stall flutter.

mode, as is evident from the negative values of the fluid-damping coefficient B'_θ in Figure 7.17b (f = frequency of vibration). In contrast, pure bending vibration is damped ($B' > 0$) for all profiles except that of the original Tacoma Narrows Bridge (profile 1, e = 12 m, d = 2.5 m); the latter exhibits negative fluid damping in the bending mode ($B'_y < 0$) for approximately $V_r \geq 8.5$ (Figure 7.54).

The critical reduced velocity for the onset of uncoupled torsional stall flutter is determined from the condition that the sum of fluid damping and mechanical damping becomes zero, $B'_\theta + B_\theta = 0$ (Equations 7.2, 7.4). With the aid of this condition, plus the definition $B_\theta \equiv 2\,I_\theta\zeta_\theta(2\pi f)$, Figure 7.17a can be transformed into a stability diagram as shown in Figure 7.17b. Herein,

$$Sc_\theta \equiv 2\,I_\theta\zeta_\theta/(\rho e^3) \tag{7.38}$$

is the mass-damping parameter for torsional vibration (ζ_θ = mechanical-damping ratio).

Information on cylinders in cross-flow with long rectangular cross-sections is given in Section 9.1 for a number of vibration modes. The rectangular profile is of interest to the hydraulic engineer since it is frequently used in trashracks. In torsional flutter of rectangular profiles, according to Nakamura et al. (1979, 1982), the phase shift between the fluid force and the body velocity plays an essential role even for small ratios of length to width.

7.3.3 *Galloping*

For a certain class of flows, distinguished by separated shear layers that do not reattach to the trailing end of the body (at least when it is at rest), movement-induced excitation (MIE) occurs *without* a phase shift between fluid force and instantaneous body velocity. The instantaneous force acting on the body in these cases is nearly the same as the force evaluated for the instantaneous flow relative to the body (quasi-steady condition). This type of MIE, called galloping, is found for (a) short bluff bodies in cross-flow and (b) gates with underflow.

(a) *Short bluff bodies.* Bluff bodies in cross-flow are susceptible to galloping if the ratio of length to height (e/d) is sufficiently small. For galloping to occur, the body needs to have only one degree of freedom (either transverse or torsional). In the following, only transverse galloping is presented in detail, although torsional and coupled-mode galloping are also possible (Sections 7.2.4, 7.5).

Figure 7.18 shows some of the profiles of cylindrical bodies that are susceptible to transverse galloping. The corresponding references are as follows: (a) Parkinson (1974), Novak (1969, 1972), Otsuki et al. (1974), Nakamura & Tomonari (1977), Washizu et al. (1978), Bokaian & Geoola (1982a, 1983); (b) Novak & Davenport (1970), Novak (1972), Bokaian & Geoola (1982a, 1983); (c, d) Bardowicks (1976); (e) Parkinson & Modi (1967), Slater (1969); (f) Bardowicks (1976), (a cruciform profile of $d/3$ thickness proved to be stable with respect to

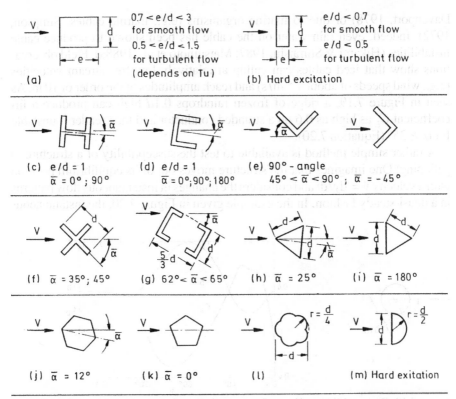

Figure 7.18. Representative profiles of cylindrical bodies susceptible to transverse galloping.

galloping according to Novak & Tanaka, 1974); (g, h) Bardowicks (1976); (i) Bokaian & Geoola (1982a, 1983); (j, k, l) Bardowicks (1976); (m) Parkinson (1963), Novak & Tanaka (1974), Bokaian & Geoola (1982a, 1983). The suscepti-bility to galloping of T-sections is illustrated in Figure 3.26.

As is shown in Sections 3.3.3 and 7.2.4, the onset of galloping can be determined theoretically with the help of either a diagram of k_a as a function of V_r (e.g., Figure 3.25) or a diagram of C_y as a function of α (e.g., Figure 7.5); the latter, in turn, can be obtained with the aid of mean-drag and lift coefficients C_D, C_L (Figure 7.4b, Equation 7.20). The uncertainty inherent in information on C_D and C_L as a function of flow-incidence angle $\bar{\alpha}$ is its dependence on a great number of factors other than cross-sectional shape (e.g., Reynolds number, free-stream-turbulence level, end effects, confinement, see Naudascher, 1991, Section 2.3). Minute changes in geometry can change the C_D and C_L character-istics so drastically that galloping may occur unpredictably. A case in point is a cable in cross-flow with a cross-section that deviates from the circular either due to ice formation (e.g., transmission lines, Richardson et al., 1965; Novak &

Davenport, 1978) or due to marine organisms (e.g., marine cables, Simpson, 1972). Indeed, even rain water on the cable has been shown to produce cable instabilities (Hikami & Shiraishi, 1987; Matsumoto et al., 1988a). Field observations show that iced cables can gallop at relatively low free-stream velocities (e.g., wind speeds of about 4.5 m/s) and reach amplitudes of the order of 10 m. As seen in Figure 7.19, a ridge of frozen raindrops $0.1d$ high can produce a lift coefficient C_L as high as 0.6 on a stranded conductor and thus render it unstable for $\bar{\alpha} \gtrsim 35°$ (Equation 7.20).

A rather simple method is available to test the susceptibility of a structure to galloping: One imagines that the structure moves out of its equilibrium position with a velocity $\dot{y} = dy/dt$ and considers the changes in instantaneous flow patterns in a quasi-steady fashion. In the example given in Figure 7.20, the instantaneous

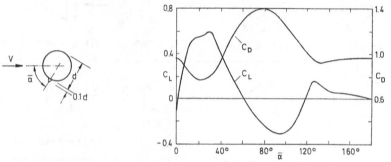

Figure 7.19. Drag and lift coefficient for a stranded conductor cable with a ridge of frozen raindrops (after Simpson, 1979).

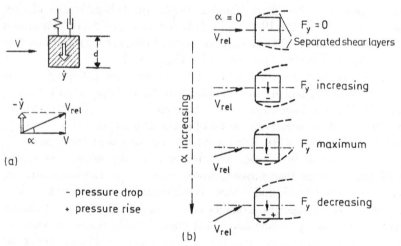

Figure 7.20. Effect of transverse body movement on separated shear layers and fluid force F_y for a square prism in cross-flow ($\bar{\alpha} = 0$, $\theta = \alpha$, Figure 7.4).

angle of incidence $\alpha = \tan^{-1}(\dot{y}/V)$ grows with $\dot{y}$, and the accompanying asymmetry in flow pattern leads to a pressure drop on the lower side until the lower shear layer reattaches to the body. This pressure change gives rise to a transverse fluid force F_y that acts *in* the direction of $\dot{y}$; hence energy is transferred to the body at the rate $F_y\dot{y}$, and movement-induced excitation is possible.

In order to determine whether vibrations will actually be excited for a given structure, one needs information on the force coefficient $C_y \equiv F_y/(d\rho V^2/2)$ as a function of α, as is shown in Section 7.2.4. Figure 7.21 depicts four basic types of C_y–versus–α characteristics along with the corresponding 'universal response' curves in which the relative response amplitude y_o/d and the reduced velocity $V_r \equiv V/(f_n d)$ are normalized with the aid of the mass-damping parameter

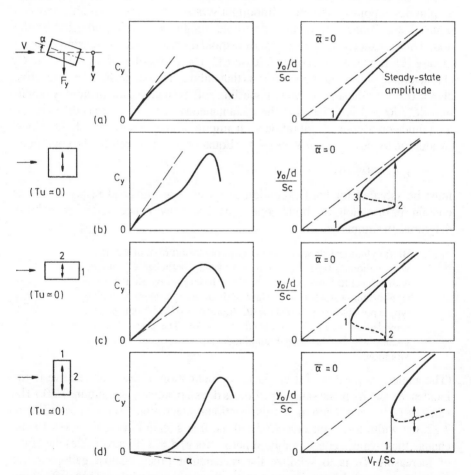

Figure 7.21. Universal response curves for cylinders undergoing transverse galloping for various C_y versus α characteristics (after Novak, 1969, 1972). (See Section 7.2.4 for explanation of symbols.)

$Sc \equiv 2m\zeta/(\rho d^2)$. Through this normalization, all response curves for a given structural geometry and flow condition (Re, Tu, etc.) collapse to a universal curve irrespective of differences in mass m and mechanical-damping ratio ζ.

The four C_y–versus–α characteristics differ mainly in the magnitudes of the slope and curvature at $\alpha = \bar{\alpha} = 0$ and in the existence of a point of inflexion. In the first two cases, instability exists with respect to *soft excitation* (from rest) exactly as predicted by Equation 7.22, i.e., with the critical value of V_r/Sc being inversely proportional to the slope $[dC_y/d\alpha]_{\alpha = \bar{\alpha}}$. In case (b), the more complex nature of the C_y characteristic produces a hysteresis in the amplitude response: y_o jumps to a higher value as the velocity increases beyond point 2 and to a lower value as it decreases below point 3. In case (c), galloping can start from rest as V_r/Sc reaches point 2; in contrast to the first two cases, however, vibration sets in with large amplitudes at point 2, and, more importantly, *hard excitation* is possible between points 1 and 2 once the relative threshold amplitude $(y_o/d)_{cr}$ indicated by the dotted line is exceeded. Whereas this threshold decreases with increasing velocity in case (c), it actually increases in case (d). The extraordinary aspect of the C_y characteristic of this 'hard oscillator' is that although the slope $dC_y/d\alpha$ is negative at $\alpha = \bar{\alpha} = 0$ (indicating stability regarding soft excitation), the instability condition $dC_y/d\alpha > 0$ is satisfied if the instantaneous value of α exceeds a certain magnitude on account of vibration with amplitudes beyond the threshold values. In addition to the condition $V_r > (V_r)_{cr}$ (Equation 7.24), therefore, the condition

$$y_o/d > (y_o/d)_{cr} \tag{7.39}$$

must be satisfied for hard excitation to occur. The universal response curves contain information on both conditions and may hence serve as stability diagrams. In summary:

> Bluff cylindrical structures are susceptible to soft MIE of the transverse-galloping type if the transverse-force coefficient C_y, plotted with respect to flow-incidence angle α, has a positive slope for the equilibrium position $\alpha = \bar{\alpha}$. Hard MIE of the transverse-galloping type is possible even if the slope $dC_y/d\alpha$ at $\alpha = \bar{\alpha}$ is zero or negative, provided $dC_y/d\alpha > 0$ is satisfied at $\bar{\alpha} + \Delta\alpha$. The larger $\Delta\alpha$, the greater will be the trigger amplitude $(y_o)_{cr}$ necessary to induce vibration. $\tag{7.40}$

(The angle α is positive if it produces an inclination of the body relative to the incidence velocity in the same direction as does a positive $\dot{y}$, see Figure 7.20.) The value of C_y is not a function of cross-sectional shape alone. As seen from Figure 7.22, the width-to-height ratio L/d and the free-stream turbulence level Tu can change these values considerably. Whereas for $e/d = 2$ (Figure 7.22a) the effect of increasing Tu is to stabilize the rectangular prism against galloping, for $e/d = \frac{1}{2}$ (Figure 7.22c) an increase in Tu decreases the stability. These effects are clearly exhibited in Figures 7.22b, d. In smooth flow, the prism with $e/d = \frac{1}{2}$ becomes unstable only in combination with large threshold amplitudes and only

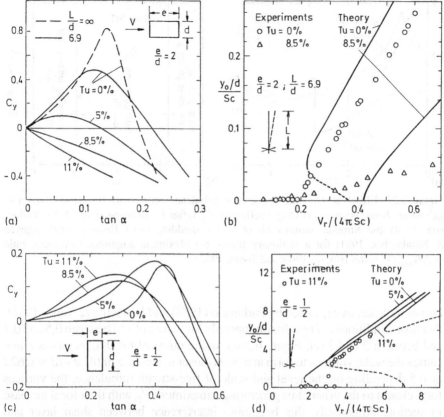

Figure 7.22. Transverse-force coefficient C_y as a function of flow-incidence angle α and corresponding universal response curves for rectangular prisms in cross-flow ($\bar{\alpha} = 0$) of different turbulence levels Tu. (a, b) $e/d = 2/1$; (c) $e/d = 1/2$, $L/d = 128$; (d) $e/d = 1/2$, $L/d = 85$ (after Novak, 1972).

for $V_r/(4\pi Sc) \geq 6$, where $V_r \equiv V/(f_n d)$; the critical value of $V_r/(4\pi Sc)$, in other words, is more than an order of magnitude larger than that for $e/d = 2$. With a turbulence level of 11%, in contrast, transverse galloping for $e/d = \frac{1}{2}$ can start from rest at a considerably reduced onset velocity. Effects of nonuniform approach velocity and mode of vibration are discussed in conjunction with Figure 3.30. The frequency of vibration f in air is very nearly equal to the natural frequency f_n, whereas in water, according to Bokaian & Geoola (1982b), it is between about 85 and 100% of f_n determined in still water (cf. Parkinson, 1985).

The most important factor influencing galloping vibrations of cylindrical structures is the size and shape of the 'afterbody', i.e., the structural part downstream of the lines of separation. In Figure 7.23, this influence is illustrated with the example of a rectangular prism of variable e/d ratio. The C_D and Sh

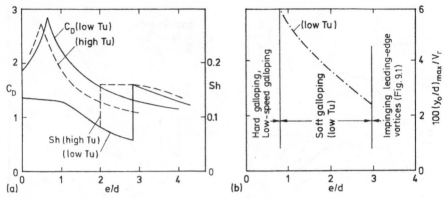

Figure 7.23. Effect of e/d and free-stream turbulence on various flow and transverse-galloping characteristics. (a) Drag coefficient C_D (after Courchesne et al., 1982; Parkinson, 1974) and Strouhal number Sh of vortex shedding (after Brooks, 1960; Nguyen & Naudascher, 1991) for a stationary prism. (b) Maximum amplitude-to-velocity ratio $(y_o/d)_{max}/V_r$ (after Brooks, 1960; and Smith, 1962).

curves indicate, as explained by Parkinson (1988) and Laneville & Yong (1983), that flow conditions change discontinuously for values of e/d between 0.5 and 0.7 and between 2 and 3 (cf. Figure 6.20a). In the range $e/d < 0.6$, an increase in e causes the wake vortices to form at a point closer to the base until, at $e/d = 0.625$ or 0.5 depending on the level and scale of free-stream turbulence, the vortices form closest to the prism thus generating a maximum C_D with their local increase in suction. Apparently, this beginning interference between shear layer and afterbody has a significant effect on the pressure loading on the prism sides, since softly excited galloping commences with $e/d \simeq 0.7$ for smooth flow and with $e/d \simeq 0.5$ for Tu = 12% (Nakamura & Tomonari, 1977). Low-speed galloping in a range of reduced velocities below $V_r = 1/Sh$ has been observed for $e/d < 0.6$ by Nakamura & Hirata (1991).

As e/d increases beyond 0.6, the longer afterbody forces the vortices to form further downstream, and this shift reduces C_D. The abrupt change in Strouhal number Sh that occurs for e/d between 2 and 3, depending on the turbulence level Tu, correlates with intermittent reattachment (or impingement) of the shear layers on the sides of the afterbody. Since reattached shear layers produce a pressure loading on the prism sides that opposes transverse movements (the shear layers can no longer affect an afterbody!), galloping is prohibited at e/d values larger than these. The fact that the maximum ratio of reduced amplitude to reduced velocity decreases with increasing e/d in the range of soft galloping (Figure 7.23b) is clearly due to the self-limiting reattachments (shown in the lowermost sketch in Figure 7.20b) which occur at lower and lower amplitudes as e/d increases. A similar decrease and eventual suppression of maximum amplitude in the range $0.7 < e/d < 3$ is produced by free-stream turbulence (Figures 7.22a, b). Flow-induced vibrations in the range $e/d > 3$ are treated in Section 9.1. Yaw effects on galloping have been addressed by Skarecky (1975).

Unusual effects of afterbody shape have been observed with the profiles shown in Figures 7.18i, m. In accordance with the preceding discussion, the D-section will not gallop from rest. As a matter of fact, relatively large amplitudes are required to trigger vibration in this case, as indicated by the dotted lines in Figure 7.24b. The influence of free-stream turbulence is dramatic; large values of Tu substantially reduce the critical velocity for onset of instability and increase the galloping steady-state amplitudes (Figure 7.24b). Similar modifications occur in three-dimensional flows around short-span cylinders or in inclined flows, $\alpha \neq 0$ (Novak & Tanaka, 1974). If the D-section is oriented with its curved surface facing upstream, it will neither gallop nor vibrate on account of vortex shedding, because it has no afterbody.

The quasi-steady assumption utilized for the predictions in Figures 7.5, 7.22, and 7.24 requires that $(V_r)_{cr}$ at which galloping commences be rather large (Section 7.2.4). If $(V_r)_{cr}$ is small, the range of galloping comes close to, or even overlaps, the resonance range of vortex-induced excitation near $V_r = 1/Sh$ (Equation 6.4), and the quasi-steady assumption no longer holds. Nevertheless, investigations with $(V_r)_{cr}$ close to $1/Sh$ seem to indicate that the theoretical galloping curves are approached asymptotically in these cases as well. For a D-section cylinder with $(V_r)_{cr} = 5.4$ and $1/Sh = 7.4$, for example, vortex-excited vibrations from rest occur in the range $4.1 < V_r < 7.9$ (Parkinson, 1963); for V_r increasing beyond 7.9, the amplitudes increased more or less as predicted by the quasi-steady galloping theory, yet triggering was necessary to obtain vibrations in tests that were started with zero amplitude in that range.

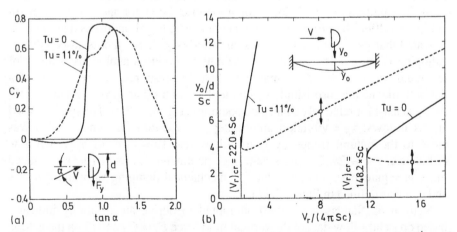

Figure 7.24. Effect of free-stream turbulence on (a) coefficient C_y (Re $= 9\times10^4$) and (b) universal galloping-response curve for D-Section cylinder constraint to vibrate with a sinusoidal mode shape normal to the flow (after Novak & Tanaka, 1974).

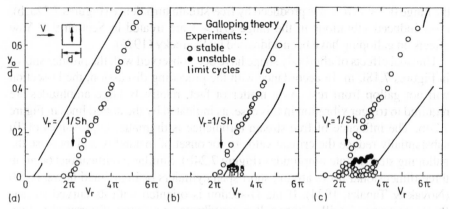

Figure 7.25. Response diagrams for square prism in low-turbulence cross-flow showing combined effects of galloping and vortex resonance (after Parkinson & Wawzonek, 1981; Parkinson, 1980). (a) $Sh(V_r)_{cr} = 0.15$; (b) $Sh(V_r)_{cr} = 1.06$; (c) $Sh(V_r)_{cr} = 1.49$.

Similar asymptotic approaches of the upper branch of the theoretical galloping curve have been found for a square prism with ratios of $(V_r)_{cr}$ to $1/Sh$ close to or even below 1.0 (Figure 7.25). It seems that one can approximate the combined effects of galloping and vortex shedding by drawing an almost straight line from zero near $V_r = 1/Sh$ to the theoretical galloping curve as indicated in Figure 7.25. More recent information on these combined effects is contained in articles by Bearman et al. (1987), Parkinson (1985), and Corless & Parkinson (1988).

(b) *Gates with underflow*. If a gate with separated flow underneath is free to vibrate in the vertical direction, its motion will produce changes in the instantaneous flow pattern (Figure 7.26a) similar to those shown for one side of the prism in Figure 7.20b. Therefore, one can expect the gate to gallop just as does a prism provided that its underside possesses an 'afterbody shape' that is susceptible to galloping (Figure 7.18) and that the vibration produces analogous patterns of separated shear layers. As a matter of fact, the response characteristics of a flat-bottom gate are surprisingly similar to those of a square prism (Figure 7.21b, right), at least for relatively small gate openings (Figure 7.26b). The symbols used are as follows: y_o = vibration amplitude; f = vibration frequency, which was close to the natural frequency f_n in still water; Tu = level of approach-flow turbulence; $Sc \equiv 2m_r\zeta$ = mass-damping parameter; $m_r \equiv m/(\rho e^2)$ = reduced mass; m = gate mass per unit width; ζ = mechanical-damping ratio; and V, a, e, s, and h_s are as defined in Figure 7.27a.

Nguyen & Naudascher (1986) attempted to derive an instability indicator based on steady-flow data of the vertical fluid force $F_y = C_y e \rho V^2/2$ on the gate as a function of gate inclination α. They obtained C_y versus α diagrams as functions of, mainly, gate-bottom shape and relative gate opening s/e and showed that these

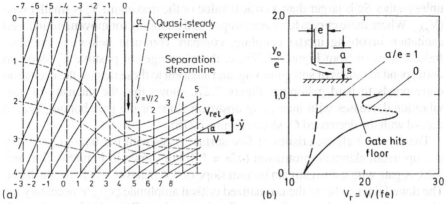

Figure 7.26. (a) Instantaneous streamline pattern for a gate moving downward at speed $\dot{y} = V/2$ (after Nguyen & Naudascher, 1986b). (b) Response characteristics of a free-surface gate at the small relative gate opening $s/e = 1.0$ (after Kanne et al., 1990). [$h_s \simeq 14$; Tu = 8%; $m_r = 95.6$; Sc = 0.134 (in water); $f_n e^2/\nu \simeq 360$.]

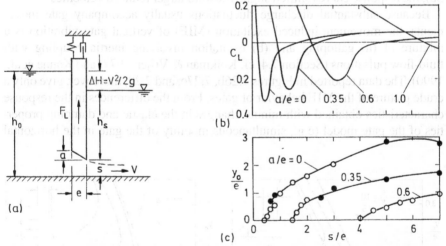

Figure 7.27. (a) Definition sketch of a free-surface gate. (b) Mean-lift coefficient for leaf gates with various underside inclinations at Tu = 1.5%; (c) Relative vibration amplitude of model gate for Tu = 5.5%, $h_s/e \simeq 14$, $f_n e^2/\nu = 490$, $V_r \equiv V/(fe) = 30$, $m_r = 67.8$; and Sc = 0.21 (in air); full symbols denote hard excitation (after Nguyen & Naudascher, 1986b).

can be replaced, as far as the indicator for instability is concerned, by diagrams of fluid-dynamic uplift coefficients C_L (or $\bar{\kappa}_B$) versus s/e. Information on $\bar{\kappa}_B (s/e)$ for a great variety of gate shapes is contained in Section 3.3 of the IAHR Monograph on Hydrodynamic Forces (Naudascher, 1991). Where the C_L versus s/e curves exhibit a positive slope, there is the danger of soft excitation due to galloping, unless either Sc is larger than a critical value or the maximum V_r is smaller than $(V_r)_{cr}$. Where these curves have zero slope, hard galloping (or movement-induced excitation involving inertia coupling, compare comment below) is possible. Indeed, as seen from Figures 7.27b, c, the s/e ranges of positive C_L slope do correspond to ranges of soft galloping, and adjacent to these ranges, the excitation corresponds to hard galloping. Figure 7.27c shows that the intensity of gate vibration decreases with increasing upstream slope of the gate bottom, a/e, in accord with the decreased C_L slope (Figure 7.27b).

The response characteristics of free-surface gates with flat bottom ($a/e = 0$) and upstream skinplate protrusion ($a/e = 1.0$) are shown in Figures 7.26b and 7.28. A gate with a downstream bottom slope of 45° behaves much like the latter. The dotted lines indicate the normalized critical amplitude $(y_o)_{cr}/e$ necessary for triggering large-amplitude vibrations (hard excitation). Remarkable are the findings that a gate with a flat bottom (Figure 7.28a) is only slightly more susceptible to small-amplitude vibration than a gate with the bottom removed from the separated shear layer (Figure 7.28b), whereas the opposite holds with respect to large-amplitude vibration. With increased relative submergence h_s/e (at constant Sc), the response curves shifted slightly toward larger reduced velocities.

Because substantial discharge fluctuations usually accompany gate movements, the movement-induced excitation (MIE) of vertical gate vibration is a mixture of (a) galloping and (b) excitation involving inertia coupling with fluid-flow pulsations (Section 7.4, cf. Kolkman & Vrijer, 1977; and Kanne et al., 1990). The data reported in Figures 7.26b, 7.27c, and 7.28, moreover, give only a crude picture of the MIE behavior of gates. From the differences in the response characteristics obtained with minute changes in the elastic and damping properties of the gate model (e.g., simultaneous mobility of the gate in the horizontal

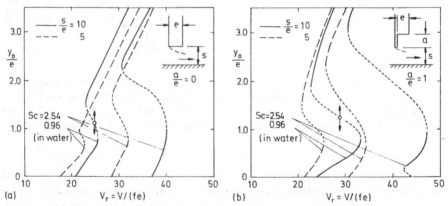

Figure 7.28. Response characteristics of free-surface gates (Figure 7.27a) with $h_s/e \approx 14$; Tu ≈ 5 to 8%; $m_r = 95.6$; $f_n e^2/\nu \approx 360$ for Sc = 0.96, and $f_n e^2/\nu = 270$ for Sc = 2.54 (after Kanne et al., 1990).

direction or changes in the nonlinear mechanical damping, cf. Kanne et al., 1990), it can be concluded that the MIE behavior of gates is strongly affected by the fine details of the mechanical system. In practice, therefore, parameters such as m_r and Sc, or the above instability criterion, are insufficient to define that system.

7.3.4 *Hysteretic changes in flow pattern*

The special feature of the excitation described in this section is a movement-induced, hysteretic change in flow pattern whereby a separated flow (jet, shear layer) either changes back and forth between two distinct patterns or periodically detaches from and reattaches to a vibrating body. The term 'hysteretic' refers to the fact that the flow patterns change in different ways depending on the direction of body motion. Since these changes are associated with variations in the fluid force acting on the body, the hysteretic nature of the change brings about a phase shift or time lag between body motion and exciting force. In many respects, the mechanism of excitation is similar to that of stall flutter (Section 7.3.2). In this case, however, the phase shift almost exclusively dominates the transfer of energy from the flow to the body.

A typical example for this type of excitation is a pair of parallel circular cylinders in cross-flow (Figure 7.29a). The various effects of interference in such arrangements of cylinders have been summarized by Zdravkovich (1977, 1984a, 1985, 1987, 1988). Although both cylinders may vibrate, the vibrations of the downstream cylinder are consistently more intense than those of the upstream one. Within the range of positions indicated by the shaded area in Figure 7.29b, a lightly damped cylinder with degrees of freedom in the x and y direction vibrates as illustrated by the loops in that figure. With just one degree of freedom in the y

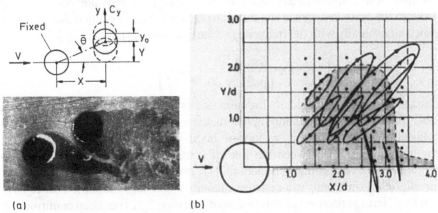

(a)   (b)

Figure 7.29. (a) Instantaneous flow pattern around twin circular cylinders in smooth cross-flow (Re = 1.5×10^4) and (b) Range and modes of vibration of downstream cylinder (after Matsumoto et al., 1988b).

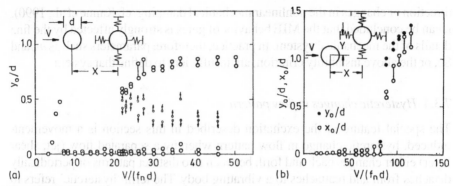

Figure 7.30. Typical response diagrams for downstream cylinder having x- and y-degrees of freedom, located a distance X downstream of a fixed, parallel cylinder of equal diameter (after Shirato, 1988). (a) Tandem arrangement: $X/d = 3$, $Y/d = 0$ [Sc $\equiv 2m\zeta/(\rho d^2) \approx 1.05$, $f_n d^2/v \approx 210$]; (b) Staggered arrangement: $X/d = 2.2$, $Y/d = 1.55$ [Sc ≈ 5.8, $f_n d^2/v \approx 810$].

direction, the range and amplitudes of vibration were found to be only slightly smaller than with two (Matsumoto et al., 1988b). Apparently, mode coupling plays a minor role in the excitation of these vibrations, commonly referred to as *interference galloping*. (The possibility of wake-induced flutter for $X/d \gg 4$ is discussed in Section 7.5.5.)

Response diagrams from an air-tunnel study on two long-span cylinders with end plates on both ends are shown in Figure 7.30. They exhibit, typically, hard excitation for the tandem and soft excitation for the staggered arrangement. A similar, extensive study has been carried out in a water flume by Rao (1989a). For staggered arrangements, Ruscheweyh (1983) deduced the following stability criterion from a quasi-steady analysis; he employed a constant phase shift φ between the cylinder motion and the quasi-steady lift and assumed that the latter varied sinusoidally with the frequency of the cylinder vibration:

$$(V_r)_{cr} \equiv \left(\frac{V}{fd}\right)_{cr} = \sqrt{8\pi} \sqrt{\frac{2m\zeta}{\rho d^2} \frac{a}{d}} \left[\frac{\partial C_y}{\partial \theta}\Big|_{\bar{\theta}} \sin \varphi\right]^{-\frac{1}{2}} \tag{7.41}$$

Here, $(V_r)_{cr}$ = reduced velocity for onset of vibration, f = frequency of vibration, $2m\zeta/(\rho d^2) \equiv$ Sc = mass-damping parameter, $a = \sqrt{X^2 + Y^2}$ = distance between cylinder axes, and $\partial C_y/\partial \theta\big|_{\bar{\theta}}$ = slope of the lift-coefficient curve evaluated at the equilibrium angle $\bar{\theta}$ between the line of centers and the mean-flow direction (Figure 7.29a). According to Equation 7.41, instability requires a *negative* slope for $\partial C_y/\partial \theta\big|_{\bar{\theta}}$ since sinφ is a negative quantity (force lags motion). This equivalent Den Hartog criterion (cf. discussion of Equation 7.22) has been confirmed by experiments of Ruscheweyh (1983) and Bokaian & Geoola (1984a, b).

From these studies and observations of similar flow systems, one may conclude:

Flow systems in which structural movements lead to more or less abrupt changes in flow pattern are susceptible to MIE if these changes are associated with substantial variations in fluid force (or moment). A necessary condition for excitation is a structural displacement that causes a marked *decrease* in the steady-state fluid force acting in the direction of displacement. (7.42)

This indicator for movement-induced excitation (MIE), incidentally, applies to MIE of the stall-flutter type as well (e.g., Figure 7.13c). With its aid one can detect dangerous flow conditions merely from an inspection of steady-state data on fluid force or moment.

The validity of statement 7.42 as an MIE indicator is illustrated in Figure 7.31. The figure shows a simplified case for which the steady-state (or static) fluid force F_y acting on the downstream cylinder is zero for negative displacements from the equilibrium position, $y < 0$, and of constant negative magnitude for positive displacements, $y > 0$. The steady-state force-displacement curve, in other words, is assumed to be step-like as shown in Figure 7.13b by the dotted line. As the downstream cylinder moves outward from the equilibrium position, $y = 0$, some flow will switch into the gap between the two cylinders (Figure 7.29a), thus generating suction along the side facing the wake. If this switch were instantaneous, the force histogram would resemble the idealized square wave given by the dotted line in Figure 7.31d. Due to fluid inertia, however, a finite time is required for the flow to react to the change in cylinder position and for the 'jet'

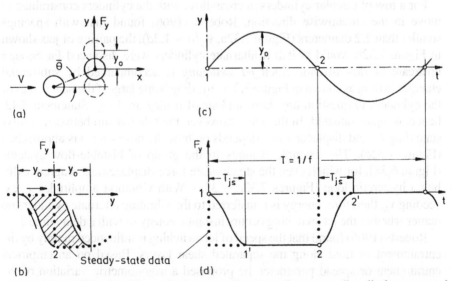

Figure 7.31. Schematic of force-displacement diagram and corresponding displacement and force histograms illustrating a simplified delay-time model for interference galloping of the downstream cylinder, placed in a position, $\bar{\theta}$, of maximum lift gradient.

to be established (Knisely, 1985b). During this delay time, denoted as T_{js}, the force histogram will assume a form like that shown by the solid line $1 - 1'$. When the cylinder returns toward equilibrium, the jet-like flow will switch out of the gap. In Figure 7.31d, this flow redistribution is presented as requiring the same amount of time, T_{js} (line $2 - 2'$). Consequently, the actual force-displacement curve forms a hysteresis loop which is traversed in the clockwise direction, indicating positive work done on the cylinder. (Loop area $= W_e =$ work done per cycle by the exciting force F_y, cf. Section 2.3.4. If f_n is very small compared to V/d, or $V_r \to \infty$, all fluid processes tend toward steady-state conditions, and the hysteresis loop disappears.)

In spite of its limitations (e.g., dynamic-overshoot and added-mass effects on F_y were disregarded in Figure 7.31), the simplified delay-time model permits an approximate prediction of the onset velocity for interference galloping by means of an empirical relationship for T_{js}/T as a function of the 'suddenness parameter' $V_r y_o/d$ (Knisely & Kawagoe, 1988). More important in the context of this monograph, however, it enables the practicing engineer, in conjunction with the indicator test 7.42, to assess systems for possible excitations. For example, dangerous arrangements of two parallel cylinders can be disclosed by merely inspecting steady-state fluid-force information like that contained in the IAHR Monograph on Hydrodynamic Forces, Figure 2.81 (Naudascher, 1991). More-over, the indicator test can also be used to determine which of the structures with bistable flow, depicted in Figure 6.38, might be susceptible to movement-induced excitation of the hysteretic-flow-change variety.

For a row of circular cylinders in cross-flow, with the cylinders constrained to move in the streamwise direction, Roberts (1966) found that with spacings smaller than 2.2 diameters (Figure 7.32a, $y_G/d \leq 1.2d$), the pairing of jets shown in Figure 7.32b would switch if alternate cylinders were displaced far enough upstream or downstream. Such *jet switching* is accompanied by substantial changes in drag as shown in Figure 7.32c, the drag being large if the near wakes of the cylinders in question are short, and small if they are long. Statement 7.42, hence, is again satisfied. In this case, however, the relationship between steady-state drag F_x and displacement x depends on how the position (x) is approached (Figure 7.32c). The example belongs to the group of bistable-flow systems (Figure 6.38), for which even the steady-state force-displacement diagram exhi-bits a hysteretic loop (Figures 7.32c, 7.33a). With vibration amplitudes x_o ex-ceeding x_A, therefore, energy is transferred to the vibrating alternate cylinders no matter whether the jet switching occurs instantaneously or with a delay time T_{js}.

Roberts (1966) found that the speed of jet switching is influenced mainly by the entrainment of fluid along the separated shear layers. Based on an empirical entrainment or spread parameter, he proposed a trigonometric variation of the drag $F_x(t)$ during jet switching as plotted in Figure 7.33d with a constant value of $\tau_{js} \equiv T_{js}V/d$ of approximately 10. The corresponding qualitative picture analo-gous to Figure 7.31 is shown in Figure 7.33. Whenever the movable alternate

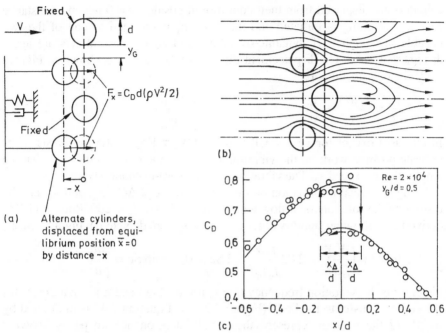

Figure 7.32. (a) Closely-spaced, infinite row of cylinders with alternate cylinders free to move in unison in streamwise direction; (b) Instantaneous flow pattern; (c) Steady-state drag coefficient as a function of cylinder displacement (after Roberts, 1966).

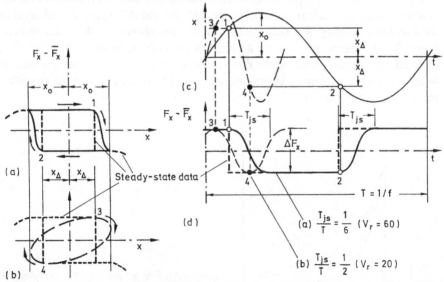

Figure 7.33. Qualitative illustration of the excitation of a system as shown in Figure 7.32a. (a, b) Force-displacement diagrams for (a) $T_{js}/T = \frac{1}{2}$ and (b) $T_{js}/T = \frac{1}{6}$; (c, d) Corresponding histograms of (c) displacement and (d) fluid force.

cylinders are displaced from their equilibrium position $\bar{x} = 0$ beyond a distance $\pm x_\Delta$, jet switching is initiated and the force F_x is changed to one of the two bistable-state values. If dynamic-overshoot and added-mass effects are again disregarded, the force-displacement diagrams take the forms shown in Figures 7.33a, b, as a function of

$$\frac{T_{js}}{T} \equiv \frac{\tau_{js}d/V}{T} = \frac{\tau_{js}}{V_r} \tag{7.43}$$

In all cases except for very small values of $V_r \equiv V/(fd)$, the hysteresis loops indicate positive work on the vibrating cylinders; the only condition is that the amplitudes x_o exceed x_Δ. The vibration is of the hard-excitation type.

Using empirical information on the magnitudes C_D, ΔC_D, x_Δ, and τ_{js}, and by restricting the solution to air flow with $y_G/d = 0.5$ and $V_r < 200$, Roberts (1966) derived the following stability criterion for the system depicted in Figure 7.32a:

$$0.137 \, \kappa \left(\frac{V}{f_n d}\right)^2 - 2.02 \frac{f}{f_n f_n d} \frac{V}{} - 2 \, \mathrm{Sc} = 0, \quad \text{with Sc} \equiv \frac{2m\zeta}{\rho d^2} \tag{7.44}$$

Herein, f is the vibration frequency ($f \approx f_n$ for air flow) and κ is a parameter for the energy transfer to the body (Figure 7.34). If Equation 7.44 is multiplied by $\pi x_o^2 \rho V^2/2$, the first term represents the work W'_e done on the body per cycle owing to the jet-switching process, the second term gives the energy W''_e dissipated per cycle fluid-dynamically, and the last term is the energy dissipated per cycle, W_d, due to work done by the mechanical damping force.

Figure 7.35 gives a graphical representation, comparable to Figure 2.6, of these contributions to the energy balance. If one adds W'_e and W''_e to get the total input of fluid-dynamic energy W_e, one obtains the curves shown by the solid lines in Figure 7.35b. The intersections of these curves with the appropriate lines for $-W_d$, denoted by circles, yield stable-limit-cycle vibrations due to the movement-induced excitation considered, and those denoted by triangles yield the condition for the onset of vibration.

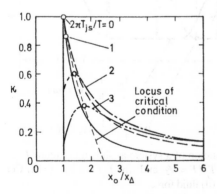

Figure 7.34. Energy parameter κ according to Roberts (1966).

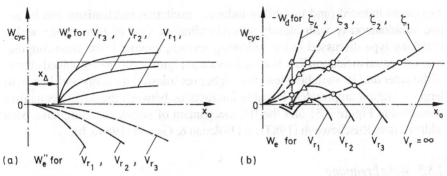

Figure 7.35. Energy diagrams for (a) fluid-dynamic energy input per cycle and (b) total energy input and energy dissipation per cycle for $\bar{x} = 0$ (after Roberts, 1966). ($V_{r1} < V_{r2} < V_{r3} < \infty$; $\zeta_1 < \zeta_2 < \zeta_3 < \zeta_4$, where ζ = mechanical-damping ratio; $W_e = W_e' + W_e''$.)

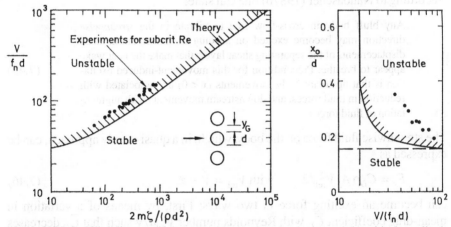

Figure 7.36. Stability diagram giving (a) critical reduced velocity and (b) critical amplitude for a row of closely spaced circular cylinders ($y_G/d = 0.5$), spring-supported in the streamwise direction (after Roberts, 1966).

In Figure 7.36, this onset condition, deduced from Equation 7.44, is compared with experimental results. The theoretically-derived stability criteria are on the safe side regarding both the critical reduced velocity and the critical amplitude that must be exceeded for onset of MIE. A significant result, displayed in this figure, is the absence of instability below a certain value of reduced velocity $V/(f_n d)$, independent of reductions in the mass-damping parameter. At relatively high frequencies f_n, apparently, the time available is not sufficient for the shear-layer entrainment to initiate the jet-switching process (Equation 7.43).

Certain elements of jet switching, as studied by Roberts, may be present in heat-exchanger tube banks (Paidoussis & Price, 1988). Actually, a number of movement-induced (and instability-induced) excitation mechanisms can be active simultaneously with closely spaced cylinders. One of them is the wake-breathing type discussed in the following section; another is the fluid-coupling type presented in Section 7.4. With reference to Figure 7.50, the combined effects of the latter *and* jet switching lead to a higher net transfer of energy and, hence, to larger vibration amplitudes. Possible interactions between the MIE mechanism presented in Figure 7.31 and the IIE mechanism of vortex shedding have been addressed by Ruscheweyh (1983) and Bokaian & Geoola (1984a, b).

7.3.5 *Wake breathing*

If a bluff body undergoes streamwise vibration, the wake region between the shear layers that separate from this body will alternately expand and contract. In the context of this monograph, this process is referred to as wake breathing. According to Naudascher (1987b) one can state:

> Any bluff body in cross-flow free to vibrate in the *streamwise* direction may become excited on account of movement-induced displacements of the separating shear layers that make the near wake appear to breathe. Precondition for this movement-induced excitation is that upstream body movements ($\dot{x} < 0$) are associated with reduction in fluid forces, and downstream movements with augmentation of fluid forces.   (7.45)

The streamwise fluid force on the body, which, in a quasi-steady approach, can be expressed as

$$F_x = C_D \rho\, A_\perp V_{\text{rel}}^2 / 2 \qquad \text{with } V_{\text{rel}} = V - \dot{x}, \tag{7.46}$$

can become an exciting force in two ways: First, by means of a variation in mean-drag coefficient $\overline{C}_D$ with Reynolds number $V_{\text{rel}} d / \nu$ such that $\overline{C}_D$ decreases with increasing relative velocity V_{rel}; and, second, by means of a movement-induced fluctuation, $C_D' = C_D - \overline{C}_D$, part of which is in phase with the body velocity $\dot{x}$. ($A_\perp$ = cross-sectional area perpendicular to the flow.)

Wake breathing of the *first type* may occur if the body has a rounded shape so that the separated flow forms a wide or narrow wake depending on the laminar or turbulent state of the boundary layer ahead of separation. It is characterized by periodic, successively-occuring laminar and turbulent separations as the body vibrates near the critical Reynolds number. According to an analysis based on an analogy with galloping by Martin et al. (1981), this type of movement-induced excitation can occur within a narrow range of reduced velocities near $V/(fd) =$ Re*$(fd^2/\nu)^{-1}$, where Re* is the Reynolds number for which the C_D versus Re curve has the greatest negative slope, $-\partial C_D/\partial \text{Re}$.

The *second type* of wake breathing is not confined to bluff bodies of rounded shape, nor is it restricted to a narrow Reynolds-number range. Here, the 'brea-

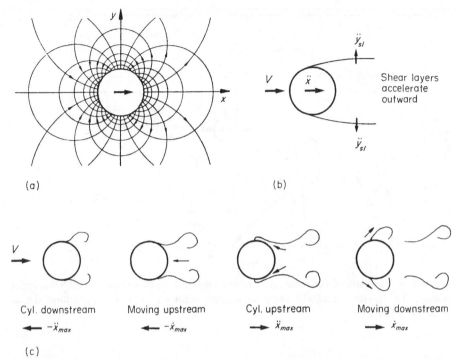

(a)   (b)

(c)

Figure 7.37. Illustration of the effect of fluctuating 'added-mass flow' for a circular cylinder vibrating freely in the streamwise direction. (a) Instantaneous streamlines of accelerated cylinder. (b) Accompanying acceleration of free shear layers. (c) Streamlines at different instants during a cycle of cylinder vibration in the range of $1.5 < V/(fd) < 2.0$ (after Aguirre, 1977).

thing' of the wake is associated with what Kolkman (1976, 1980) has termed 'added-mass flow': As the body accelerates downstream (Figures 7.37a, b) it displaces fluid which not only causes an added-mass force $A'\ddot{x}$ (Section 3.2), but also changes the instantaneous flow pattern. During the half of the vibration cycle for which $\dot{x} > 0$, this added-mass flow has the tendency to displace the separating shear layers outward and, thus, to widen the wake. During the other half of the cycle, $\dot{x} < 0$, this tendency is reversed. Since the fluid force F_x acting on the body is strongly affected by the wake width, this process produces a fluctuation in F_x that is either in phase with $\dot{x}$ (exciting force) or 180° out of phase with $\dot{x}$ (damping force) depending on the relative magnitudes of two opposing effects: the first related to $C_D = \bar{C}_D + C'_D$, and the second related to $V_{rel} = V - \dot{x}$ (Statement 7.45). As shown by Naudascher (1987b), the in-phase component can be approximated by a simple, linear relation between C_D and $\dot{x}$, at least near the onset of vibrations from rest. But the breathing-type of excitation is so strongly intermixed with instability-induced vortex formations (Section 9.1) that it appears impossible to deduce a simple stability criterion or indicator test.

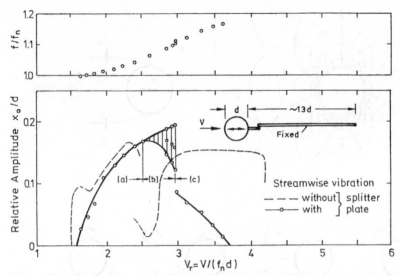

Figure 7.38. Response of a circular cylinder vibrating freely in the streamwise direction (after Aquirre, 1977). (Re = 2 to9×10³; equivalent body density $\rho_e = 0.94\rho$; logarithmic decrement $\delta = 0.157$.)

A verification of the mechanism of wake breathing is represented in Figure 7.38. By means of a fixed splitter plate of roughly $13d$ length, Aguirre (1977) eliminated the shedding of alternate vortices as well as the feedback from the far wake, which are known to be the essential ingredients of vortex-induced cylinder vibration. According to the amplitude obtained with and without splitter plate, the cylinders behave nearly alike in the midportion of the first excitation range 'a', and the amplitudes attained near the end of range 'a' are slightly larger if alternate vortex shedding is prevented. One may conclude, therefore (Naudascher, 1987b), that the movement-induced wake breathing is responsible for the large-amplitude vibration in this range, and instability- or vortex-induced phenomena are simply augmenting the vibration near the beginning and disturbing it near the end of range 'a'. The sudden drop of amplitude for the cylinder without splitter plate near $V_r = 1/(2Sh) = 2.5$ (Sh $\equiv f_o d / V$, Figure 6.12) is due to the incompatibility of two vortex-shedding modes: the symmetric one that is an integral part of wake breathing (Figure 7.37c) and that dominates in the first excitation range $1.2 < V_r < 2.5$; and the antisymmetric one of frequency f_o that dominates in the second excitation range $2.5 < V_r < 4.0$. The nearly constant amplitude for the cylinder without splitter plate in range 'c' appears to be associated with a combined action of wake breathing [decreasing in strength with V_r according to the full-line curve] and alternate vortex shedding [increasing in strength as V_r increases beyond the resonance point $V_r = 1/(2Sh)$]. (Further explanations are given in connection with Figure 6.13.)

Intensified investigation of streamwise cylinder vibrations (e.g., King et al., 1973; Hardwick & Wootton, 1973; King & Johns, 1976; Griffin & Ramberg, 1976) was prompted by severe problems with marine piles encountered during the construction of a deep-water oil terminal in an estuary (Wootton et al., 1974). Most likely these problems were exacerbated by the fact that the two types of wake breathing were occurring simultaneously (Martin et al., 1981). At any rate, combined effects of several excitation mechanisms should be carefully assessed during design if a structure is to be free of dangerous vibrations. Square and rectangular prisms free to vibrate in the streamwise direction, for example, may be excited by so many mechanisms that they are treated separately in Section 9.1.2.

7.3.6 *Excitation of slender bodies subjected to axial flow*

In contrast to the rigid-body motions of the bodies described in Sections 7.3.1 through 7.3.5, the motions of flexible bodies such as rods, tubes, pipes, and elastic barges (Figure 7.39) all exhibit an undulation such that the amplitude and phase of the body displacement vary with distance from its leading end. For the typical configuration sketched in Figure 7.40a, one may distinguish between the forces

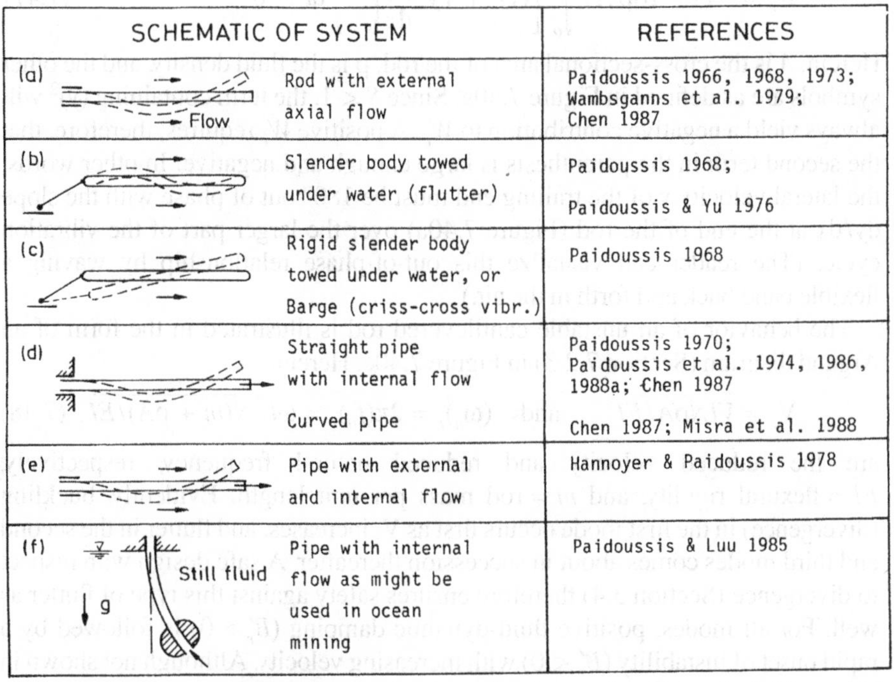

SCHEMATIC OF SYSTEM		REFERENCES
(a)	Rod with external axial flow	Paidoussis 1966, 1968, 1973; Wambsganns et al. 1979; Chen 1987
(b)	Slender body towed under water (flutter)	Paidoussis 1968; Paidoussis & Yu 1976
(c)	Rigid slender body towed under water, or Barge (criss-cross vibr.)	Paidoussis 1968
(d)	Straight pipe with internal flow Curved pipe	Paidoussis 1970; Paidoussis et al. 1974, 1986, 1988a; Chen 1987 Chen 1987; Misra et al. 1988
(e)	Pipe with external and internal flow	Hannoyer & Paidoussis 1978
(f)	Pipe with internal flow as might be used in ocean mining	Paidoussis & Luu 1985

Figure 7.39. Basic configurations of slender bodies with external or internal axial flow which are susceptible to movement-induced excitation.

F_S and F_N acting on the cylindrical surface of a deflected body element and the forces F_D and F_L acting on the trailing end of the body. In the case of bodies with bluff trailing ends, the flow separates and the forces associated with the 'trailing-end stall' dominate the surface forces. At sufficiently high flow velocities, these end forces lead to a number of fluid-elastic instabilities: divergence in the form of buckling in the first mode, and single-mode flutter in successively higher modes as the velocity is increased. Divergence (Section 3.4) sets in if the lift F_L at the trailing end surpasses the flexural force that tends to restore the equilibrium position of the cantilevered structure. Flutter occurs due to, first, a reduction of the effective flexural rigidity of the system on account of centrifugal loading, and, second, positive work done by the end forces on the body.

(a) *Cantilevered rods.* For cantilevered rods in axial flow, the work done by the end forces depends strongly on the taper S of the trailing end (Paidoussis, 1968). For an ideally streamlined end, $S = 1$, the form drag is zero, and only coupled-mode flutter is possible (similar to the case of a rod pinned or clamped at both ends, Paidoussis, 1973). For the general case, $S < 1$, single-mode flutter is possible if the work done by the forces at the end of the rod over one cycle of vibration, W_e, is positive:

$$W_e = -(1 - S)\rho \, AV \int_o^T \left[(\dot{y})^2 + V\dot{y} \, \frac{dy}{dx} \right]_{x = L} dt > 0 \qquad (7.47)$$

Herein, A is the cross-sectional area of the rod, ρ is the fluid density, and the other symbols are as defined in Figure 7.40a. Since $S < 1$, the term containing $(\dot{y})^2$ will always yield a negative contribution to W_e. A positive W_e requires, therefore, that the second term in the parenthesis is large enough and negative. In other words, the lateral velocity $\dot{y}$ of the trailing end must be 180° out of phase with the slope dy/dx at the end of the rod (Figure 7.40a) over the larger part of the vibration cycle. (The reader can visualize this out-of-phase relationship by waving a flexible band back and forth in the air.)

The behavior of an unstable cantilevered rod is illustrated in the form of an Argand diagram (Section 7.2.3) in Figure 7.40c. Herein,

$$V_r = VL\sqrt{\rho A/EI} \qquad \text{and} \qquad (\omega_n)_r = 2\pi(f_n)_r = \omega L^2 \sqrt{(m + \rho A)/EI} \quad (7.48)$$

are the reduced velocity and reduced natural frequency, respectively, EI = flexural rigidity, and m = rod mass per unit length. Evidently, buckling (divergence) in the first mode occurs first as V_r increases, and flutter in the second and third modes comes about in succession thereafter. A safe design with respect to divergence (Section 3.4) therefore ensures safety against this type of flutter as well. For all modes, positive fluid-dynamic damping ($B'_r > 0$) is followed by a rapid onset of instability ($B'_r < 0$) with increasing velocity. Although not shown in Figure 7.40c, stability is regained in the first mode before the second-mode instability sets in. The regaining of stability in the second mode, according to that

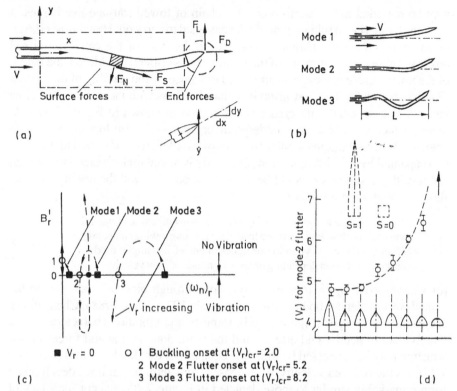

Figure 7.40. Instabilities of a cantilevered rod subjected to axial flow. (a) Surface and end forces on rod; (b) Mode shapes; (c) Argand diagram showing variation of reduced damping B'_r with reduced natural frequency $(\omega_n)_r$ as the reduced velocity V_r is increased for a particular rod; (d) Effect of trailing-end taper on V_r for second-mode flutter onset (after Paidoussis, 1973).

figure, occurs after onset of instability in the third mode. Hence, the second and third modes may exist simultaneously in an uncoupled way. The effect of the degree of taper on the critical reduced velocity for second-mode flutter is shown in Figure 7.40d; if the end of the rod becomes blunter and the afterbody length decreases, $(V_r)_{cr}$ increases. In other words, instability grows with increased streamlining of the trailing end. In summary, one may state:

> A flexible cantilevered slender structure with a streamlined trailing end subjected to external axial flow of sufficiently high velocity is susceptible to flutter in one of its (single) modes. This flutter is usually preceded by divergence in the first mode. (7.49)

(b) *Slender bodies towed under water.* The flexible slender body shown in Figure 7.39b, with zero cross-sectional area A at its leading end and fastened to a cable, may be regarded as an idealization of a chain of towed sausage-like barges, a Dracone barge (i.e., a rubber-coated tube designed to carry oil, Hawthorne, 1961); or an anchored airship (Munk, 1924). As in the case of the cantilevered rod discussed above, the geometry of the trailing end is critical; in addition, the length of the towing cable can exert a substantial influence on the stability of the system (Paidoussis, 1968). The first instability that develops with increasing axial-flow velocity is associated with criss-cross vibration as shown in Figure 7.39c. At higher velocities, a body may undergo divergence and, with further increase in velocity, flutter vibrations similar to those of cantilevered rods; the latter can be accompanied by a whirling motion if the body is towed under water. For optimal stability, the trailing end should be as blunt as possible and the towing cable as short as possible. In summary:

> A flexible slender body in axial flow with streamlined leading-end and bluff trailing-end, fastened to a relatively long towing cable, may develop criss-cross vibration or flutter in one of its (single) modes, the type of motion depending on the magnitude of flow velocity.   (7.50)

(c) *Cantilevered pipes.* In many ways, flow through pipes is analogous to the axial flow past slender bodies (Paidoussis et al., 1974, 1986, 1988a; Chen, 1987). For cantilevered pipes (Figure 7.39d), stability and vibration characteristics are mainly determined by end effects, and the work done by the end forces on the structure may be expressed by the same equation as for slender bodies (Equation 7.47). As the reduced velocity V_r increases from zero, instabilities develop in a form remarkably similar to those depicted in Figure 7.40c, except for a lack of divergence in the case of pipe flow. At small values of V_r, the natural frequency decreases with increasing V_r and the pipe experiences increased fluid damping; at sufficiently high V_r values, this trend reverses and negative damping is possible. An important parameter affecting pipe stability is the mass ratio $\rho A / (m + \rho A)$, as illustrated in Figure 7.41. The sudden increases in reduced velocity and frequency, V_r and ω_r, near $\rho A / (m + \rho A) = 0.295$ (and also near 0.67 and 0.88) correspond to a temporary recovery of stability (Gregory & Paidoussis, 1966).

An important conclusion that can readily be derived from Equation 7.47 is that for $V < 0$, with the flow from the free towards the fixed end of the cantilevered pipe (Figure 7.39f), instability occurs for vanishingly small V; stabilization in this case is obtained only through friction with the external fluid or by means of artificial augmentation of this external form of dissipation (Paidoussis & Luu, 1985). Three-dimensional motions of a system fitted with an inclined nozzle or caused by other geometrical peculiarities have been examined by Lundgren et al. (1979) and Sethna & Gu (1985). Further aspects of cantilevered-pipe instabilities are discussed in reviews by Paidoussis & Issid (1984), Paidoussis (1987), and Chen (1987). Investigations on the stability of curved pipes have been summarized by Paidoussis (1987), Chen (1987), and Misra et al. (1988).

Based on the available information, one may conclude:

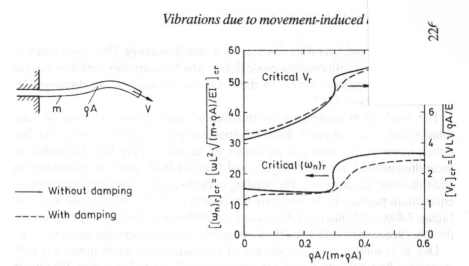

Figure 7.41. Theoretical prediction for onset of second-mode flutter of a cantilevered pipe with and without internal damping in the tube material (after Gregory & Paidoussis, 1966). (Corresponding experimental data are not shown.)

A flexible cantilevered pipe conveying fluid is susceptible to flutter in one of its (single) modes, much as is a cantilevered slender rod with external axial flow. If the fluid flow is reversed, as in ocean mining systems, flutter sets in at extremely small flow velocities. (7.51)

Except for the last-mentioned case of a cantilevered pipe aspirating fluid (Figure 7.39f), the movement-induced excitations discussed in this section are of limited practical significance. Most hydraulic structures are sufficiently stiff that unusually high flow velocities would be required for these excitations to occur. Dangerous vibrations may arise, however, if the axial flow is bounded by coaxial walls or if the flow occurs through a cluster of slender bodies placed side by side; they are caused by hydrodynamic coupling as explained in Section 7.6. The closer the coaxial walls or the tighter the cluster, the lower is generally the critical velocity for onset of vibration.

7.3.7 *Ovalling or flutter of shells*

Pipes with very thin walls (tubes) may undergo circumferential-mode (n) instabilities as well as the lateral-mode (m) instabilities, addressed in the previous section. The shell-type flutter or 'breating' vibrations to which they give rise were investigated both for clamped-clamped and cantilevered shells (Paidoussis & Denise, 1972; Weaver & Unny, 1973; Weaver & Myklatun, 1973). Their onset can be strongly influenced by viscous forces, especially for thin-shell configurations (Paidoussis, Chan & Misra, 1984, 1985). Short cantilevered tubes (Figure 7.42b) have a large number of circumferential modes that can be excited sequentially with increasing flow velocity. For long tubes (Figure 7.42c), the circumferential and lateral modes can exist simultaneously. This coexistence is not to be confused with coupled-mode flutter, which occurs only for tubes fixed at both ends. Quantitative ranges of these various instabilities are discussed by Paidoussis & Denise (1972).

The mechanism underlying circumferential-mode flutter is essentially the same as that of lateral-mode flutter of pipes conveying fluid (Section 7.3.6). For flutter, however, the flow can no longer be treated as plug flow but rather as three-dimensional and 'centrifugal', and Coriolis loads must be considered in addition to the flexural ones. The bending of the free end of the tube away from its equilibrium position in the upstream direction (i.e., $\partial y/\partial x$ out-of-phase with $\dot{y}$ in Figure 7.40a and Equation 7.47) causes circumferential distortions of the end of the tube, because its wall is not thick enough to maintain a circular cross-section.

Due to nonlinear effects, variation of circumferential-mode number n with increasing flow velocity V may not occur in a well-ordered sequence. Paidoussis & Denise (1972) have demonstrated that, as V is increased, a short cantilevered tube (Figure 7.42b) first undergoes $n = 2$ mode flutter vibration of increasing amplitude; then higher-mode vibrations ($n = 3$ and 4) set in; and, finally, a switch back to an $n = 2$ mode vibration of extremely large amplitude is observed, corresponding to a nonlinear limit-cycle oscillation.

A special type of 'flapping' vibration may occur in addition to circumferential-mode flutter if the fluid-conveying tubes have a non-circular cross-section (Figure 7.42d). The mechanism for this flapping vibration is similar to that observed for two adjacent plates with flow between them, and it has, in fact, been treated on this basis (Weaver & Paidoussis, 1977).

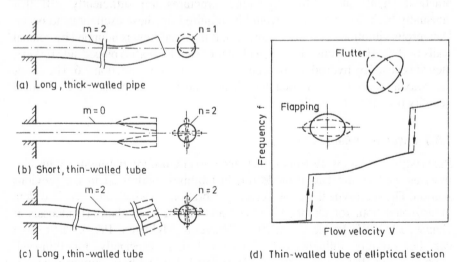

(a) Long, thick-walled pipe

(b) Short, thin-walled tube

(c) Long, thin-walled tube

(d) Thin-walled tube of elliptical section

Figure 7.42. Possible lateral and circumferential modes of flutter vibration of cantilevered pipes or tubes conveying fluid. (a-c) Pipes or tubes with circular cross-section; (d) Tube with elliptical cross-section.

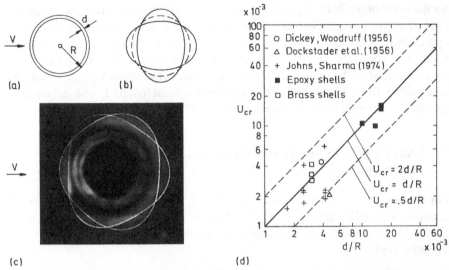

Figure 7.43. Ovalling vibration of cylindrical shells in cross-flow. (a) Definition sketch; (b, c) cross-sectional view of circumferential mode, *n*, showing the two extreme positions of the shell: (b) *n* = 2, (c) *n* = 3; (d) Critical dimensionless velocity (Equation 7.52) for onset of ovalling as a function of thickness-to-radius ratio (after Paidoussis & Helleur, 1979).

Circumferential-mode flutter of cylindrical shells can also be excited by cross-flow (Figures 7.43b, c). It is known as ovalling and was first observed with thin-walled, tall, metallic chimney stacks in wind. Dickey & Woodruff (1956) and Dockstader et al. (1956) describe some field observations, and Johns & Allwood (1968) report a case of ovalling due to a typhoon which eventually led to the collapse of a chimney. Several Japanese publications (e.g., Katsura, 1985; Uematsu & Uchiyama, 1985) were inspired by the anticipation of ovalling in silos as their construction becomes progressively lighter with the use of high tensile-strength steels.

In a recent review, Paidoussis, Price & Ang (1988) summarize how our concept of the mechanism of ovalling vibrations has evolved from that of subharmonic vortex excitation to that of single-mode flutter. An analytical model based on the flutter concept (Paidoussis & Wong, 1982) yielded, in its original form, good qualitative predictions of modal orientation with respect to flow direction as well as sequence of vibration modes with increasing flow velocity. With the improved flutter analysis, satisfactory quantitative agreement has also been reached, at least for shells clamped at both ends and placed in a uniform stream. A crude design aid, proposed by Paidoussis & Helleur (1979), utilizes the dimensionless velocity

$$U \equiv \frac{V}{\sqrt{E_s/[\rho_s(1 - \nu_p^2)]}} \qquad (7.52)$$

for the onset of ovalling, U_{cr}, which can be correlated with the thickness-to-radius ratio d/R over remarkably large ranges of d/R and L/R, in spite of differences in approach-flow conditions (Figure 7.43d). Here, E_s = Young's modulus for the shell, ρ_s = shell density, v_p = Poisson's ratio, L = length of shell. Hence, for a given material of the shell and a likely maximum wind speed V_{max}, U_{max} can be determined from Equation 7.52, and safety against ovalling can be ensured through observance of the relationship $0.45\ d/R > U_{max}$.

7.4 EXCITATION INVOLVING COUPLING WITH FLUID-FLOW PULSATIONS

For some flow systems, movement-induced excitation (MIE) develops on account of coupling between a body vibration and a fluid oscillation, the latter arising in conjunction with pulsations in flow rate and/or head produced by the vibrating body.

> All flow systems with structural movements that are accompanied by
> appreciable variations in flow rate and/or head should be assessed   (7.53)
> for possible MIE.

Whether or not the potential for MIE leads to vibration must be found from more detailed considerations, as shown in the following. In this context, one should differentiate between 'press-open' and 'press-shut' devices, which are devices controlling the flow through small openings such that the fluid force tends to press them either open or shut.

7.4.1 *Valves and turbines*

For the press-open device shown in Figure 7.44a, Den Hartog (1985) proposed the following arguments for the case of flow through inlet D at a constant rate, $\bar{Q}$, while the outflow from E varies depending on the position of the valve m. If the valve vibrates as indicated in Figure 7.44b, the outflow will fluctuate by an amount Q' about the mean, and Q' will be negative during the first two quarters and positive during the second two quarters of the vibration cycle. The greater the excess volume dV that is stored within this chamber, the higher will be the pressure $p = \bar{p} + p'$ in the valve chamber of volume V. Since the inflow exceeds the outflow during the first two quarter cycles of vibration, the pressure rises. Thus, when the valve reaches again the equilibrium position at $t = T/2$, moving to the left, the pressure is at a maximum. In a similar way, when the valve is moving to the right at $t = T$, the pressure is a minimum. Since $A \gg A_v$ (Figure 7.44a provides definition of terms), the pressure in excess of the mean, p', tends to push the valve to the left with a force $p'A$, where A denotes the cross-sectional area of the valve. Consequently, the excess pressure is more or less in phase with the velocity $\dot{x}$ of the valve and, hence, does positive work on the latter.

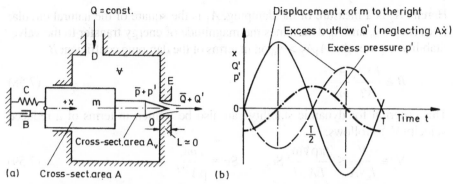

Figure 7.44. (a) Schematic of lightly-damped fuel-injection valve vibrating axially due to leakage-flow pulsations Q'; (b) Corresponding histograms under the assumption that effects on Q' other than changes in outflow area are negligible (after Den Hartog, 1985).

If changes in volume produced by the in-and-out motion of the valve, $A\dot{x}$, are considered and if pressure variations are assumed not to affect the outflow velocity (an assumption that does not hold for low frequencies), then the excess rate of outflow Q' may be expressed as:

$$Q' = -V\frac{A_v}{l}x - A\dot{x} \tag{7.54}$$

where $V = \bar{Q}/\bar{A}_{gap}$ = mean outflow velocity, A_v = base area of the valve cone (Figure 7.44a), and l = length of valve cone. The energy transfer to the valve is determined by the part of Q' that is in phase with x (the only part represented in Figure 7.44b).

The pressure variations accompanying the changes in V are obtained from the definition of the fluid modulus of elasticity, $E = dp/(dV/V)$, with the aid of $dV/dt = -Q'$ and Equation 7.54:

$$\frac{\dot{p}}{E} = -\frac{Q'}{V}; \qquad \dot{p} = +\frac{E}{V}\left(V\frac{A_v}{l}x + A\dot{x}\right) \tag{7.55}$$

The equation of motion for the valve of mass m is

$$m\ddot{x} + B\dot{x} + Cx = -pA \tag{7.56}$$

Eliminating p from Equations 7.55 and 7.56 by means of differentiating the latter, one obtains

$$\dddot{x} + A_2\ddot{x} + A_1\dot{x} + A_0x = 0 \tag{7.57}$$

$$A_2 = \frac{B}{m}; \quad A_1 = \omega_N^2 = \frac{C}{m} + \frac{A^2E}{mV}; \quad A_0 = \frac{AEVA_v}{mVl}$$

Herein, A_2 is a measure of the damping, A_1 is the square of the natural circular frequency ω_N, and A_0 determines the magnitude of energy transfer to the valve. Stability requires that $A_1 A_2 > A_0$ or, in terms of the damping coefficient B

$$B \geq \frac{EVA_v}{Vl} \frac{A}{\omega_N^2} \qquad (7.58)$$

This criterion for dynamic stability can also be written in terms of a reduced velocity V_r as follows:

$$V_r \equiv \frac{V}{f_n \sqrt{A}} \leq \frac{2\pi\rho V \omega_N^2}{EA_v/l} Sc, \qquad Sc \equiv \frac{2m\zeta}{\rho A^{3/2}} \qquad (7.59)$$

where $f_n = \sqrt{C/m}/2\pi$ and $\zeta = B/(4\pi m f_n)$. If V_r exceeds the critical value given by Equation 7.59, according to this crude consideration, the valve will be excited to vibrate.

Equation 7.57 is a cubic differential equation, rather than the quartic that one would have obtained for a system of two coupled oscillators. Actually, the system in Figure 7.44a does comprise two oscillators: (a) the valve and (b) a compressibility-governed Helmholtz resonator (Figure 4.3a). The latter consists of a strong 'fluid spring' (the chamber of volume V) and a negligibly small fluid mass (the fluid within the neck of length L). To demonstrate the susceptibility to movement-induced excitation in this case, only the *fluid-spring* coupling had to be accounted for. One may thus state:

> Valves controlling the outflow from a Helmholtz resonator of negli-
> gible fluid inertia are susceptible to movement-induced excitation if   (7.60)
> they act as press-open devices.

In most cases of movement-induced excitation involving fluid-flow pulsations, the dominant part in the excitation mechanism is played by *fluid-inertia*. This situation occurs particularly for 'press-shut devices', i.e., devices controlling flow through small openings such that the fluid force tends to press them shut. A typical example is given in Figure 7.45. If high-pressure steam or water leaks into a chamber J due to a difference in head ΔH across the valve or the turbine block m, any axial displacement of m changes the rate of leak Q as well as ΔH. This change decelerates or accelerates the mass m_{fl} of the fluid in pipe E by means of a variation of the pressure in J, which in turn affects the motion of m.

The mass m_{fl} in the pipe can oscillate both in a Helmholtz-resonator mode on account of chamber J with volume V and in an organ-pipe mode if the pipe length L is larger than one-quarter wave length of a standing acoustic wave λ. In most practical cases, V is negligible and $L \ll \lambda/4 = c_a T/4$ (Section 4.3.1), so that only the *inertial* part of the fluid oscillator needs to be considered in this case. When m moves to the right, the rate of flow decreases and m_{fl} tends to be decelerated. This deceleration is associated with a pressure reduction in J or an increase in ΔH

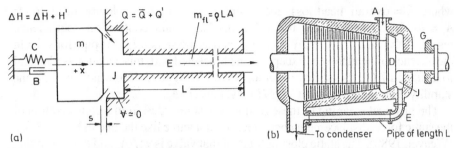

Figure 7.45. (a) Schematic of plug valve in near-closed position or idealization of the turbine in Figure 7.45b. (b) Sketch of steam turbine with dummy piston D, thrust bearing G, and equilibrium pipe E (after Den Hartog, 1985).

which, in turn, causes an increased fluid force on m in the direction of motion. Movement-induced excitation therefore is possible provided the work done by this force exceeds the energy dissipated by damping. If, on the other hand, V is large and the fluid is easily compressible (as in the case of steam), a fluid-spring action results that reduces pressure variations, and the system will most likely be stable. In fact, cures can be effected by the insertion of a slightly larger chamber J in a steam turbine (Figure 7.45b).

The system sketched in Figure 7.45a, say a plug valve, may be unstable both statically and dynamically. For a simple case with $V = 0$, $L \ll c_a T/4$, $b\bar{s}/A \ll 1$, and $\rho A_v L/m \gg 1$, the following criterion indicates absence of divergence or *static* stability (Kolkman, 1976):

$$\frac{2\gamma\Delta\bar{H}}{C/\sqrt{A_v}} \le \frac{\bar{s}}{\sqrt{A_v}} \tag{7.61}$$

Herein, $\bar{s}$ = initial, zero-load valve opening; b = effective valve width; A_v = effective cross-sectional area of valve plug; m = valve mass; A = cross-sectional area of pipe downstream from valve; L = length of downstream pipe; $\Delta\bar{H}$ = hydrostatic head across valve; C = stiffness of valve support; and $\gamma = \rho g$ = specific weight of fluid.

Recently, D'Netto & Weaver (1987) used a nonlinear theory, modified on the basis of experiments, to show that the critical *dynamic* stability threshold for movement-induced excitation of a plug valve is identical to that of divergence. Their finding is significant since it reveals the basic mechanism to be associated with negative fluid stiffness (Section 3.4) and valve vibrations to come about, essentially, because of system nonlinearities. Their criterion for static and dynamic stability reads:

$$\frac{2\gamma\Delta\bar{H}}{C/\sqrt{A_v}} \le \frac{\bar{s}}{\sqrt{A_v}} \left[\frac{8\psi}{27} C_d^2 A_r^2 \right], \qquad \psi = 1 + \xi - \frac{3}{C_c A_r} + \left(\frac{3/2}{C_d A_r} \right)^2 \tag{7.62}$$

where ξ = overall head-loss coefficient related to the velocity head in A; $A_r = b\bar{s}/A$; C_d and C_c are the valve discharge and contraction coefficients, respectively; and $\sqrt{2g\Delta\bar{H}}$ may be interpreted as the velocity in the pipe for $s = \bar{s}$ in the absence of head losses. Except for the bracketed part, this equation is identical to Equation 7.61. Its validity is demonstrated in Figure 7.46b, which presents the stability criteria in the form $C/(\gamma\Delta\bar{H}A_v)$ versus $\bar{s}/\sqrt{A_v}$.

The physical meaning of these results is best revealed from a consideration of the static force-displacement characteristic of a valve like the one in Figure 7.46a (Weaver, 1980). The static closing force of that valve is $\gamma\Delta\bar{H}A_v$ and the restraining force is C times the linear extension of the spring holding the valve. At an initial, zero-load opening, these two forces are balanced,

$$C\bar{s} = \gamma\Delta\bar{H}A_v \qquad (7.63)$$

For a spring constant C smaller than that given by this equation, the valve will be held closed, whereas for a larger value of C, the valve will remain open under static conditions. With fluid flow, however, the head difference ΔH is dynamic because of the flow past the plug and the inertia effects resulting from plug movements, so that dynamic instability develops near the static characteristic (Figure 7.46b). Within a certain region below the static characteristic, the static head is sufficient to hold the valve closed; but sudden closure creates pressure waves that cause the valve to move off the seat far enough to reestablish the flow, and so the valve continues to open. In a certain region above the static characteristic, the fluid inertia associated with a reduction in discharge creates sufficient dynamic head to close the valve; as the fluid comes to rest, ΔH decreases and the spring can open the valve. (Fuller details are given in Griffiths, 1969, and Weaver et al., 1978.)

The function of fluid inertia in this process is to maintain higher flow rates

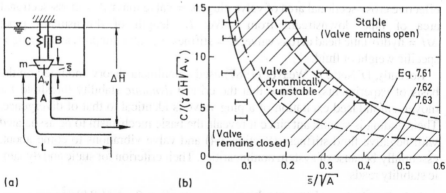

Figure 7.46. (a) Schematic of plug-valve system. (b) Stability diagram of plug valve for large fluid inertia; horizontal lines = experimental stability threshold $L/\sqrt{A_v} = 221$ (after D'Netto & Weaver, 1987).

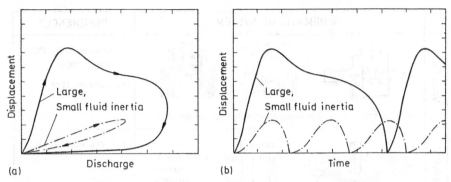

Figure 7.47. (a) Displacement-discharge behavior and (b) corresponding displacement histograms, typical for valves, seals, and gates operating at small openings (after Weaver & Ziada, 1980).

while the valve is closing and to delay the reestablishment of flow while the valve is opening. In this way, a phase shift in the fluid force is brought about, and this shift is the precondition for energy transfer to the structure. Figure 7.47a illustrates this phase shift via the displacement-discharge relationship. The corresponding histograms in Figure 7.47b show that the opening times, being dominated by the valve restraint, are about the same for large and small fluid inertias, whereas the closing times differ considerably, independent (!) of the valve's natural frequency in still fluid. Herein lies the key also to an understanding of why a relative increase in restraining force, compared with the dynamic fluid closing force, can result in a *decrease* in actual vibration frequency and an increase in amplitude (Weaver & Ziada, 1980). In summary:

> Devices controlling flow through small openings are susceptible to movement-induced excitation if they are press-shut devices and exposed to large fluid-inertia effects from the adjacent flow passage. (7.64)

7.4.2 *Leakage-flow channels*

An interesting variety of this excitation mechanism occurs in connection with boundaries of leakage-flow channels (Figures 7.48h, i, m, n). If the elastic response of the structure, of which that boundary is a part (e.g., a wall or plate), can alter the leakage-flow rate in a way equivalent to that of the press-shut devices discussed in the foregoing paragraphs, then movement-induced excitation is possible. Most hydraulic- and energy-system components have suffered from this type of excitation (Paidoussis, 1980; Parkin, 1980; Mulcahy, 1983). As a general rule, a leakage-flow channel behaves like a press-shut device if its flow is controlled such that the fluid force tends to pull it shut. The presence of (a) diverging boundaries, (b) a substantial local head loss near the entrance, (c) a

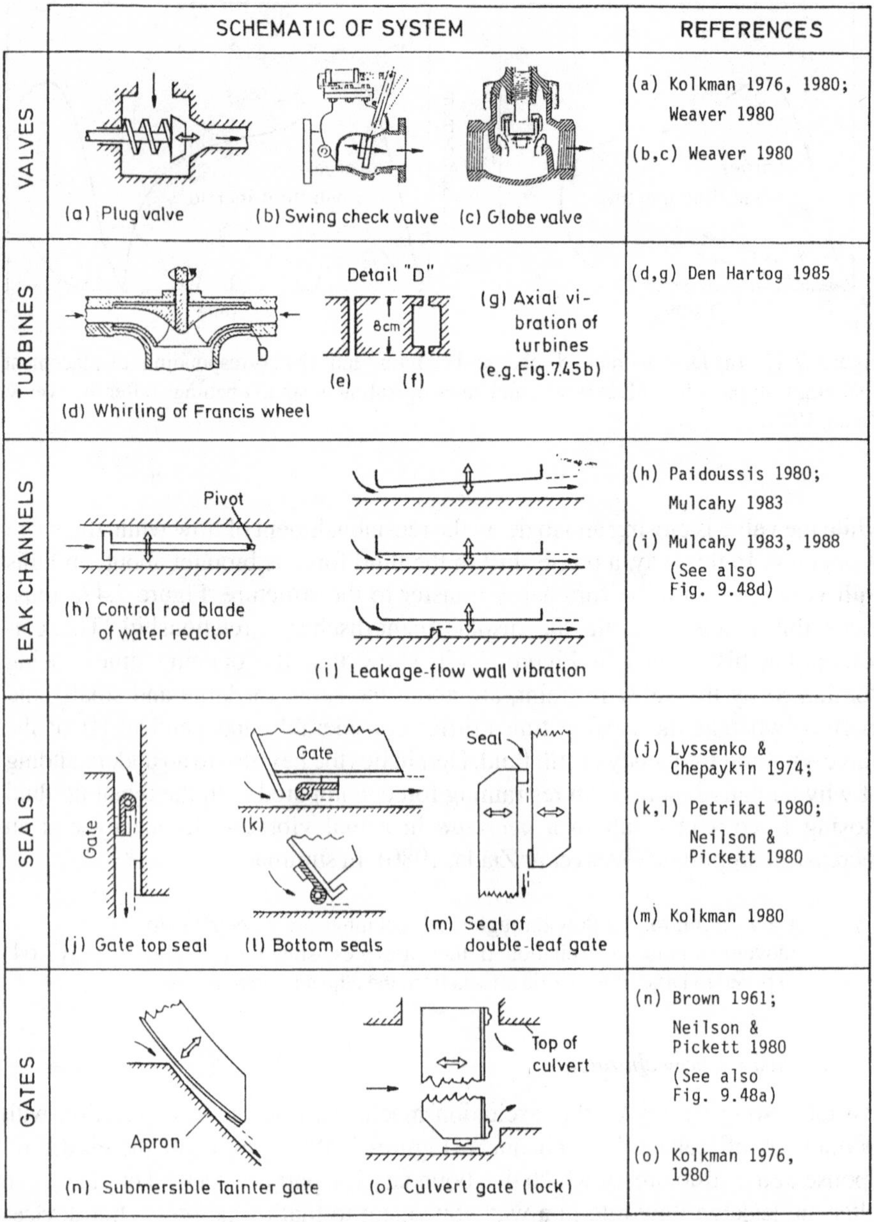

	SCHEMATIC OF SYSTEM	REFERENCES
VALVES	(a) Plug valve (b) Swing check valve (c) Globe valve	(a) Kolkman 1976, 1980; Weaver 1980 (b,c) Weaver 1980
TURBINES	Detail "D" 8cm (e) (f) (g) Axial vibration of turbines (e.g.Fig.7.45b) (d) Whirling of Francis wheel	(d,g) Den Hartog 1985
LEAK CHANNELS	Pivot (h) Control rod blade of water reactor (i) Leakage-flow wall vibration	(h) Paidoussis 1980; Mulcahy 1983 (i) Mulcahy 1983, 1988 (See also Fig. 9.48d)
SEALS	Gate (k) (j) Gate top seal (l) Bottom seals Seal (m) Seal of double-leaf gate	(j) Lyssenko & Chepaykin 1974; (k,l) Petrikat 1980; Neilson & Pickett 1980 (m) Kolkman 1980
GATES	Apron (n) Submersible Tainter gate Top of culvert (o) Culvert gate (lock)	(n) Brown 1961; Neilson & Pickett 1980 (See also Fig. 9.48a) (o) Kolkman 1976, 1980

Figure 7.48. Examples of systems susceptible to movement-induced excitation involving coupling with fluid-flow pulsations.

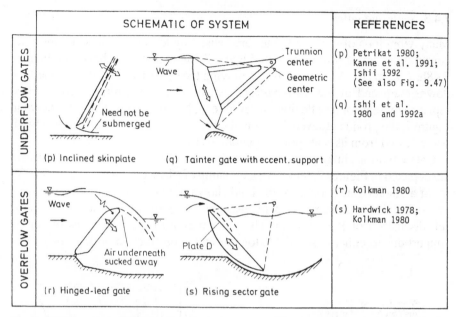

SCHEMATIC OF SYSTEM	REFERENCES

Figure 7.48 (continued). Examples of systems susceptible to movement-induced excitation involving coupling with fluid-flow pulsations.

sharp-edged entrance, or (d) a flow constriction upstream of the channel midpoint (Figure 7.48i) is thus sufficient to produce the potential for movement-induced excitation. Every leakage-flow channel should therefore be checked individually for stability (Mulcahy, 1988). In fact, even whirling-type vibrations of turbines have been traced back to leakage-flow problems. The whirling of a Francis wheel, for example, was found to be excited by periodic variations of the labyrinth width (D in Figure 7.48d) due to lateral movements of the wheel. The excitation was suppressed by changing the D-detail from (e) to (f), thereby creating a wide inner chamber for which the width variations and, hence, the pressure fluctuations were negligible.

Lateral vibrations of valves with an annular flow (e.g., Figure 7.48c) or of any cylindrical body in a narrow duct conveying fluid (Mateescu & Paidoussis, 1987; Paidoussis et al., 1990) are caused by a similar mechanism. The center location of a plug in an orifice, for example, is potentially unstable. Any disturbance from this position will create an increased velocity on the side with the smaller gap and a decreased velocity on the other side, thus producing a fluid force which tends to increase the eccentricity. Instrumental for a phase lag in force fluctuations and, thus, a potential for movement-induced excitation is again the fluid inertia.

7.4.3 *Seals and gates*

Many vibrations of seals and gates are caused by leakage flows with upstream constrictions in accordance with Figure 7.48i (examples shown in Figures 7.48h, j, k, m, n). A cure consists in moving the constriction of flow toward the downstream end of the flow passage. As an alternative, rubber seals can be replaced by seals of less flexible material (e.g., Figure 9.48g). The gate shown in Figure 7.48o produced severe horizontal vibrations of the plug-valve type when it was released from its seats prior to operation (Kolkman 1980, p. 383).

Gate vibrations involving flow-rate pulsations of a more complex mechanism are depicted in Figures 7.48p-s. An inclination of a leaf-gate plate or an eccentricity of a Tainter-gate support of the kind shown in (p) and (q) turn the gates into press-shut devices and make them susceptible to movement-induced excitation (cf. discussion of Figure 9.47). The latter may arise even in cases of centric support due to either unavoidable tolerances of construction or deformations of

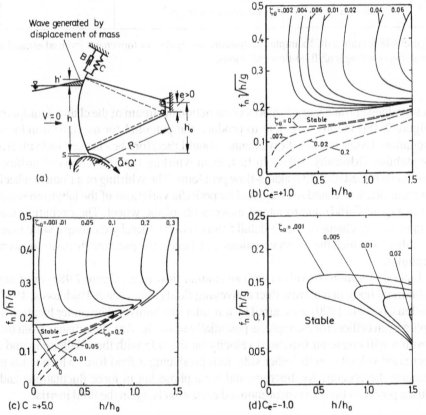

Figure 7.49. Stability diagrams for a Tainter gate with eccentric trunnion releasing flow from an infinite reservoir in near-close positions (after Ishii et al., 1977).

the supporting arms under load. Prompted by the complete failure of a Tainter gate in Japan, Ishii et al. (1977, 1980, 1992a) developed stability criteria as shown in Figure 7.49 and verified them experimentally. Here, $f_n = \sqrt{C_\theta/I_\theta}/2\pi$ = natural frequency; $C_\theta = CR^2$ = torsional spring constant per unit gate width; I_θ = moment of inertia of gate mass and added mass per unit gate width; $\zeta_\theta = B_\theta/(4\pi f_n I_\theta)$ = damping ratio including mechanical and fluid damping; $C_e = C^*e/(RI_r)$ = eccentricity and mass parameter; $I_r = I_\theta/(\rho R^4)$ = reduced mass; C^* = geometric parameter; and the other symbols are as explained in Figure 7.49a. Within the range of practical significance, $f_n\sqrt{h/g} > 0.2$, Tainter gates are stable if designed as press-open devices ($C_e \leq 0$). If designed as press-shut devices ($C_e > 0$), i.e., with trunnion support above the geometric center, Tainter gates become unstable over the entire practical range of $f_n\sqrt{h/g}$ values unless $f_n\sqrt{h/g}$ is very large or $\zeta_{s\theta}$ exceeds a critical value (Figures 7.49b, c). Since I_θ and ζ_θ increase with increasing h/h_o due to growing contributions of added mass and fluid damping, respectively, the curve separating stable and unstable ranges in Figure 7.49 must be determined individually for any given Tainter-gate system. Increasing the relative gate opening s/h has a stabilizing effect on the gate (Ishii et al., 1980).

The two overflow gates in Figures 7.48r, s have in common, first, substantial underpressures near the crest which tend to pull them closed and, second, relatively high tailwater levels. The excitation mechanism is most likely mixed, involving fluctuations in flow contraction due to 'added-mass flow' analogous to Figure 7.37 and, possibly, unstable flow reattachment as discussed in Section 6.6 (Kolkman, 1980). Vibrations were suppressed by providing sufficient aeration to keep air under the nappe (Figure 7.48r) and by perforating plate D to reduce the underpressure in that zone (Figure 7.48s), respectively.

7.4.4 *Closely-spaced bodies*

Another example of a mixed type of excitation, involving jet switching in addition to fluid-flow pulsations, is illustrated in Figure 7.50. If the flexibly-mounted cylinders depicted in Figure 7.32a vibrate about an equilibrium position upstream of the row of fixed cylinders, $\bar{x} < 0$, and if they are closely spaced, they will produce discharge and/or head variations much like a press-shut device and become subject to excitation of the kind described in this chapter. If they vibrate about an equilibrium position downstream, $\bar{x} > 0$, the same mechanism will lead to fluid damping. The combined effects of this and the jet-switching mechanism described in Section 7.3.4 will modify the energy diagram of Figure 7.35b approximately as shown in Figure 7.50b. For a case of $\bar{x} < 0$ and relatively low mechanical damping, $\zeta = \zeta_1$, soft excitation with a build-up of large amplitudes from rest will be possible, signified by point 1. For $\zeta = \zeta_2 > \zeta_1$, the curve for energy dissipated per cycle due to the mechanical damping, $-W_d$, has three intersections with the curve representing work done by the fluid forces, W_e: point 2 corresponds to soft excitation due to the flow-pulsation mechanism alone; point

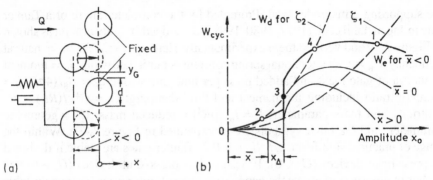

Figure 7.50. (a) Closely spaced, infinite row of staggered cylinders with alternate cylinders free to move in unison in streamwise direction; (b) Approximate effect of equilibrium position $\bar{x}$ on energy diagram for V_r = const (after Roberts, 1966; $\zeta_2 > \zeta_1$).

4 corresponds to hard excitation due to additional jet switching; and point 3 determines the 'threshold amplitude'. Depending on whether disturbance amplitudes are smaller or larger than this threshold value, the cylinder will vibrate with an amplitude given by solution 2 or 4, respectively.

7.5 EXCITATION INVOLVING MODE COUPLING

7.5.1 *Basic features of mode coupling*

A body free to vibrate in more than one degree of freedom may be excited by movement-induced mechanisms even if each mode by itself is free of excitation. The most important prerequisite for such coupled-mode excitation is coincidence or merging of the individual mode frequencies; primarily in this way does vibration in the one mode produce fluid-force components that transfer flow energy to the body in the other mode(s). Frequency merging occurs if mode frequencies vary with flow velocity in a way that two of them merge to a common value as the velocity is increased. Examples for this situation are given in Figures 7.3d, e.

Figure 7.51 presents an overview of systems susceptible to the mode-coupling type of movement-induced excitation. For most of these systems, single-mode flutter is possible only for certain body configurations and orientations, usually associated with flow separation (stall), and for specific conditions of body support (Sections 7.3.1, 7.3.2, 7.3.6, 7.3.7). With one or more additional degrees of freedom, movement-induced excitation can occur for a wider range of geometric and flow conditions, and it will typically lead to more intense vibrations.

Quite universally, one may state:

SCHEMATIC OF SYSTEM	REFERENCES
(a) Slender Bodies in Cross-Flow 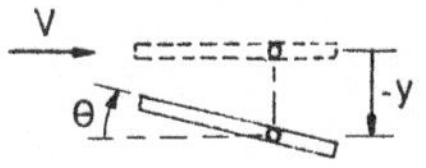 Flutter of foils, vanes, and blades: 1. Airfoils 2. Hydrofoils 3. Turbomachinery blading	1. Bisplinghoff et al. 1955 2. Abramson 1969 3. Carta 1978 Recent overviews: Garrick et al. 1976, 1981; Carta 1978; Dowell et al. 1989; McCroskey 1982
(b) Long Bluff Bodies in Cross-Flow 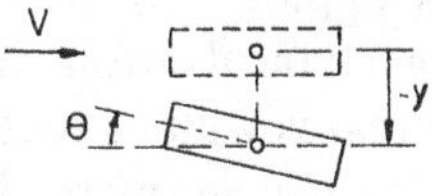 Bridge-deck flutter	Scanlan & Tomko 1971; Simiu & Scanlan 1978
(c) Short Bluff Bodies in Cross-Flow 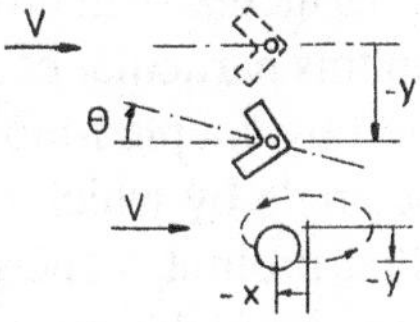 1. Coupled-mode flutter or galloping 2. Wake-induced flutter 3. Stranded cable flutter	1. Parkinson & Modi 1967; Blevins 1977 2. Simpson 1971; Wardlaw et al. 1973; Wardlaw 1980 3. Simpson 1979
(d) Rods and Pipes with Axial Flow 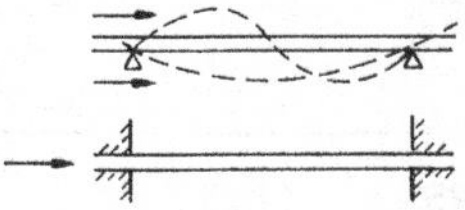 Pinned-pinned or clamped-clamped rod or pipe flutter	Paidoussis 1973; Paidoussis & Issid 1974; Weaver 1974a; Paidoussis 1987
(e) Shells with Axial Flow 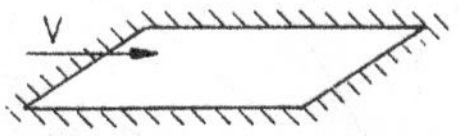 Plate and panel flutter	Dowell et al. 1989
(f) Long-Span Gate with Small-Gap Underflow 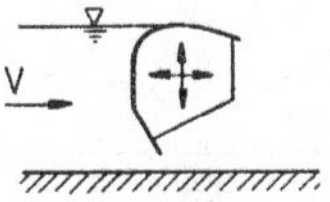 With or Without Overflow	Ishii & Knisely 1992 Ishii et al. 1993 (See also Fig. 9.49)

Figure 7.51. Examples of systems susceptible to movement-induced excitation involving mode coupling.

Structures free to vibrate in two modes are susceptible to coupled-mode movement-induced excitation if the natural frequencies of the two modes are close to each other or tend to merge with increasing flow velocity, and if the conditions provide for energy transfer between the modes. (7.65)

7.5.2 Slender bodies in cross-flow

The system most thoroughly investigated is the *unstalled rigid foil* or blade free to vibrate in both the bending (y) and the torsional (θ) modes (Figure 7.51a). The variations of natural circular frequency (ω_n) and fluid damping (B') of these modes with flow velocity (V) are presented in Figure 7.3 for a typical airfoil under nonstall conditions. Whereas ω_n in the torsional mode decreases with increasing V, that of the bending mode increases, until at $V = V_{cr}$ the frequencies merge into that of coupled-mode vibration. As V increases from zero, moreover, B' increases initially in both modes until, at sufficiently high velocities, B' for torsion decreases rapidly toward zero while B' for bending continues to increase. In other words, only the torsional mode leads to instability initially in this particular case, but this suffices to drive the structure into the more dangerous coupled-mode flutter on account of frequency merging provided the phase angle by which the torsional mode leads the bending mode takes on a value that promotes energy transfer. Although some structures exhibit instability in *both* modes at the onset of flutter, this finding can be generalized as expressed in statement 7.65.

Figure 7.52 illustrates the effect of three important parameters affecting the

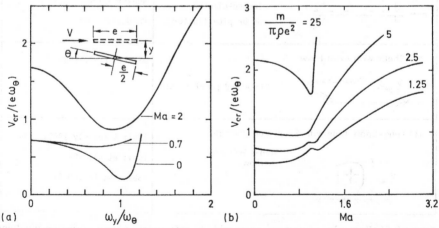

Figure 7.52. Effects of ratio of natural frequencies at zero velocity, ω_y/ω_θ, Mach number, Ma, and mass ratio, $m/(\pi\rho e^2)$, on the critical velocity, V_{cr}, for the onset of rigid-blade, nonstall flutter (after Bisplinghoff et al., 1955). (Near Ma = 1, the possibility of shock-wave formation can completely alter the flutter behavior.)

onset of rigid-blade, nonstall flutter: the ratio of the natural circular frequencies at zero velocity, ω_y/ω_θ, the Mach number, Ma, and the mass ratio, $m/(\pi\rho e^2)$, where m = blade mass. The closer the ratio ω_y/ω_θ to unity, the lower is the velocity V_{cr} required for onset of flutter in incompressible flow, Ma $\simeq 0$ (Figure 7.52a). At a high subsonic Mach number, Ma $= 0.7$, the sensitivity of V_{cr} to ω_y/ω_θ is smaller, while at Ma $= 2$, it is greater. With increasing mass ratio, the flutter onset velocity increases irrespective of Mach number (Figure 7.52b). If the foil or blade is relatively light compared to the surrounding fluid, according to this result, it is more susceptible to flutter. Particularly low values of $m/(\pi\rho e^2)$ are encountered in marine engineering, where the flutter of hydrofoils is of interest (Abramson, 1969).

Additional factors the designer should consider in securing stability with respect to coupled-mode flutter are: finite-span and sweep effects, proximity to a free surface, cavitation, and the relative locations of fluid-dynamic center, elastic axis, and center of gravity (cf. discussion of Figure 7.2). Regarding these relative locations, Pines (1958) and Carta (1978) stress the following basic criteria:

> For a slender, rigid body in cross-flow, coupled-mode, nonstall flutter cannot occur if the center of gravity lies upstream of the elastic center. In the event that it lies downstream of the elastic center, flutter can occur only with the fluid-dynamic center lying ahead of the elastic axis *and* the zero-velocity frequency ratio ω_y/ω_θ remaining below a specified value dependent on geometry and radius of gyration. (7.66)

Strictly speaking, these criteria apply only to cases of unseparated, incompressible fluid flow past rigid blades. The importance of Mach number is illustrated in Figure 7.52. Its effect on the location of the fluid-dynamic center is dramatic: if Ma < 1, it is located at the quarter-chord point, but with Ma > 1, it moves near midchord.

A *stalled rigid foil* or blade is shown in Section 7.3.2 to be susceptible to single-mode flutter (e.g., Figure 7.14a). As can be inferred from Figure 7.53, stalled foils may undergo coupled-mode flutter as well. The counterclockwise loop of C_M versus α in Figure 7.53a indicates negative work done by the fluid on the foil during pure torsional motion and hence positive damping; only for light or deep stall (Figures 7.53b, c) are there regions of negative damping (cross-hatched parts of loop, $B'_\theta < 0$, Equation 7.9), but they are outweighed by the contributions to positive damping. However, the variations in lift (C_L) and drag (C_D) produced by the torsional motion, also depicted in this figure, show that there are negative contributions to the fluid-coupling coefficients $B'_{y\theta}$ (Equation 7.8) and $B'_{x\theta}$, respectively; therefore positive work on the foil can be expected, and hence excitation, if the foil is free to move also in the y and x directions.

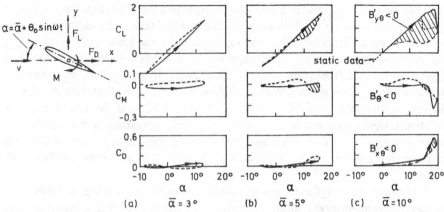

Figure 7.53. Effect of stall regime (Figures 7.9a, b) on the variation of lift (C_L), moment (C_M), and drag (C_D) coefficients with incident angle α during forced torsional vibration with amplitude $\theta_o = 10°$ and $V/(f_s e) = 31.4$ (after McCroskey & Pucci, 1981). (a) No stall; (b) Light stall; (c) Deep stall.

7.5.3 *Long bluff bodies in cross-flow*

For massive *long bluff bodies* in air flow, such as bridge decks, Equations 7.8 and 7.9 may be simplified as follows:

$$F_y(t) = -B'_y\dot{y} - B'_{y\theta}\,\dot{\theta} - C'_{y\theta}\theta$$
$$M(t) = -B'_{\theta y}\dot{y} - B'_\theta\,\dot{\theta} - C'_\theta\theta \qquad\qquad (7.67)$$

From the variation of the remaining 'added' coefficients with reduced velocity in Figure 7.54, obtained from free-vibration experiments by Scanlan & Tomko (1971), it is evident that coupled-mode flutter is possible ($B'_{y\theta} < 0$) for a profile close to that of the original Tacoma Narrows Bridge (profile 1) in addition to flutter in the pure torsional or bending mode ($B'_\theta < 0$ or $B'_y < 0$, cf. discussion of Figure 7.17). Further information on long bluff bodies in cross-flow is given in Section 9.1.

7.5.4 *Coupled-mode galloping*

Movement-induced excitation of *short bluff bodies* in cross-flow (Figure 7.18) has the special feature, independent of the mode, that the instantaneous fluid forces can be expressed by their static counterparts (Section 7.2.4). The difficulty in the case of the coupled transverse/torsional mode lies in the formulation of the effective incidence angle of flow, α, since the instantaneous linear velocity at a given point of the body is a function of its distance from the elastic axis. Following the approach of Sisto (cf. Dowell et al., 1989, or Blevins, 1977), one may define α for a 'representative distance' R_1 from the elastic axis. Thus,

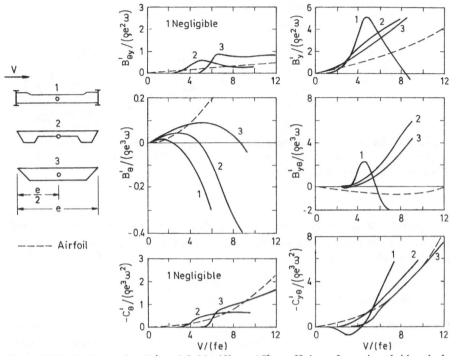

Figure 7.54. Fluid-damping (B') and fluid-stiffness (C') coefficients for various bridge-deck sections in smooth cross-flow (after Simiu & Scanlan, 1978).

$$\alpha = \bar{\alpha} + \theta - R_1\dot{\theta}/V - \dot{y}/V \tag{7.68}$$

where $\bar{\alpha}$ = incidence angle for the body at rest; θ = angular deflection; $R_1\dot{\theta}/V$ = effect of the body's angular velocity $\dot{\theta}$; $\dot{y}/V$ = effect of the body's transverse velocity $\dot{y}$. Based on the quasi-static assumption, the transverse force F_y and moment M on the body become, for small variations in α,

$$F_y = \frac{\rho}{2}V^2 e C_y, \qquad C_y = -a_1\alpha$$
$$M = \frac{\rho}{2}V^2 e^2 C_M, \qquad C_M = -a_2\alpha \tag{7.69}$$

Herein, a_1 and a_2 represent the slopes of the curves for C_y and C_M versus α at $\alpha = \bar{\alpha}$, respectively, determined from steady-flow experiments (e.g., Figure 7.55). Substituting α in Equations 7.69 by Equation 7.68, one obtains

$$F_y(t) = \frac{\rho}{2}V^2 e[-a_1(\theta - R_1\dot{\theta}/V - \dot{y}/V)]$$
$$M(t) = \frac{\rho}{2}V^2 e^2[-a_2(\theta - R_1\dot{\theta}/V - \dot{y}/V)] \tag{7.70}$$

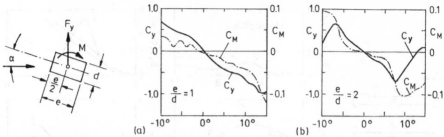

Figure 7.55. Steady-state force and moment coefficients, as defined in Equation 7.69, for rectangular prisms in smooth cross-flow (after Nakamura & Mizota, 1975). ($Ve/\nu = 2.1 \times 10^5$; $L/e = 10.7$, with end plates.)

Comparison of these equations with Equations 7.8 and 7.9 yields the added coefficients, which are all independent of reduced velocity in this case of coupled-mode galloping.

7.5.5 *Wake-induced flutter*

From the many types of flow-induced *cable* vibrations (reviewed by Simpson, 1979), two are selected in the following because of their intriguing excitation mechanisms. One of these involves conductor cables that are suspended in bundles of two or more conductors spaced 10 to 20 diameters apart (e.g., Figure 7.56a), the bundle geometry being maintained by rigid spacers that divide each span between towers into subspans 45 m to 60 m long. Above a critical wind speed, the cable span in the wake of an upstream cable may be excited to vibrate, typically in an elliptical orbit (Wardlaw et al., 1973; Simpson, 1971).

Like all coupled-mode-flutter phenomena, wake-induced flutter occurs only within a limited range of *x-y*-frequency coalescence. Only if the horizontal and vertical frequencies differ by a small amount at zero velocity, surprisingly enough, is there a position for the leeward cable where, at a suitable velocity, fluid stiffness will enforce frequency coalescence. The range of flutter (Figure 7.56c), as well as the vibration amplitudes, are highly dependent on background turbulence. At a turbulence level of 10%, for example, the system depicted in Figure 7.56 was stable. In low-turbulence flow, the typical situation for wind over undeveloped ground, amplitudes can grow so large that conductors clash, producing cable and spacer damage.

With a wind velocity of 10 m/s, a cable diameter of 30 mm, and a frequency of 1 Hz, a typical reduced frequency is $fd/V = 0.003$. Because of this extremely low value, the excitation mechanism can be interpreted on purely quasi-steady grounds as shown in Figure 7.57: The two force-displacement diagrams, constructed on the basis of static drag and lift data (e.g., Figure 7.56b), illustrate clearly that the vibrating cable is brought into the high-drag zone on its down-

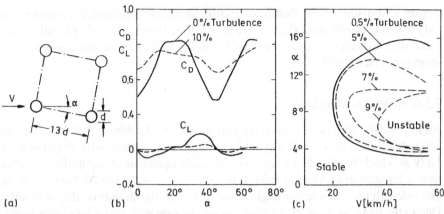

Figure 7.56. (a) Typical arrangement of conductor bundle; (b) Steady-state drag (C_D) and lift (C_L) data for leeward cable at Re $= 1.8\times10^4$; (c) Corresponding stability diagram for a given system (after Wardlaw, 1980).

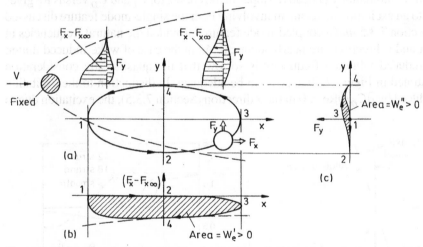

Figure 7.57. (a) Sketch of typical orbit of a circular cylinder with x and y degrees of freedom in the wake of another. (b, c) Force displacement diagrams for (b) streamwise and (c) transverse fluid-dynamic forces based on quasi-steady considerations.

stream path (1-2-3) and into the low-drag zone on its upstream path (3-4-1) only by means of the coupled transverse motion. The areas W'_e and W''_e signify positive work done on that cable.

Vibration can be attenuated by (a) flexible spacers that incorporate damping elements; (b) bundles rigged at incidence to the windstream so that no cable lies in the wake of another; (c) large conductor separation and staggered (small) subspan lengths (Simpson, 1979; Wardlaw, 1980). Increased mechanical damping has only a minor stabilizing effect (Price & Piperni, 1988). The role played by the wake-flutter mechanism in tube-bundle vibrations has been addressed by Paidoussis & Price (1988).

7.5.6 *Stranded-cable flutter*

An important difference between smooth and stranded cables is that the latter can generate lift near the critical Reynolds number, Re_{cr}, defined as the magnitude of Vd/ν at which the boundary layer on the cable upstream of separation changes from a laminar to a turbulent state. If the approach flow is yawed, the strands on one side will present a larger angle to the flow than the strands on the other side. Due to the increase in relative roughness on the side with larger incidence angle, the boundary layer on that side will become turbulent as it approaches Re_{cr} before the boundary layer does on the opposite side. The corresponding asymmetry in separating streamlines produces the lift shown in Figure 7.58.

The combination of negative slopes in the curves for C_L and C_D versus Re gives rise to an excitation mechanism involving both the single-mode feature discussed in Section 7.3.5 *and* a coupled-mode feature, provided the natural frequencies in the x and y directions are nearly identical. As in the case of wake-induced flutter, the reduced vibration frequency is so low that the quasi-steady consideration illustrated in Figure 7.59 appears to be adequate. In distinction to the excitation produced by $\partial C_D/\partial Re < 0$ in the x direction (Section 7.3.5), the excitation due to

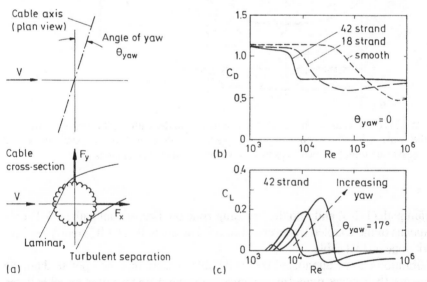

Figure 7.58. (a) Sketches of a stranded cable in yawed approach flow; (b, c) Steady-state drag (C_D) and lift (C_L) coefficients for a typical stranded cable (after Simpson, 1979).

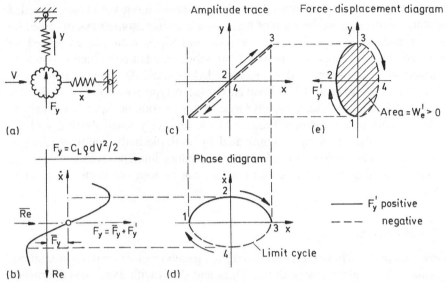

Figure 7.59. Sketches showing role of lift forces in the coupled-mode excitation of a stranded cable in yawed approach flow.

$\partial C_L/\partial \text{Re} < 0$ (Figure 7.58c) can only act with simultaneous body movements in the x and y directions. If the body moves with a velocity $\dot{x}$, its relative approach-flow velocity, and hence the effective Reynolds number, is reduced in proportion to $\bar{V} - \dot{x}$ (Figure 7.59b). Based on this relationship one can find an orbit, or amplitude trace (Figure 7.59c), for which the corresponding force-displacement diagram (Figure 7.59e) indicates maximum work W'_e done per cycle on the body by the quasi-steady lift fluctuations F'_y. The actual orbit of vibration is determined by the maximum of the sum of work done by lift *and* drag, $W'_e + W''_e$, in accordance with the minimum-energy principle.

The violent vibration of the CEGB Severn-Crossing transmission line in 1959 with vertical amplitudes of at least 8.5 m at a wind speed close to 50 km/h (Re $\simeq 4.2{\times}10^4$) was probably caused by a mechanism of this sort (Simpson, 1965, 1979). For underwater cables, typically 25 mm in diameter, the critical Reynolds number is associated with $V \simeq 1.5$ m/s, a normal velocity in tidal flows. Because of the reduction in critical range from about 2.5 m/s in air to about 0.2 m/s in water, however, the danger of severe vibration in marine applications is rather limited.

7.5.7 *Rods and pipes with axial flow*

Single-mode vibrations of rods and pipes subjected to external and internal flow are discussed in some detail in Section 7.3.6. Suffice it here to point out that cantilevered rods and pipes may also undergo coupled-mode vibrations, preceded by static divergence in the form of first-mode buckling, sometimes followed by second-mode buckling. In cases of rods and pipes with pinned-pinned or clamped-clamped supports (Figure 7.51d), vibrations can be excited only due to coupled-mode instabilities; they are preceded by static divergence (Paidoussis, 1973; Paidoussis & Issid, 1974; Weaver, 1974a). A typical example of this kind is illustrated in Figure 7.3e. Coupled-mode flutter of rods or pipes is unlikely to endanger hydraulic structures because of the relatively small flexibility of the latter, unless the problem is aggravated by multiple-body coupling (Section 7.6.3). However, bellows or rubber-type isolators interconnecting pipes may undergo buckling so violently that anchors can be torn out from the concrete floor.

7.5.8 *Panel flutter*

Flow past plates whose edges are fixed (i.e., panels) can give rise to instabilities qualitatively similar to those of thin pipes and shells with fixed ends (Kornecki, 1974). In incompressible flow, static divergence (buckling) precedes the onset of flutter, while in supersonic flow only flutter can occur. Panel flutter is of interest not only in aerospace technology (NASA, 1972); vibrations of taut canvas or plastic tarps, as well as flags, are associated with essentially the same excitation mechanism.

Figure 7.60a shows a simplified representation of a continuous elastic panel made up of three hinged plates with two spring supports. The system has two degrees of freedom (y_1 and y_2), and the equations of motion are coupled in a

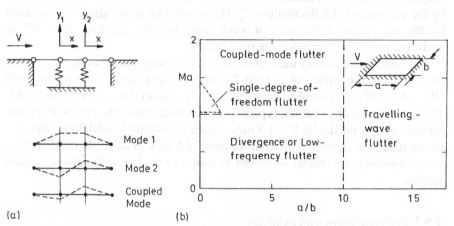

Figure 7.60. (a) Simulation of coupled-mode panel flutter by a series of discrete, rigid plates (after Dowell et al., 1989). (b) Flutter regions of a panel as a function of length-to-width ratio *a/b* and Mach number Ma (after Dowell, 1970).

fashion similar to that described with Equations 7.5 through 7.9, in that case for a body with y and θ movements. By employing similar techniques as for airfoil flutter (Sections 7.3.1, 7.5.2), one can determine the conditions for onset of flutter as well as the flutter modes. The important features, exhibited by the elastic panel and its hinged-plate model, are as follows. Flutter occurs in the coupled mode with a maximum amplitude near the trailing edge (Figure 7.60a); it sets in upon coalescence, i.e., merging, of the natural frequencies of the first and second modes. This coalescence is attainable because the natural frequencies vary with the velocity. The flutter frequency lies between those for the first and second modes. Because of structural nonlinearities, elastic-panel flutter does not usually lead to rapid catastrophic failure with increased flow velocity as it does in cases of airfoil or vane flutter, but leads rather to long-time fatigue failure due to vibrations with amplitudes of the order of the plate thickness (Dowell, 1970, 1974).

The basic regimes of flutter for a flat elastic plate, fixed at all four edges, as a function of length-to-width ratio and Mach number are shown in Figure 7.60b. For very long plates, $a/b > 10$, only travelling-wave flutter has been observed; it is characterized by modal lines that move streamwise along the plate rather than remaining stationary as for the other flutter modes (Dowell, 1966). Among the parameters that determine the actual flutter characteristics are structural damping, mass ratio, plate curvature, effects of an adjacent cavity, and boundary-layer effects (Dowell, 1970, 1974). An additional factor in hydrodynamic applications is the presence of a free surface (Weaver & Unny, 1972): For sufficiently small depths of flow, the free-surface has a stabilizing effect. An unusual feature of coupled-mode panel flutter, important for the design, is the possible destabilization of the system arising from structural damping (Benjamin, 1963). More details about this feature have been presented by Dowell (1970). The stability criterion of parallel-plate fuel assemblies is briefly discussed in Section 7.6.3.

7.6 EXCITATION INVOLVING MULTIPLE-BODY COUPLING

7.6.1 *Basic features of multiple-body coupling*

If two or more neighboring bodies vibrate, the motion of each affects those of the others because of fluid-dynamic coupling. The added mass, added damping, and added rigidity of a given body (Sections 3.1, 3.2) are then affected not only by the motion of that body but also by the motion of its neighbors. This coupling is clearly illustrated in Figure 7.61. If one of four tubes in a four-tube cluster, say 'a', is forced to vibrate, *all* tubes respond, and this response depends strongly on the space between the tubes. Each tube in this experiment has the same mechanical characteristics and, hence, the same resonance frequency in isolation. The magnification factor in Figure 7.61 is defined as the ratio of the response amplitude x_o or y_o (observed at the excitation frequency $f = \omega/2\pi$) to a reference

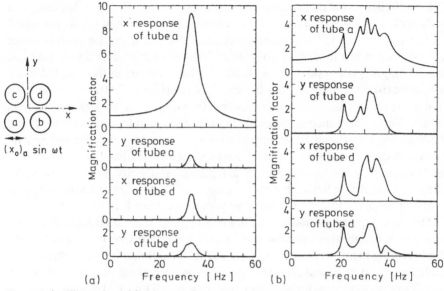

Figure 7.61. Illustration of fluid-dynamic coupling effects for a cluster of four parallel tubes in quiescent liquid (after Chen, 1975): (a) Widely-spaced cluster; (b) Closely-spaced cluster.

deflection; the latter is the x deflection of tube 'a' that occurs if it is subjected to a static load of the same magnitude as the dynamic-load amplitude.

For a *widely-spaced cluster* (Figure 7.61a), the x-response characteristic of tube 'a' is similar to that of a solitary tube (cf. Figure 2.4a), but significant responses can also occur for that tube in the y direction, and for tube 'd' in both the x and y directions, all being near the excitation frequency. For a *closely-spaced cluster*, on the other hand (Figure 7.61b), resonance peaks occur at a number of frequencies. In fact, as shown in Figure 7.62d, $2N$ natural frequencies exist for a cluster of N bodies with 2 degrees of freedom each, and, the closer the body spacings are, the further apart these frequencies will be. Moreover, the total amplitude response for each of the deflections y_a, x_d, and y_d, given by the areas under the response curves in Figure 7.61, is greater than their counterparts for a widely-spaced cluster. According to these observations, the effect of multiple-body coupling is to make the system receptive to disturbances over a broader frequency range (an important fact for extraneously and instability-induced excitations) and to enhance its susceptibility to movement-induced excitation.

If the bodies are distributed masses like tubes, then there exist $2N$ natural frequencies for each of their axial modes (e.g., Figure 7.62b). In Figures 7.62c, d, the natural circular frequencies ω_n are given in the dimensionless form

$$\omega'_n = \omega_n \left[\frac{EI}{(m + \rho A)L^4} \right]^{-1/2} \tag{7.71}$$

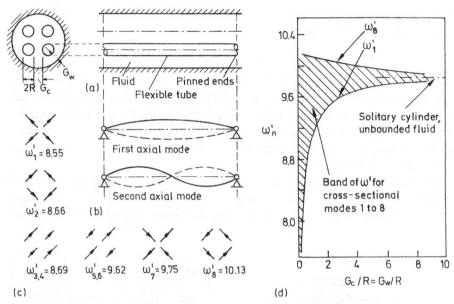

Figure 7.62. (a) A typical arrangement of four identical flexible tubes in a rigid enclosure filled with fluid. (b) First and second axial modes of tube with pinned-pinned supports. (c) Cross-sectional mode patterns and corresponding ω_n' for first axial mode and for $G_c/R = G_w/R = \frac{1}{4}$. (d) Dimensionless natural frequencie ω_n' for first axial mode as a function of relative tube spacing. (All data for 'standard case', after Paidoussis, 1979, and Paidoussis et al., 1977.)

where L, m, A, and EI are the length, mass per unit length, cross-sectional area, and flexural rigidity of the tube, respectively. The cross-sectional shape of the rigid enclosure has only a minor effect on the magnitudes of ω_n', but it affects the mode shapes considerably (Figure 7.62c). The latter shapes are distinguished by different directions and phase shifts. (Some of the modes and ω_n' values are identical because of the geometric symmetry of the system.)

Further insight into interbody fluid coupling is obtained from the equations of motion in which the fluid forces are expressed by 'added' coefficients (Section 7.2.1). Considering for simplicity just the role of added mass with respect to the x direction fluid force F_{x_b} on tube 'b' in Figure 7.61 due to motion of the tubes 'a' and 'b' in the x and y direction, one obtains

$$F_{x_b} = -A'_{x_b x_b} \ddot{x}_b - A'_{x_b x_a} \ddot{x}_a - A'_{x_b y_b} \ddot{y}_b - A'_{x_b y_a} \ddot{y}_a \tag{7.72}$$

Herein, e.g., $A'_{x_b y_a}$ is the added-mass contribution of the y_a motion to the x_b force, which can be determined theoretically as shown by Chung & Chen, 1977 (cf. Chen, 1987). Equation 7.72 describes the increased effect of added mass in the x direction that tube 'b' experiences on account of its closeness to tube 'a'. If fluid-dynamic coupling involves added damping and rigidity as well, but only x_b and x_a motions are possible, then

$$F_{x_b} = -A'_{x_b x_b} \ddot{x}_b - B'_{x_b x_b} \dot{x}_b - C'_{x_b x_b} x_b - A'_{x_b x_a} \ddot{x}_a - B'_{x_b x_a} \dot{x}_a - C'_{x_b x_a} x_a \quad (7.73)$$

A similar equation is obtained for the fluid force F_{x_a} on tube 'a' in the x direction. The equations have the same form as Equations 7.8 and 7.9, which describe the coupling of two modes (i.e., y and θ) rather than of two bodies. In both cases, the energy exchange between the coupled motions, and hence the onset of vibration, depends strongly on the phase angle between these motions (y and θ, in one case, and $\dot{x}_b$ and $\dot{x}_a$ in the other).

The interbody phase angle plays a dominant role also regarding the stability of multiple-body systems. Figures 7.63a, b show the effect of the phase angle between forced torsional motions of adjacent blades, called interblade phase

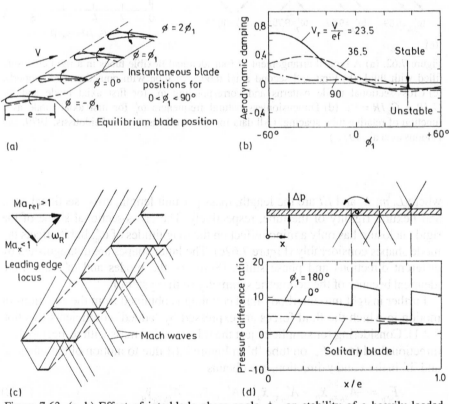

Figure 7.63. (a, b) Effect of interblade phase angle, ϕ_1, on stability of a heavily loaded cascade, derived from forced torsional-vibration tests ($\theta_o = 2°$) in a wind tunnel (after Carta & St. Hilaire, 1978). (c, d) Effect of Mach-wave reflections on pressure-difference distribution along a typical blade in a supersonic cascade performing torsional vibration at $V_r \equiv V/(ef) = \pi$ according to computations of Verdon & McCure, 1975 (after Carta, 1978).

angle ϕ_1, on the induced aerodynamic damping for a heavily loaded cascade without observable stall. One can deduce from Figure 7.63b that results for solitary blades are completely inadequate as a basis for estimating the stability of this particular multiple-blade arrangement. The situation is similar to that of stalled cascades, which become unstable at lower reduced velocities V_r than do unstalled cascades (Yashima & Tanaka, 1978).

An additional complication in cases of supersonic flows arises from the reflection of shock waves from adjacent bodies. As shown in Figure 7.63d, such wave reflections can make the pressure loading of cascade blades completely different from those for the incompressible, solitary-blade counterpart. Again, use of solitary-airfoil theory would strongly underestimate the danger of excitation, and the interblade phase angle ϕ_1 has a significant effect on stability.

In summary:

> A system of N closely-spaced, identical bodies, each with 2 degrees of freedom and surrounded by fluid, possesses $2N$ interbody mode patterns and natural frequencies. As a consequence, the system is receptive to extraneous and instability-induced excitations over a broad frequency range, and its susceptibility to movement-induced excitation of any kind is greatly enhanced. $\qquad(7.74)$

As shown in the following sections, the significance of multiple-body coupling lies more in the increased danger for excitations of all kinds than in a unique excitation mechanism distinctly dependent on such coupling.

7.6.2 *Multiple bodies in cross-flow*

For a cluster of flexible cylinders in cross-flow, one can distinguish between fluid-damping- and fluid-stiffness-controlled instabilities, characterized by fluid-forces which act predominantly as velocity-dependent damping and displacement-dependent stiffness, respectively (Chen, 1987; Paidoussis & Price, 1988). In the latter case, the instability is essentially due to fluid-inertia effects on the phase relationship between fluid force and body vibration. In this case, multiple-body coupling is decisive. In general, however, the two mechanisms are superimposed on each other.

Figure 7.64a shows the stability diagram for a typical example of multiple-body coupling: a row of N identical flexible cylinders in uniform crossflow. Herein, m, ζ^*, and f_n^* are the mass per unit length, the mechanical-damping ratio, and the natural frequency of a cylinder, respectively, all measured in vacuum, and $\bar{V} = V/(1 - d/s)$ is the gap velocity. In contrast to the system described in Figures 7.32-7.36, there are $2N$ modes of vibration in this example rather than one. For low mass-damping parameters Sc* (of the order of 1), the instability is fluid-damping controlled, according to Chen & Jendrzejczyk (1983), and highly affected by instability-induced excitation (IIE) due to vortex shedding; vibrations

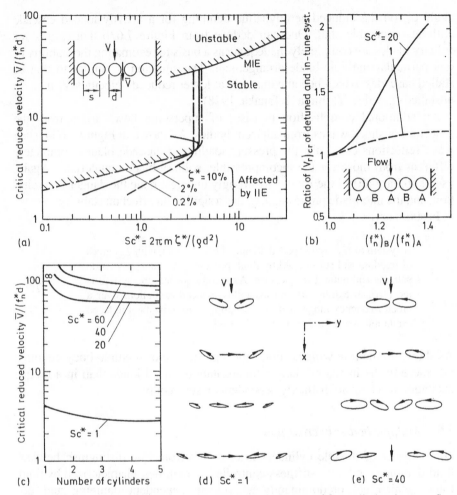

Figure 7.64. (a) Lower bounds for onset of vibration of a row of five cylinders in cross-flow at $s/d = 1.33$. (b) Effect of detuning of alternate cylinders on lower stability threshold. (c) Lower stability threshold as a function of the number of cylinders for $\zeta^* = 0.02$. (d, e) Schematic of vibration modes for rows of cylinders with two, three, four, and five tubes at Sc* = 1 and 40 (after Chen & Jendrzejczyk, 1983).

set in even if only one cylinder is flexible and the others rigid. If Sc* exceeds about 3 to 5, the instability is controlled by fluid stiffness, and the critical reduced velocity is greater by roughly an order of magnitude; in this case, at least two flexible cylinders are necessary for instability. (Figure 7.64a shows only the threshold for soft excitation, i.e., excitation from rest.) Whereas two parameters, ζ^* and $m/(\rho d^2)$, are required to descibed the effects of mass ratio and damping in the low-Sc* range, the two can be combined in one parameter, Sc*, in the high-Sc* range. In both ranges, the critical reduced velocity $(V_r)_{cr}$ decreases with the number of neighboring flexible bodies (Figure 7.64c). This destabilizing effect is typical of multiple-body coupling.

Figures 7.64d, e show the mode shapes associated with the stability threshold for two, three, four, and five flexible cylinders in a row. For the case controlled by fluid stiffness (Sc* = 40), the orbits have both x and y components. *Frequency detuning* of alternate cylinders has a stabilizing effect on the system in most cases, as evident from Figure 7.64b. This stabilization is far more effective in the range of large Sc* where multiple-body coupling has been proved to be essential for excitation.

Among the multiple-body systems, the heat-exchanger tube bundle is one of the most thoroughly investigated because numerous failures have led to frequent shut downs of power stations (Paidoussis, 1980). Most of these failures are due to vibration-induced fretting at the tube supports; such fretting occurs either in regions of high-velocity cross-flows (e.g., in entrance or exit nozzles and near baffle plates) or in regions with lowered tube stiffness (e.g., in U-bends). In the most dramatic of these cases, the failures are caused by fluid-elastic instability (MIE) which sets the tubes in violent whirling vibration (Figure 7.65f) if the reduced velocity $V_r \equiv \bar{V}/(f_n d)$ exceeds a critical value (Figure 7.65e). In this case,

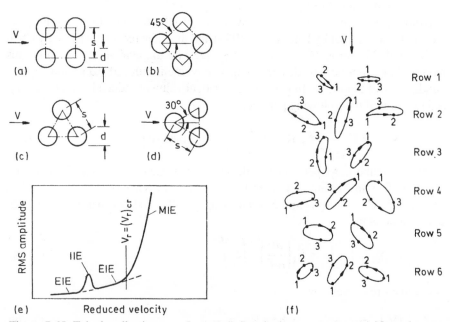

Figure 7.65. Tube bundles in cross-flow. (a-d) Standard arrangements: (a) Normal square; (b) Parallel triangle; (c) Rotated square; (d) Normal triangle. (e) Typical response of a tube in the bundle. (f) Schematic of orbits of whirling tubes (MIE) for a normal-triangular tube bundle of $s/d = 1.375$ (after Zdravkovich & Namork, 1979).

$\bar{V} = V/(1 - d/s)$ and f_n = tube natural frequency in still fluid. The standard arrangements of tube bundles are depicted in Figures 7.65a-d.

The mechanism behind this whirling excitation is not completely understood, but it probably includes versions of the mechanisms described in Sections 7.3.4, 7.3.5, and 7.5.5, at least in widely-spaced tube bundles. At low values of mass damping Sc, the instability is dominated by mechanisms controlled by fluid damping, whereas at larger Sc values, fluid-stiffness-controlled mechanisms come into play, similar to those of wake-induced flutter (Paidoussis & Price, 1988). For closely-spaced bundles, a mechanism involving movement-induced flow redistribution in the 'flow channels' on either side of a vibrating tube, and an appropriate phase lag between tube motion and flow redistribution, permits a remarkably good prediction of the stability threshold for a parallel triangular array (Lever & Weaver, 1982, 1986). This mechanism belongs to the group of leakage-flow-induced vibrations (Section 7.4.2). Multiple-body coupling seems not to be a prerequisite for excitation for most tube-array geometries over the common range of Sc; however, it is a factor in lowering the stability threshold and an agent of amplification in the post-critical region in which tube amplitudes become very large. The reduction of the stability threshold results from the increase in added mass associated with the relative motion of closely spaced tubes and, hence, a reduction in their fluid-coupled natural frequency f_n. This latter effect is only appreciable in liquid flows, i.e., for small values of Sc.

Among the several reviews of flow-induced vibrations of tube bundles, the most extensive is that by Paidoussis (1983) and the most recent are those by Chen (1987) and Weaver & Fitzpatrick (1988). Despite the availability of numerous theories, the approach to design against movement-induced excitation in tube bundles is still empirical. In general, the critical value of reduced velocity, $(V_r)_{cr}$, is a function of the following parameters:

$$(V_r)_{cr} = F\left[\frac{m'}{\rho d^2}, \delta_a, \text{Re, Tu,} \frac{L_v}{d}, \frac{s}{d}, \text{array pattern}\right] \qquad (7.75)$$

Here, m' = tube mass per unit length plus added mass, δ_a = logarithmic decrement of tubes in still air, and Tu and L_v are level and scale of the approach-flow turbulence. The relationship between $(V_r)_{cr}$ and the first two parameters can be approximated by

$$(V_r)_{cr} \equiv \left(\frac{\bar{V}}{fd}\right)_{cr} = K_1\left(\frac{m'}{\rho d^2}\right)^{C_1}\left(\delta_a\right)^{C_2} \qquad (7.76a)$$

or, even more crudely, by

$$(V_r)_{cr} = K(\text{Sc}')^C, \quad \text{where Sc}' \equiv \frac{m'\delta_a}{\rho d^2} \qquad (7.76b)$$

Figure 7.66 provides a rough guide on the lower-bound curves for the critical reduced velocity in single-span tube bundles in the form of Equation 7.76b. Even

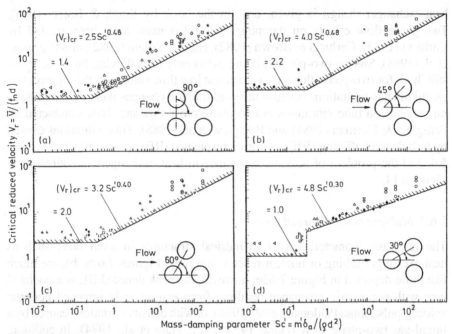

Figure 7.66. Lower bounds for onset of whirling vibration (MIE) in single-span tube bundles of various arrangements with cross-flow (after Weaver & Fitzpatrick, 1988). (a) Normal-square, (b) Rotated-square, (c) Normal-triangular, and (d) Parallel-triangular arrangements.

though $(V_r)_{cr}$ increases with increased pitch ratio s/d for a given bundle geometry (Chen, 1984), the scatter of data is so large that one need not account for this effect.

From the several coupled natural frequencies for single-span tube bundles in liquid (Section 7.6.1), the lowest one is usually excited first; this minimum value, therefore, should be used in evaluating $(V_r)_{cr}$. Because the direction of approach flow is essential, moreover, the lower of the two possible bounds (Figures 7.66a, b or c, d) are to be used if the flow direction is in doubt (Yeung & Weaver, 1983).

In many modern applications, the tubes have multiple spans of uneven length; thus, the tube supports are closer together in regions of high velocity and provide extra stiffness. In such cases, the threshold velocity from Figure 7.66 must be modified to account for uneven flow distribution and for the fact that this velocity may not be associated with the lowest natural tube frequency (Weaver & Goyder, 1990; Weaver & Parrondo, 1990).

In practical applications of Figure 7.66, the greatest difficulty is posed by the prediction of damping. Some authors use δ_a 'in still fluid' rather than in still air. Semi-empirical formulas for estimating δ_a in heat-exchanger tube bundles for design purposes have been presented by Pettigrew et al. (1986). Information on heat-exchanger design is given, e.g., in the book by Singh & Soler (1984). Two-phase flow effects on damping and added mass have been treated by Carlucci (1983), Carlucci & Brown (1982), Hara & Kohgo (1982), and Pettigrew et al. (1988). Since a two-phase mixture exists only for a flowing fluid, the δ_a 'in still fluid' for two-phase flows is evaluated at low flow velocities; this approach is useful but has limitations because the void fraction changes with V_r. Among the rare studies on tube vibrations in two-phase cross-flow are those conducted by Pettigrew & Gorman (1981) and Pettigrew et al. (1988). Tube vibrations caused by turbulence buffeting (EIE) or vortex resonance (IIE) are discussed in Section 5.4, and the problem of acoustic resonance in heat exchangers is addressed in Figure 8.14.

7.6.3 Multiple bodies in axial flow

The response characteristics of cylindrical structures in axial flow, such as heat-exchanger tubing or nuclear-reactor fuel rods (Figures 7.67a, b), are much like those depicted in Figure 7.65e; however, the peak denoted IIE is associated here with parametric resonance resulting from extraneously imposed periodic velocity pulsations (Paidoussis et al., 1980), or with density variations caused by a liquid-gas two-phase flow (Hara, 1973, 1977; Hara et al., 1974). In addition, vibrations can be extraneously and movement-induced (EIE and MIE). EIE due to turbulence is strongly intensified by interbody coupling just as in cross-flow cases (Paidoussis & Curling, 1985). Nevertheless, most important for structural safety is again the avoidance of MIE in the form of fluid-elastic instabilities, which are in many ways similar to those of pipes conveying flowing fluid (Section 7.5.7). The effect of increasing flow is generally to diminish the natural frequencies to the point at which buckling, or divergence, sets in (Figure 7.3e); at still greater velocities of flow, the tubes or rods begin to flutter (Paidoussis, 1981, 1983). Figure 7.67c shows a typical stability diagram for a solitary flexible tube with internal and unbounded external flow. The flow velocities are expressed in a nondimensional form equivalent to a reduced velocity,

$$U_i \equiv \frac{V_i L}{\sqrt{EI/\rho A}} \, , \qquad U_e \equiv \frac{V_e L}{\sqrt{EI/\rho A}} \tag{7.77}$$

where L, A, and EI are as defined in conjunction with Equation 7.71. The discrepancy between theory (full lines) and experimental results (circles and triangles) in Figure 7.65c calls for the application of a safety factor to account for imperfections in the actual system.

For multiple-tube systems (Figure 7.67b), the stability characteristics are similar to those of the solitary tube except that the critical dimensionless velocities are reduced with decreasing body spacings (Figure 7.67d). As the velocity exceeds the critical values given in Figure 7.67d, the system buckles successively in all its modes (Figure 7.62c), and then it flutters. This flutter exhibits complex

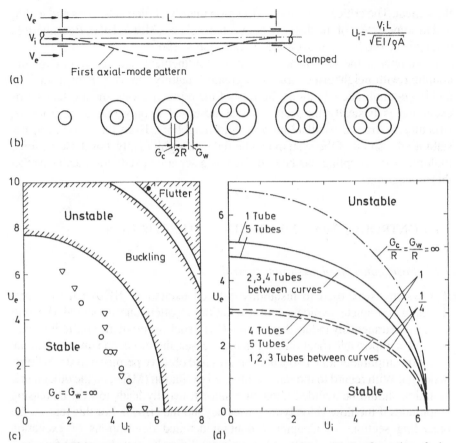

Figure 7.67. (a) Flexible tubes with internal and external axial flow. (b) Configuration of tube clusters considered in Figure 7.67d. (c) Stability map for solitary tube with clamped ends (after Hannoyer & Paidoussis, 1978). (d) Stability map for tubes as shown in Figure 7.67b, with clamped ends, all for 'standard case' (after Paidoussis & Besançon, 1981).

coupled modes because the system is unstable with respect to a progressively larger number of modes at higher velocities (Paidoussis, 1979; Paidoussis & Gagnon, 1984). It is common practice, therefore, to design the system in such a way that it stays safely below the lowest stability threshold presented in Figure 7.67d. Since this threshold relates to *static* stability, safe design is simply a matter of providing sufficient flexural rigidity EI (Equation 7.77); damping is not important in that case.

A situation similar to the cluster of tubes prevails for the flat-plate equivalent, the parallel-plate fuel assembly. The decisive stability criterion for such multi-plate assemblies is again that of buckling. Safety against collapse requires that the maximum dimensionless flow velocity U, defined in Equation 7.77, stays below the critical. The critical velocity is a function mainly of plate thickness and width, and to some extent of the distance between the plates (Miller, 1960; Rosenberg & Youngdahl, 1962; Scavuzzo, 1965; Wambsganss, 1967).

In summary, the two factors (a) flow confinement and (b) fluid-dynamic coupling with neighboring structures greatly magnify the danger for instability and large-amplitude vibrations. In most of the reported cases, the mechanism of excitation is basically the same as that observed if only one of a group of structures is free to oscillate among rigid neighbors. Even then, however, the stability threshold of the system can be reduced significantly. For these reasons, multiple-body coupling has been treated as a separate excitation mechanism in this monograph.

7.7 CONTROL OF MOVEMENT-INDUCED EXCITATION

7.7.1 *Basic concepts of excitation control*

If a system is exposed to instability-induced excitation (IIE) such as vortex shedding, complete avoidance of excitation is often uneconomical (Section 6.8.2); if extraneously induced excitation (EIE) such as turbulence buffeting takes place, it is impossible (Section 5.3). In these cases, therefore, one must determine vibration amplitudes and associated stresses and observe permissible stress limits in design. With regard to movement-induced excitation (MIE), on the other hand, vibration should be avoided altogether since it usually leads to ever increasing amplitudes if the flow velocity increases beyond the critical. For this reason, the preceding sections of Chapter 7 contain detailed descriptions of excitation mechanisms and methods of predicting instabilities. Knowledge of the former is indispensable in assessing a system for potential sources of MIE, and application of the latter is the minimum prerequisite for vibration-free design.

One can control MIE in three ways:

(a) Eliminate the source of MIE by modifying the flow field (fluid-dynamic solution);
(b) Ensure MIE stability by modifying the structural components (structural-dynamic solution);
(c) Failing both, employ a passive or active damper to reduce the excitation to a tolerable level.

The first mentioned method is the most effective and most economical. It requires thorough knowledge of the basic flow-structure interactions that give rise to MIE; and it involves, typically, a modification of the geometry of the structural flow boundaries or the use of appropriate fittings. Depending on the excitation mechanisms, it may also entail, for example, a change of the excitable mode to avoid frequency merging (in mode-coupling MIE) or a detuning of neighboring bodies (in multiple-body-coupling MIE, cf. Figure 7.64b).

The second, or structural-dynamic, method aims generally at stiffening the endangered structural component and increasing its lowest natural frequency f_n. The goal is to keep the relevant reduced velocity $V_r \equiv V/(f_n d)$ below the critical value obtained from the appropriate stability diagram for all flow conditions encountered in the prototype. The increasing of f_n can be difficult, however, since it requires changes of structural stiffness *and* mass, i.e., parameters that frequently cannot be adjusted independently of each other. In some cases, moreover, structural stiffening reduces the vibration frequency (cf. discussion of Figure 7.47).

If the critical reduced velocity pertains to dynamic rather than static stability, MIE can be avoided by an increase in structural or mechanical damping so that the mass-damping parameter Sc is larger than that associated with the threshold of stability. As explained in Section 5.3, increased damping can best be achieved in practice by altering the details of structural support. The designer must be aware, however, that in very rare cases such as coupled-mode panel flutter, increased damping can destabilize the system (Section 7.5.8).

The third method, controlling MIE by means of passive or active dampers, is usually a last resort, after all other methods have failed, or if excessive vibrations occur in a system already in operation. Most common among the passive damping devices is a tuned mass damper (TMD), or dynamic vibration absorber, which consists of a comparatively small vibratory system m, B_2, C_2 attached to the structural mass M as shown in Figure 7.68a. If excited by vibrations of the structure, the TMD will vibrate at the same frequency but with a phase shift thereby reducing the amplitude of the structural vibrations. An illustration is the system response to a periodic force $F_o \sin \omega t$ acting on the mass M (Figure 7.68b). The equations of motion for negligibly small mechanical damping of M are

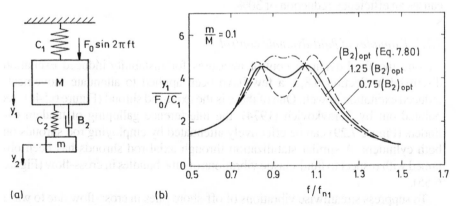

Figure 7.68. (a) Schematic of tuned mass damper. (b) Response of mass M to an excitation $F_o \sin 2\pi f t$ for $m/M = 0.1$ and for different values of damping B_2.

$$M\ddot{y}_1 + C_1\dot{y}_1 + C_2(y_1 - y_2) + B_2(\dot{y}_1 - \dot{y}_2) = F_o \sin 2\pi ft$$

$$m\ddot{y}_2 + C_2(y_2 - y_1) + B_2(\dot{y}_2 - \dot{y}_1) = 0 \tag{7.78}$$

where y_1 and y_2 are the deflections of M and m, respectively. Optimum damping for a given mass ratio m/M is obtained by adjusting the natural frequency f_{n2} of the absorber to

$$f_{n2} = \frac{1}{2\pi}\sqrt{\frac{C_2}{m}} = \frac{1}{1 + m/M} \tag{7.79}$$

and its damping coefficient to

$$(B_2)_{opt} = 4\pi m f_{n1} \sqrt{\frac{3m/M}{8(1 + m/M)^3}} \tag{7.80}$$

(Den Hartog, 1985; Snowdon, 1968). By means of this 'tuning', one reduces the maximum response amplitude of M to $\sqrt{1 + 2\,M/m}$ times the static deflection $(y_1)_{stat} = F_o/C_1$, which means an effective damping ratio of

$$\zeta_{eff} = \frac{1}{2}\left[1 + \frac{2}{m/M}\right]^{-\frac{1}{2}} \tag{7.81}$$

A limiting factor in most practical cases is the space available for the deflections of the two masses, i.e.,

$$y_1 - y_2 = \sqrt{y_1(y_1)_{stat}}\left[2\,\frac{m}{M}\,\frac{f}{f_{n1}}\,(B_2)_{opt}\right]^{-\frac{1}{2}} \tag{7.82}$$

Since optimal tuning (Equations 7.79, 7.80) is difficult to maintain under all circumstances, allowance must be made for deviations in tuning. Maintenance of control is particularly important for *frequency* tuning where a 6% deviation causes an efficiency reduction of 30%.

7.7.2 *Examples of fluid-dynamic control*

Among the fluid-dynamic control measures for instability-induced excitation discussed in Section 6.8.2, a few have been applied to attenuate movement-induced excitation as well. One of these is the axial-rod shroud (Figure 6.54c). As pointed out by Zdravkovich (1974), the interference galloping of tandem cylinders (Figure 7.29) can be effectively attenuated by employing rod shrouds on both cylinders. A similar stabilization through axial-rod shrouds has been obtained with respect to fluid-elastic vibrations of tube bundles in cross-flow (Figure 6.55).

To suppress streamwise vibrations of off-shore piles in cross-flow due to wake breathing (Section 7.3.5), Wootton et al. (1974) proposed radial fins of $0.1d$ extending over 15% of the piles' wetted length and mounted at angles of $\theta = 45°$ and 135° from the stagnation point. Dickens (1979) found that an arrangement of fins at $\theta = 135°$ is about as effective as the twin arrangement. Both studies, of course, apply to unidirectional flow only.

An unusual fluid-dynamic attenuation of excitation related to stall flutter was reported by Melbourne & Cheung (1988). Noting that the major contribution to the wind loading of cantilevered grandstand roofs stems from the high negative pressures on the top surface near the leading edge, they investigated the effect of venting that area. With a slot near the leading edge with a width of $0.05\,L$ (L = cantilever length), they achieved a 25% reduction in the dynamic response of the roof. Similar venting by means of perforating the endangered structure can attenuate galloping and vortex-excited vibrations. Again, however, there are exceptions. For example, an I-beam in a cross-flow, the beam flanges placed normal to the flow, galloped at lower reduced velocities if the web was perforated (Figure 7.73c).

Figure 7.69 shows an example of fluid-dynamic control in a case of galloping. The two essential principles employed are: first, a reduction in distance between the separating streamlines and the sides of the prism (Figure 7.69a) that minimizes the movements of these streamlines relative to the prism (Figure 7.20); and, second, a generation of lateral forces on the added side plates (Figures 7.69a, b) that counteract the exciting force. Additional advantages of the measure lie in the simultaneous decrease in drag (Roshko & Koenig, 1978) and the reduced susceptibility to vortex excitation (Cooper, 1988). A control measure similar to Figure 7.69a proved successful also in cases of rectangular and cruciform-section prisms in cross-flow (cf. Figure 9.9 and Ruscheweyh, 1988).

For unidirectional flow with a mean angle of incidence $\bar{\alpha} = 0$, the fluid-dynamic damping can be optimized by selecting a specific ratio for a/d or s/d as a function of free-stream turbulence. In Figure 7.69e, either the positive slope $[dC_y/d\alpha]_{\alpha=\bar{\alpha}}$, or the range in which this slope is positive, or both, have been reduced for most of the shapes investigated, in comparison to that for the bare prism. The simultaneous reduction in drag coefficient C_D is demonstrated in Figure 7.69f for shape 'c'. A successful application of this particular MIE control measure has been made to the pylons of a cable-stayed bridge (Shiraishi et al., 1988, and Figure 9.9). However, these benefits have their price. For example, the cross-flow response to turbulence buffeting increases substantially for shape 'a' at small incidence angles due to the increased slope in the C_y versus α curve (Cooper, 1988); and the double-valued C_D and C_L results for shape 'c' at incidence angles approaching 45° are a clear sign of bistable states of flow (Section 6.6).

Curing one vibration problem can thus create a different one. This observation can be generalized. It is imperative, therefore, never to limit considerations of vibration control to one particular excitation mechanism. If fittings are applied (Figures 7.69a, b), moreover, they must be designed not only to stabilize the structure but also to be free of vibration themselves. One may conclude, therefore, that exhaustive investigations are necessary before measures like those reported herein can be applied safely in a particular engineering system.

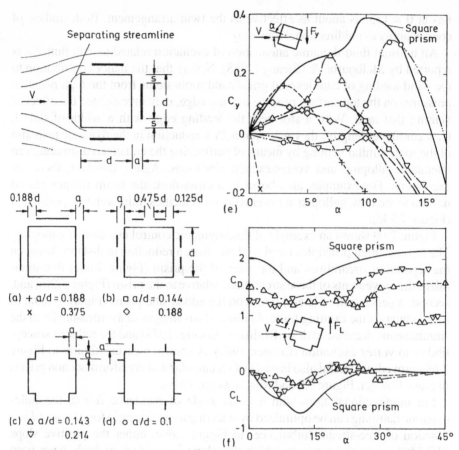

Figure 7.69. Illustration of principle of fluid-dynamic damping in case of a galloping square prism. Cross-sectional shapes investigated are given in (a-d). (e) Steady-state transverse-force coefficient for a mean incidence angle $\bar{\alpha} = 0$. (f) Steady-state drag and lift coefficients for configuration c. (Re = 1.06×10^5, Tu = 0.1%; after Naudascher et al., 1981.)

7.7.3 Examples for added damping devices

The damping of a structure can be increased by several means. Some of these involve damping devices placed near the support of the structure as shown in Figure 5.11. Supplementary dampers can be effective also without fixed supports. Figure 7.70 depicts three examples for the addition of *dissipative dampers* at locations at which relative motion occurs between adjacent structures or structural parts. The arrangement shown in Figures 7.70a, b, for example, can serve to control interference galloping (Figure 7.29), and that of Figure 7.70c can attenuate the dynamic response of a building to a variety of excitations (Feld, 1971; Casparini et al., 1980).

Additional damping elements can also be used, of course, if excessive vibrations occur at a structure already completed. An example is shown in Figure 7.71. Damping cylinders of this kind will attenuate any type of excitation occurring at small gate openings including those referred to in Figures 7.48p, q.

Dissipative dampers extract kinetic energy from the vibrating structure either by means of viscous, viscoelastic, and frictional forces (Blevins, 1977), or with the aid of a body moving through a mass of pellets (Langer, 1969), or by means of fluid sloshing back and forth through high-resistance passages. The latter method stems from spacecraft technology where nutation dampers, i.e., torus-shaped, hollow rings partially filled with liquid, are used to control low-frequency oscillations. Dampers of this kind can also be used to suppress galloping and vortex-excited vibrations (Modi & Welt, 1984, 1988; Modi et al., 1988). Nutation dampers with perforated inside tubes, baffles, horizontal layers, and floating pieces of wood provide the highest dissipation rates.

In contrast to these dissipative devices, the *tuned mass damper* (TMD) attenuates vibrations of a structure (*M*) by setting up opposite movements of a secondary system (*m*) attached to the structure near a point of maximum amplitude (Figure 7.72). The restoring force in that secondary system can be

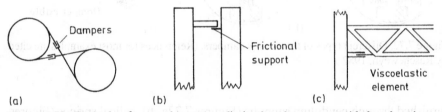

Figure 7.70. Examples for supplementary dissipative dampers provided at locations of relative motion between adjacent structures or structural parts (after Ruscheweyh, 1982).

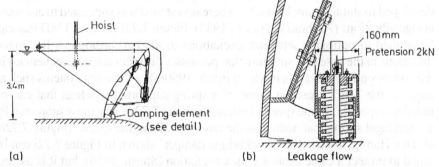

Figure 7.71. Damping cylinder fixed at the center of a 22.5 m-span reversed Tainter gate. Gate had vibrated during leakage flows at gate openings between 8 and 25 mm (after Petrikat, 1980).

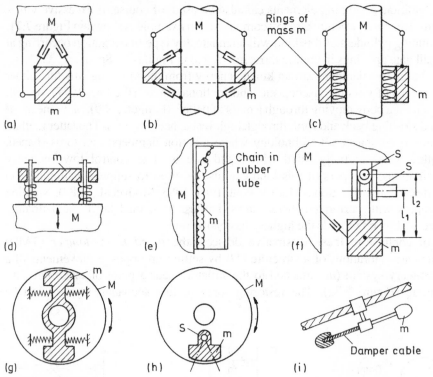

Figure 7.72. Different types of tuned mass dampers. (References for most examples are cited by Ruscheweyh, 1982.)

furnished through pendulum supports (Figures 7.72a, b), helical springs (Figures 7.72c, d, g), or torsional springs (*S* in Figures 7.72f, h). In Figure 7.72c, the helical springs provide both restoring and damping forces, and the pendulum is used merely as a support (Hirsch, 1980). Systems a, b, c, and e were originally developed to stabilize smoke stacks, whereas system d was suggested to attenuate bridge vibrations (Aschrafi & Hirsch, 1983). Figure 7.72f shows a TMD that can be applied to optimally attenuate excitations in the two orthogonal horizontal directions in the case of a structure that possesses different natural frequencies in the corresponding vibration modes (Hirsch, 1980). For torsional systems such as engines, the TMD takes the shape of a spring-supported flywheel that can be tuned to a specific engine speed of rotation (Figure 7.72g), or a mass supported by a centrifugal spring that maintains the tuning of all engine speeds (Figure 7.72h, cf. Den Hartog, 1985). The 'Stockbridge damper' shown in Figure 7.72i can be tuned to protect a cable against vortex excitation (Sturm, 1936), but it is useless with respect to such low-frequency excitations as mentioned in Sections 7.5.5 and 7.5.6.

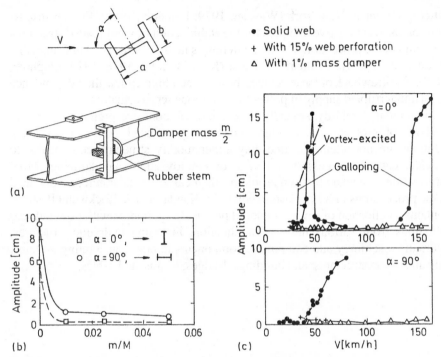

Figure 7.73. (a) I-beam with vibration absorber. (b) Effect of mass ratio m/M on maximum amplitude of I-beam ($a = 81$ cm, $b = 71$ cm, $L = 34$ m) for wind speeds up to 170 km/h. (c) Typical response of I-beam ($a = 36$ cm, $b = 25$ cm, $f_x \simeq f_y \simeq 5$ Hz) as obtained from wind-tunnel tests with sectional model (after Wardlaw & Cooper, 1978).

Figure 7.73 summarizes results on wind-induced vibrations of I-beam bridge-deck hangers. They are typical of any cylindrical structure in a cross-flow. As shown in Figure 7.73b, little advantage is gained by increasing the damper mass beyond about one percent of the structural mass. For flow parallel to the web ($\alpha = 90°$), the beam was found to be excited in both the bending and torsional modes. Torsional vibration amplitudes, originally 28°, were virtually eliminated by the one-percent TMD shown in Figure 7.73a. The response diagrams from a similar study in Figure 7.73c show, moreover, that a one-percent TMD of slightly different form effectively attenuates both vortex-excited and galloping vibrations. The problem with the use of a piece of rubber in place of a spring, however, is the gradual change in tuning due to aging.

In contrast to the passive dampers discussed above, *active dampers* use auxiliary systems to produce a control force that opposes the exciting force in magnitude and phase. The concept of active dampers stems from aircraft design (Sensburg et al., 1979). One of the best-known applications of that concept to wind problems in structural engineering is the damper system of the 728 m high

Citicorp Center in New York (Wiesner, 1979; Leipholz, 1980). Further discussions on the use of active dampers for the stabilization of wind-loaded structures are found in the proceedings of the conferences held in 1978/1979 in Gainsville, Florida, and Waterloo, Ontario, Canada (Klein & Salhi, 1978, 1979; Leipholz, 1980). The drawback of these devices, besides their high cost, is their dependence on an uninterrupted supply of power to the various servo actuators.

In summary, fluid-dynamically acceptable solutions are preferable to other methods of protecting a structure against flow-induced excitations. If such solutions are not attainable, one may either modify structural components to provide, e.g., the necessary stiffness, or employ additional damping. More information on actual vibration problems, their cause and their cure, is contained in the proceedings of a symposium edited by Naudascher & Rockwell (1980). It contains a collection of several hundred practical experiences with flow-induced vibrations from various fields of application, including hydraulic structures, hydraulic machinery and equipment, components in power-generating systems, ship and ocean structures, and buildings, bridges, beams, and cables.

CHAPTER 8

Vibrations due to excitation of fluid oscillators

8.1 SOURCES OF EXCITATION AND DAMPING

The various types of excitation mechanisms, summarized in Figure 1.1, can produce oscillations of a *fluid oscillator* in a manner analogous to the excitation of a body oscillator. Most fluid oscillations of this kind are undesirable as they give rise to pressure pulsations that produce fluctuating forces on adjacent structures. If one of these structures constitutes a body oscillator, these forces can cause substantial structural vibrations even without a direct flow-induced excitation of the structure, especially if the resonant frequency of the fluid oscillator is close to that of the body oscillator. An essential design task, therefore, is an assessment of not only the resonant frequencies of the fluid oscillators in the system (Section 4), but also the possible fluid-oscillator responses.

Flow-related mechanisms that may excite oscillations of a fluid oscillator can be classified (Section 1.2) in terms of extraneously induced (EIE), instability-induced (IIE), and movement-induced excitation (MIE). Examples involve: penstock resonance due to turbine runner blades passing guide vanes (EIE); cavity resonance caused by an unstable shear layer along the cavity (IIE); and tube resonance due to an unsteady shock wave (MIE). The distinguishing characteristics of the three types of excitation are much the same as in cases of body oscillators: an excitation independent of the oscillator response in case of EIE; a response-dependent amplification of excitation and possible locking of exciting frequency to oscillator frequency in case of IIE; and a mechanism by which oscillator movements withdraw energy from the flow in case of MIE.

Insight into the basic characteristics of the response of a fluid oscillator can be obtained by considering excitation of the oscillator by a fluctuating pressure source near the oscillator and examining the resultant fluctuations of fluid displacement, x, fluid pressure, p, of fluid velocity, v. In practice, these fluctuations may be generated by pulses from a rotating cascade of blades (EIE), from impinging shear layers (IIE), or from perturbations in the location of a shock wave (MIE). In all cases, the response is a function of the oscillator damping in much the same manner as the response of a body oscillator was seen to depend on

269

its damping. In cases of EIE and IIE, moreover, the response of the fluid oscillator is also affected by the ratio of the frequency of excitation f to the natural frequency f_R of the oscillator, again in analogy to body oscillators.

For the damping of fluid oscillators, one may distinguish between wave-radiation damping (predominant in cases of Helmholtz or open-basin resonators with short necks), damping due to frictional and flow-separation losses (predominant in duct or pipe resonators), and damping due to energy absorption at reflecting boundaries (predominant in resonators with two- or three-dimensional standing waves). These basic classes of damping are illustrated by typical examples in Figure 8.1, involving fluid oscillators driven by EIE. In many practical cases, of course, the damping is mixed.

Radiation damping. Under the condition that the moving mass of fluid $\rho A l_{eff}$ in the relatively short neck of the Helmholtz resonator of Figure 8.1b radiates sound waves into the surrounding fluid in much the same manner as does a piston in an infinite baffle, the equation of forced oscillation becomes

$$[\rho A(l + l')]\ddot{x} + \left[2\pi \frac{\rho f^2 A^2}{c_a}\right]\dot{x} + \left[\frac{\rho c_a^2 A^2}{V}\right]x = A\tilde{p}_{io} \cos (2\pi ft) \qquad (8.1)$$

where $l + l' = l_{eff}$ is the effective length of the oscillating mass including the added mass (Kinsler & Frey, 1962). The symbols in this equation were defined in conjunction with Equation 4.10, except for f and $\tilde{p}_{io}$, which are the frequency and pressure amplitude of the extraneous excitation (a sound wave impinging on the resonator opening in this case). The radiation damping is described by the coefficient of $\dot{x}$ as seen from comparing Equation 8.1 with the equation of forced vibration of a body oscillator (Figure 8.1a)

$$m\ddot{x} + B\dot{x} + Cx = F_o \cos (2\pi ft) \qquad (8.2)$$

The solution of Equation 8.1, derived in analogy to the solution of Equation 2.13, is presented in Figure 8.1b. If one compares the damping term, encircled by a dashed line, with the corresponding term for the body-oscillator system in Figure 8.1a, one finds that the ratio $B/B_{cr} = 2\zeta$, for the case of radiation damping, is a function of the excitation frequency f. Consequently, the form of the amplification or response curves $x_o = x_o(f/f_R)$ will differ from those of the corresponding body oscillator shown in Figure 2.4a. For high frequencies, the 'added' length l' in Equation 8.1 is a function of the frequency f much as the added mass A' in the case of the vibrating piston represented in Figure 3.12.

Frictional damping. If a Helmholtz resonator has a long neck, the damping due to pipe friction may overshadow the radiation damping. The example depicted in Figure 8.1c stands for a pipe terminated by a valve cavity through which a small amount of water flows due to leakage (Streeter & Wylie, 1967, p. 135). Since the speed of sound c_a in water is large, l is much smaller than c_a/f, and the column of

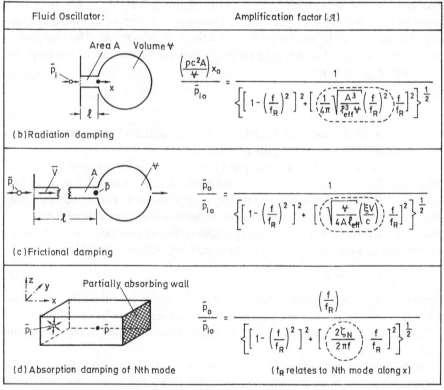

Figure 8.1. Effect of different types of damping on amplification factor for (a) Discrete-mass body oscillator, (b, c) Discrete-mass type fluid oscillators, (d) Distributed-mass type fluid oscillator. ($\tilde{p}_{io}$ = input pressure amplitude; $\tilde{p}_o$ = response pressure amplitude.)

fluid within the pipe will oscillate as a discrete mass when excited by an input pressure pulsation $\tilde{p}_i = \tilde{p}_{io} \cos(2\pi ft)$. The frictional force opposing the oscillatory flow per unit area can be approximated by the linear expression $\xi(\rho\overline{V}/2)\dot{x}$, where $\dot{x}$ = fluctuating velocity of the fluid mass, $\overline{V}$ = average velocity in one direction during a half period, and ξ = loss coefficient obtained from estimating the average frictional loss. According to the solution for the response pressure amplitude $\tilde{p}_o$ in Figure 8.1c, the form of the amplification curves $\tilde{p}_o = \tilde{p}_o(f/f_R)$ is the same as that of the corresponding body oscillator (Figure 2.4a).

Absorption damping. Fluid oscillators of the distributed-mass type, as shown in Figure 8.1d, are not really analogous to body oscillators. Instead, the response equation has to be deduced from the wave equation accounting for the absorption of energy at the reflecting walls. The corresponding damping can be represented in two ways. Following Beranek (1954), the response amplitude $\tilde{p}_o$ obtained with a pressure source in the oscillator enclosure of strength K_o, $\tilde{p}_{io} = K_o/(\pi f_R)$, is as given by the equation in Figure 8.1d; it represents an average in time and space over the interior of the enclosure for the Nth oscillation mode. The damping is expressed in terms of a damping ratio ζ_N, defined by the maximum response amplitude at resonance as

$$(\tilde{p}_o)_{max} = \frac{\tilde{p}_{io}}{2\zeta_N/\omega}, \qquad \omega = 2\pi f$$

In contrast to the response of discrete-mass type oscillators, the response amplitude $\tilde{p}_o$ decreases to zero as f/f_R approaches zero.

An alternative to the foregoing approach is to relate the damping of distributed-mass type fluid oscillators to a reflection coefficient K_r, which has a value less than unity. The resultant wave system can be approximately expressed as the sum of an undamped incident wave (pressure or surface-displacement amplitude a_{io}) and a reflected wave that was attenuated at the reflecting boundary by a factor K_r. The amplitude of the standing wave at the partially reflecting boundary is then

$$a_o = a_{io}(1 + K_r) \tag{8.3}$$

(Ippen, 1966). The amount of wave energy dissipated by such other causes as friction, flow separation, and turbulence can be accounted for by an adjustment of either ζ_N or K_r as long as it is small compared to the absorption damping (Figure 8.10).

In cases for which damping cannot be determined theoretically, it can be obtained experimentally by means of the quality factor Q or the resonance width Δf as explained in Section 2.4.1 (Figure 2.4c). Extensive information on the damping of pipeline and surge-tank systems is given by Streeter & Wylie (1967) and Jaeger (1949). An example of the determination of the damping characteristics for a U-tube oscillator in a navigation lock is given in the IAHR Monograph on Hydrodynamic Forces (Naudascher, 1991, Section 2.2.5). Discussions of the performance of acoustic resonators as affected by porous materials or screens at the resonator mouth, damping material along the cavity walls, adjacent resonators, etc. have been presented by Ingard (1953), Kinsler & Frey (1962), Beranek (1954), and Rayleigh (1945).

8.2 EXTRANEOUSLY INDUCED EXCITATION OF FLUID OSCILLATORS

Among the most common sources of extraneously induced excitation (EIE) of fluid oscillators are:
- – Incident waves;
- – Turbulence;
- – Two-phase flows (e.g., entrained air leading to blow-out);
- – Cavitating flows;
- – Vibrating structural parts in the system;
- – Or indeed any pressure or velocity pulsation generated independent of the fluid oscillator.

Streeter & Wylie (1967), for example, list turbine runner blades passing guide vanes, hunting governors, reciprocating pumps, oscillating valves, vapor cavities and/or vortex shedding in draft tubes as possible sources for the excitation of resonating pipelines. Whether an exciting flow instability must be classified as instability-induced excitation (IIE) or as EIE depends on whether or not that excitation is a function of the fluid-oscillator response. Only in the latter case (EIE) can the response be derived by means of amplification factors such as presented in Figure 8.1. In cases of IIE, one must take into account the effect of the fluid-oscillator response on the excitation (Section 8.3). With movement-induced excitation (MIE), the oscillator response is self-excited (Section 8.4). Not included in this monograph are discontinuous, or transient, excitations of free oscillations, such as those generated in a surge tank by the sudden closure of a valve.

Among the most intensively studied resonance problems are pressure oscillations in *piping systems* resulting from fluctuations of pressure or discharge induced by
- – a surface wave in a reservoir near a pipe entrance,
- – a reciprocating pump in the system, or
- – the controlled variations of a valve opening.

Whenever the frequency f of these fluctuations matches one of the resonant frequencies f_R of the piping system, resonance gives rise to a strong acoustic standing wave in the pipeline. Figure 8.2 shows examples of pressure-amplitude distributions for pipelines with vibrating valves at the end, and Figure 8.3 depicts a pipeline with a closed end section excited by surface waves. In either case, a pressure node occurs at the pipeline entrance and an antinode at its downstream end (Figures 8.2a, b). The corrresponding resonant frequencies f_R are obtained from Equation 4.2 in conjunction with Equations 4.4, 4.18, and 4.19.

For compound piping systems as shown in Figure 8.2c, one has to distinguish between two oscillation frequencies (Camichel et al., 1919). If the system is excited abruptly, it will respond with a fundamental-mode frequency equal to

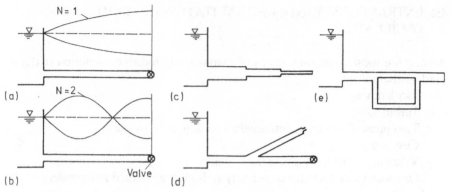

Figure 8.2. (a, b) Simple piping system with resonating acoustic standing waves in (a) fundamental and (b) first harmonic mode. (c-e) Complex piping systems: (c) Series system; (d) Branching system; (e) Parallel system.

$$(f_R)_{th} = \frac{1}{T_{th}}, \qquad T_{th} = 4 \sum_{k=1}^{k_s} \frac{L_k}{(c_a)_k} \qquad (8.4)$$

where T_{th} = theoretical period of the system, k_s = number of pipes connected in series, L_k = length of pipe k, and $(c_a)_k$ = corresponding speed of sound (Equation 4.4). If, on the other hand, the system is excited periodically, it will respond in the frequency f of excitation, and the fundamental-mode resonant frequency will be the inverse of the apparent period T_a, which follows from the impedance theory. The difference between T_a and T_{th} is due to the partial reflections of the pressure waves at each change in cross-sectional area of the piping. In a branching or a parallel system (Figures 8.2d, e), T_{th} can be taken as the largest value of $\sum 4L_k/(c_a)_k$ between the location of the disturbance and the terminal section. In general, the harmonic-mode periods in complex systems are not related to T_{th} or T_a by integers as in the case of a simple pipeline. For a series system with small step changes (Figure 8.2c), use of T_{th} in place of T_a yields acceptable approximations.

A convenient tool for determining the response characteristics of complex systems is the impedance method, described in detail by Streeter & Wylie (1967) and Defant (1961). [Impedance can be expressed as the ratio of the complex amplitudes of head to flow-rate, or pressure to velocity, fluctuations.] This method has the advantage that the complicated wave reflections that occur at pipe junctions do not need to be considered in detail. Figure 8.3 depicts the result of an analysis in which this method was used. The example presented is a series pipeline with a closed end section *excited by a surface wave* in the reservoir upstream of the pipeline. The frequency response in Figure 8.3b is presented in terms of the amplification factor $\mathcal{A}$, defined here as the ratio of the pressure-head amplitude at the closed end to the pressure-head amplitude applied at the

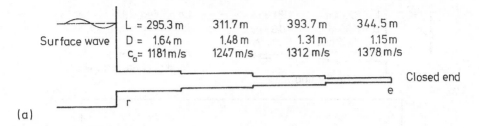

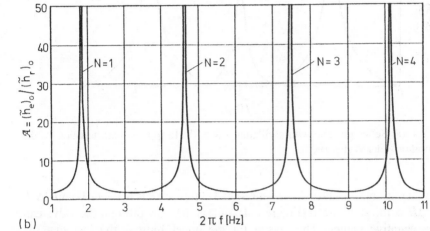

Figure 8.3. (a) Series pipeline with closed end, excited by a surface wave. (b) Response characteristics for the closed end (after Streeter & Wylie, 1967).

reservoir. The resonant frequencies for the fundamental and the three higher modes, which can be identified from the peak values, follow from the equivalent simple-pipe values

$$f_R = \frac{2N-1}{4} \left(\frac{c_a}{L}\right)_{th} = \frac{2N-1}{4} \sum_{k=1}^{k_s} \frac{(c_a)_k}{L_k}, \qquad N = 1, 2, 3, 4$$

only approximately.

In the case of an *oscillating valve* at the end of a series pipeline (Figure 8.4a), the response analysis is complicated by the fact that the forcing function is neither a specified fluctuation of pressure nor one of discharge. Generally, discharge and pressure head at an oscillating valve can be described as

$$v_* = \xi_d \sqrt{h_*}, \qquad \xi_d = \frac{C_d A}{(C_d A)_o} \tag{8.5}$$

where v_* and h_* are the dimensionless velocity and head at the valve as a function of time, and ξ_d is the normalized product of discharge coefficient C_d and area of opening A at the valve as a function of time. For a periodic variation of ξ_d, the

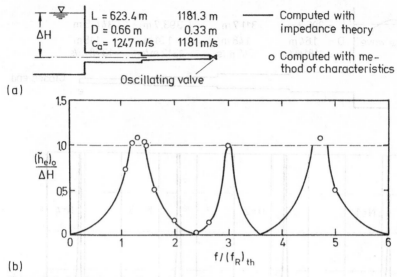

(a)

(b)

Figure 8.4. (a) Series pipeline with oscillating valve. (b) Response characteristics for the valve location (after Wylie, 1965).

maximum possible variation in pressure-head at the valve is between 0 and $2\Delta H$, where ΔH is the static head (Figure 8.4), unless the flow through the valve can assume negative values. The reason for the upper limit is the fact that the discharge decreases to nearly zero in the resonating condition. [In reality, of course, the peak pressure-head amplitude $(\tilde{h}_e)_o$ is reduced by viscous effects to a value slightly less than ΔH.] The response characteristics shown in Figure 8.4b are for a sinusoidal valve movement of frequency f. Predictions of the impedance theory are shown for values of $(\tilde{h}_e)_o/\Delta H$ up to unity; they are not valid for larger head variations because Equation 8.5 was linearized in this approach. The fundamental resonant frequency deviates from the theoretical value in accordance with Equation 8.4 because of variations in pipe diameter D and sound speed c_a.

As a final example, Figure 8.5 illustrates how drastically the response of a system can differ for different locations of excitation. In the experiments, the excitation was caused by a rotating butterfly valve. With this valve at the plenum exit (Figure 8.5b), the Helmholtz-resonator and the duct-resonator modes of pressure oscillations become excited as the valve frequency f approaches the respective resonant frequencies. With the valve located in the duct exit (Figures 8.5c, d), on the other hand, the major response occurs in the duct mode. [The ordinates in these figures denote the presure amplitude $(\tilde{p})_o$ at the blower (b) and plenum (p), normalized by the steady-state pressure difference Δp created by the valve in the fixed 'open' and fixed 'closed' positions.] Similarly drastic changes in the response of a T-branched piping system as a function of excitation location have been reported by Merkli (1978).

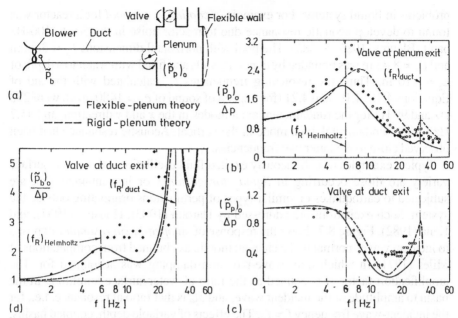

Figure 8.5. (a) Blower-duct-plenum system excited by an oscillating valve; (b, c) Response characteristics for plenum location; (d) Response characteristics for blower location (after Goldschmied & Wormley, 1977; Δp = steady pressure drop through the valve at its equilibrium setting).

Figure 8.6 presents an example of extraneously induced excitation for an acoustic distributed-mass type fluid oscillator, an *air-filled enclosure* with orthogonal rigid walls ($L_x = 10$ ft, $L_y = 15$ ft, $L_z = 20$ ft). The pressure spectrum in Figure 8.6b shows marked peaks at exactly the resonant frequencies according to Equations 4.22 and 4.5. One may distinguish three groups of standing waves, each having different damping characteristics: the axial or one-dimensional waves are more readily damped by the wall surfaces perpendicular to their axis; the tangential or two-dimensional waves are primarily absorbed by the four walls at right angles to their plane; and the oblique or three-dimensional waves, which strike all six walls, are influenced by the absorption characteristics of all the walls (Kinsler & Frey, 1962). At several frequencies (e.g., at 66 Hz), one of the two modes can be excited, depending upon the perturbation wave entering the enclosure. In a rectangular enclosure, all possible standing waves attain a maximum pressure amplitude in the corners. With the loudspeaker and microphone in diagonally opposite corners, as in the example, the response pressure amplitudes were up to 20 db (i.e., about a factor of 10) higher than without the resonator enclosure.

The type of fluid oscillator shown in Figure 8.6a may also lead to resonance problems in liquid systems. For example, a *cooling water pool* for a reactor was found to develop acoustic resonance due to reactor noise in the 50- to 100-Hz band (Blevins, 1986, p. 383). The pool with horizontal dimensions $L_x = 3.28$ m and $L_y = 8.43$ m was bounded by concrete walls and filled with water to a depth of $L_z = 8.10$ m. Thus, the resonating frequencies f_R, calculated with the aid of Equations 4.18, 4.19, and 4.21 (for a speed of sound of $c_a = 1480$ m/s), were 230 Hz and 99 Hz for the fundamental axial modes in the x and y direction, and 45.7 Hz for the fundamental axial mode in the vertical. Acoustic resonance had been observed close to the latter two frequencies.

Typical examples of extraneously excited fluid oscillations of the free-surface variety are those occurring in *liquid storage tanks* or in containers that are subjected to earthquakes or similar types of pertubations originating outside the system. Such excitations are addressed by Haroun (1983), Housner (1957), and Kana (1982). Figure 8.7 shows the response of an *open-basin resonator* exposed to incident waves normal to the channel mouth, as obtained from computations of Miles (1971) in which only wave-radiation damping was accounted for. The amplification factor $\mathcal{A}$ is defined as the ratio of water-surface amplitude in the basin to amplitude of the incident wave, and $\mathcal{A}_R$ is that ratio at resonance, i.e., for the incident-wave frequency $f = f_R$. The effects of variable depth, coupled basins, and energy losses due to entrance separation and friction are treated by Miles & Lee (1975).

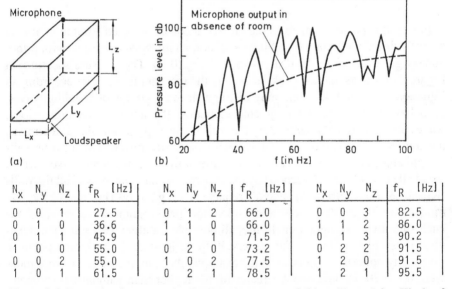

N_x	N_y	N_z	f_R [Hz]	N_x	N_y	N_z	f_R [Hz]	N_x	N_y	N_z	f_R [Hz]
0	0	1	27.5	0	1	2	66.0	0	0	3	82.5
0	1	0	36.6	1	1	0	66.0	1	1	2	86.0
0	1	1	45.9	1	1	1	71.5	0	1	3	90.2
1	0	0	55.0	0	2	0	73.2	0	2	2	91.5
0	0	2	55.0	1	0	2	77.5	1	2	0	91.5
1	0	1	61.5	0	2	1	78.5	1	2	1	95.5

Figure 8.6. Example of an acoustic distributed-mass type fluid oscillator (after Kinsler & Frey, 1962): (a) Air-filled enclosure; (b) Response measured with microphone-loudspeaker arrangement shown on the left; (c) Resonant frequencies according to Equations 4.22, 4.5.

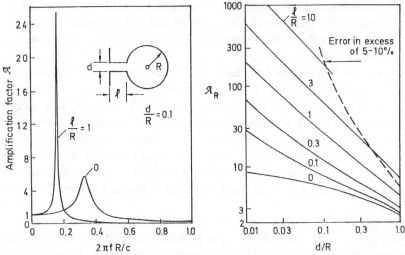

Figure 8.7. Response characteristics of an open-basin resonator (harbor) after computations of Miles (1971). (a) Harbor-mode response to incident sea waves of frequency f; (b) Corresponding amplification factor at resonance.

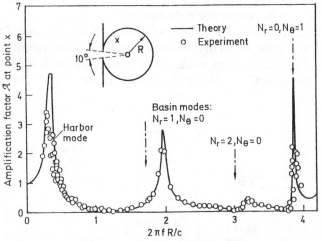

Figure 8.8. Response at point x of a circular harbor to incident sea waves of frequency f (after Lee, 1971). Basin-mode resonant frequencies have been computed using Equation 4.24 and Figure 4.10.

Numerical solutions, derived from the 'arbitrary-shape harbor' theory by Lee (1971) for a related open-basin configuration, the *circular harbor*, are compared with experimental results in Figure 8.8. The diagram includes the response in the basin modes (Figure 4.10) with one and two nodal diameters ($N_\theta = 1, 2$) and with one nodal circle ($N_r = 1$) in addition to the harbor mode (Figure 4.4). The shifts in resonant frequencies from the predictions according to Equation 4.24 are due to the basin opening. The agreement between theory and experiment is remarkably good except for the heights of the resonant peaks; these have been overestimated because, again, only radiation damping was included in the theoretical approach.

If the energy dissipation inside the harbor due to separation at the entrance and friction along the walls is small, its effect may be accounted for by a reflection coefficient K_r (Equation 8.3). As shown in Figure 8.9, the effect of such energy dissipation is concentrated around the resonance points. Figure 8.9b depicts the amplification factor at resonance as obtained from theory and experiments for a *fully open harbor* of variable length L. All experimental values for $\mathcal{A}_R$ are located between the theoretical curves with $K_r = 0.9$ and 0.8. (Reflection coefficients of this order of magnitude have a very small effect on the resonant frequency as seen from Figure 4.17.)

A summary of the resonant characteristics for the first and second modes of a *rectangular harbor* is presented in Figure 8.10 in the form of design curves. For a given aspect ratio of the basin, B/L, and a given opening ratio of the entrance, d/B, one obtains points of intersection in each part of Figure 8.10; these yield the resonant frequencies f_R (abscissa values) and the amplification factors at resonance $\mathcal{A}_R$ (ordinate values) for the first and second modes of fluid oscillation. The curves of constant d/B approach values of $f_R L/c = \frac{1}{4}$ and $\frac{3}{4}$ for the first and second mode, respectively, as the harbor becomes narrower.

An example involves a rectangular harbor that is 900 m long and 600 m wide and has an opening of 150 m at the center of the shorter side. In this case, values of $2\pi f_R L/c$ for the first and second resonant modes, as read from Figure 8.10, are 0.83 and 3.47 for the corresponding values $B/L = 0.67$ and $d/B = 0.25$. If the

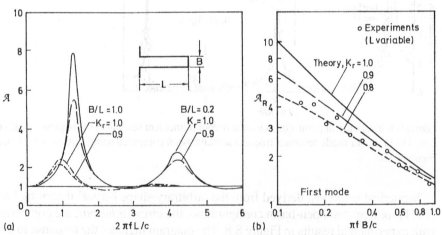

Figure 8.9. Response characteristics of a fully open harbor with different energy dissipation (after Ippen & Goda, 1963). (a) Response curves; (b) Amplification factor at resonance.

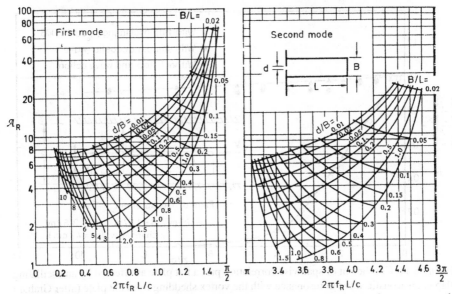

Figure 8.10. Resonant characteristics of a rectangular harbor with symmetric entrance and reflection coefficient $K_r = 1.0$ (after Ippen & Goda, 1963).

water depth in the harbor is $h = 10$ m and $c = \sqrt{gh} = 9.9$ m/s, then the resonance periods $T_R = 1/f_R$ are 11.5 and 2.75 minutes, respectively. The corresponding resonant amplification factors, not including the effect of energy dissipation inside the harbor, are 4.8 and 2.4 (Ippen & Goda, 1963).

8.3 INSTABILITY-INDUCED EXCITATION OF FLUID OSCILLATORS

In cases of instability-induced excitation (IIE), the source of excitation is a flow instability. Some of the most common flow instabilities are described in Chapter 6. What distinguishes IIE from extraneously induced excitation (EIE), discussed in the previous section, is that the force of excitation becomes affected by the oscillation it generates. An effect of this kind is illustrated in Figure 8.11 for a typical flow instability: the vortex shedding from a blunt trailing edge of a flat plate placed within a rectangular duct.

The vortex shedding from a stationary cylinder or plate is typically three-dimensional. With appropriate oscillation of either the entire two dimensional body or a part of it, however, the uniform disturbance over the span of the body triggers the vortex shedding so that it becomes more uniformly distributed, or correlated, over the span (cf. Section 6.2.3). A similar two-dimensional triggering can be obtained with a vortex-excited, two-dimensional oscillation of a fluid oscillator such as an acoustic standing wave. In either case, the maximum

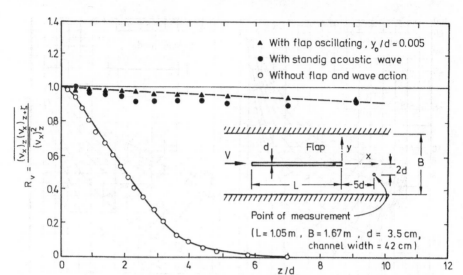

Figure 8.11. Coefficient of spanwise correlation past a flat plate as affected by an oscillating flap or an acoustic wave in resonance with the vortex shedding from the plate (after Graham & Maull, 1971).

correlating effect occurs at resonance, i.e., if the frequency of the body oscillator or the fluid oscillator coincides with the vortex-shedding frequency. As shown in Figure 8.11, this effect is almost identical in the two cases; the vortex shedding becomes highly organized by both the resonating body and fluid oscillator. Consequently, the exciting forces (IIE) will be substantially amplified and enhanced in periodicity in both cases. Actually, experiments have verified that the enhanced spanwise correlation jumps from a low to a high value as the resonant range is approached, and that the shedding locks on to the oscillator frequency (f_n or f_R) from −7% to +7% of the reduced velocity $V/(f_n d)$ at resonance for the oscillating flap and from −3% to +5% of $V/(f_R d)$ at resonance for the acoustic wave. (An explanation of the correlation coefficient R_v is given in Section 2.1.2 of the IAHR Monograph on Hydrodynamic Forces, Naudascher, 1991.)

A variety of other configurations, including corner vanes, three-dimensional grids, cascades of flat plates, and annular cascades, are susceptible to excitation by vortex shedding from trailing edges (Parker & Stoneman, 1989). An example of a typical response of a fluid oscillator to this kind of excitation is depicted in Figure 8.12. Again, the fluid oscillator consists of an acoustic standing wave between a flat plate and the walls of a rectangular duct. In the present case, however, the plate ends in a rigid circular cylinder, and a higher resonant mode was excited (i.e., the mode equivalent to that in Figure 4.13c). Note the lock-in of vortex shedding to the resonant frequency f_R in the range of roughly $0.95 < \mathrm{Sh}V/(f_R D) < 1.05$ (Sh $\equiv f_o D/V$ = Strouhal number; f_o = vortex shedding frequency unaffected by resonator). In other words, contrary to the conditions of forced oscillations (EIE), the frequency of excitation adjusts itself to the oscillation frequency which coincides closely with f_R even away from the resonance point. (For cases of EIE, in contrast, the frequency of oscillation always coincides with the frequency of excitation.)

Figure 8.13a shows a gravity-governed counterpart to the plate in a duct. In this particular case of rigid piers in an open channel, vortex-shedding from the piers

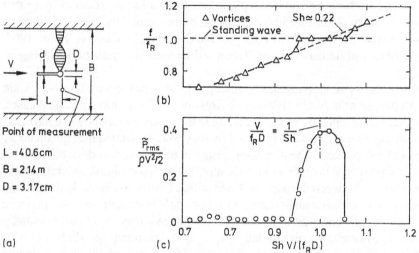

Figure 8.12. Response of a standing acoustic wave to vortex excitation (after Gaster, 1971).

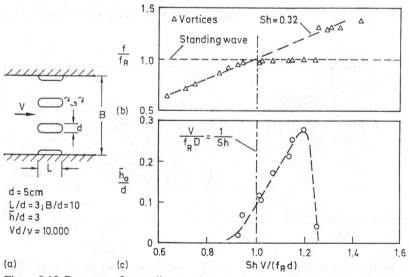

Figure 8.13. Response of a standing gravity wave between piers in an open channel to vortex excitation (after Crausse, 1939).

excited the first axial mode of a standing gravity wave as described in Figure 4.16. The response diagrams for this type of fluid oscillator depicted in Figures 8.13b, c are strikingly similar to those for a typical body oscillator shown in Figure 3.23. In both instances, the lock-in range starts as the velocity V reaches the resonant point $[V_r \equiv V/(f_R d) = 1/\text{Sh}]$, the frequency of vortex-shedding is reduced within that range, and the oscillation amplitudes subside abruptly near the end of lock-in ($\tilde{h}_o$ = amplitude of free-surface fluctuations, $\bar{h}$ = mean depth of flow). Moreover, both cases are characterized by response amplitudes far in excess of those that would have been obtained without the effect of the oscillators. (Related observations of waves in a channel with one pier are described by Clays & Tison, 1968. The possibility of subharmonic excitation is discussed in conjunction with Figure 8.31.)

Among the systems particularly susceptible to acoustic standing waves are heat-exchanger *tube banks* (Blevins & Bressler, 1987). In most practical cases, such waves are excited by vortex shedding in the upstream tube rows of the bank. In the arrangement shown in Figure 8.14a, they were generated in stagnant air by means of loudspeakers. The tube-bank casing in this figure was devised in such a way that it permitted investigation of the amplification conditions under the action of the interface between the parts with and without tubes, as a non-ideal reflecting boundary. The experimental results in Figure 8.14b indicate that the pressure amplitude in the outer part beyond the tube banks may decay exponentially ($N_x = 1, 2, 3$) or change sinusoidally ($N_x = 4, 5$) depending on whether f_R/f' is smaller or larger than 1.0, respectively (f' = frequency of the corresponding one-dimensional resonant wave in the outer duct). The response amplitude $\tilde{p}$ in the outer part of the duct may even increase above the amplitude $\tilde{p}_{\Cellipse}$ within the center part if the local amplification conditions are favorable. For the mode $N_x = 5$

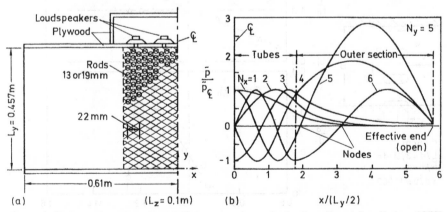

Figure 8.14. Resonant acoustic standing waves in a tube-bank casing (after Parker, 1978). (a) Experimental arrangement; (b) Theoretical variation of relative pressure amplitude in the x direction ($N_y = 5$, i.e., $\lambda_y = 2L_y/5$; $c_{\text{eff}}/c = 1.23$, Equation 4.7).

with a node very close to the interface between parts with and without tubes, the amplitude outside the tube bank exceeds $\tilde{p}_{\mathcal{C}}$ by nearly a factor of three. (The exaggerated change in wavelength from the part of the duct with tubes to the one without is related to the change in wave celerity c described by Equation 4.7.) The example clearly emphasizes the importance of fluid resonance: Under certain conditions, vibrations of duct walls induced by oscillations of the fluid oscillator, as well as deafening noise, may pose an even bigger problem than tube vibration or the vortex shedding that excited the fluid oscillator in the first place. A summary of design guidelines regarding vibration problems in tube banks, including acoustic resonance, is given by Paidoussis (1983) and Weaver & Fitzpatrick (1988).

As explained in Chapter 6, some sources of instability-induced excitation are characterized by more than one frequency of excitation. One such source is a shear layer impinging on an edge. In the example shown in Figure 8.15a, the shear layer is generated by an axisymmetric jet, and the impingement edge is formed by the round hole in a normal plate. The frequency of excitation (equivalent to f_o above) depends on the number n of wavelengths λ of the vortex pattern in the shear layer occurring between the point at which the jet exits and the plate; according to Morel (1979) it can be described in the same way as for planar shear layers (Rossiter, 1964), namely

$$\frac{f_o L}{V} \equiv Sh_L = \frac{n - C_1}{V/V_c + Ma}, \qquad n = 1, 2, 3 \ldots \tag{8.6}$$

Here, V = nominal jet velocity; V_c = disturbance convection velocity; $C_1 \approx 0.25$; $Ma \equiv V/c_a$ = Mach number; and c_a = speed of sound. For the conditions depicted in Figure 8.15b, Morel (1979) found the velocity ratio to be a function of the mode n: $V_c/V = 0.72$ for $n = 1$; 0.60 for $n = 2$; 0.57 for $n = 3$. The fluid

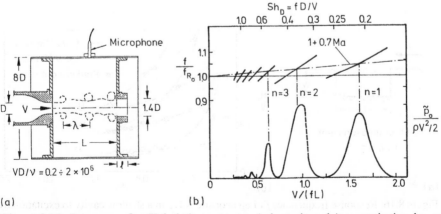

Figure 8.15. Response of a Helmholtz resonator (schematic only) to excitation by an impinging jet (after Morel, 1979).

oscillator in this example is a Helmholtz resonator with a frequency f_R that increases slightly with Mach number, $f_R/f_{R_o} \simeq 1 + 0.7\,\text{Ma}$, where f_{R_o} = resonant frequency at $\text{Ma} = 0$ (in the experiments, $\text{Ma} < 0.1$).

With increasing jet velocity V, the chamber pressure p_o in Figure 8.15a goes through a sequence of oscillations the spectra of which display a peak at a frequency f close to f_R. For each range of oscillation, f increased slightly with V (upper diagram in Figure 8.15b), and the amplitude $\tilde{p}_o$ of the pressure fluctuations first increased, reaching a maximum near the respective resonance point $V/(fL) = 1/\text{Sh}_L$, and then decreased (lower diagram in Figure 8.15b). The spectral peak of the next lower mode of oscillation often appeared before the previous peak had disappeared. The highest values of $\tilde{p}_o/(\rho V^2/2)$ occurred for mode $n = 2$ (actually values of up to 5.6!), which is in accordance with the finding of Chan (1974) that if a round jet is subjected to a periodic forcing with frequency $f = \text{Sh}_D D/V$, periodic flow is developed in the range $0.2 < \text{Sh}_D < 1.0$ with maximum amplitude near $\text{Sh}_L = \text{Sh}_D L/D = 0.92$.

If a fluid oscillator possesses more than one oscillation mode, such as a standing wave in a rectangular cavity (Figure 4.9), a multiple response is likely to occur, involving several resonant frequencies f_R. Figure 8.16a represents a shallow cavity of length L, excited by an impinging shear layer along its upper boundary. Since this fluid oscillator can be approximated by a closed cavity with an open end downstream (because of the intensive mass exchange occurring there according to Heller & Bliss, 1975), one can estimate f_R of the longitudinal standing waves with the aid of Equations 4.2 and 4.19 as

$$\frac{f_R L}{V} \simeq \frac{2N-1}{4}\frac{c_a}{V} = \frac{2N-1}{4\,\text{Ma}}, \qquad N = 1, 2, 3 \ldots \tag{8.7}$$

Thus, resonance in the case represented in Figure 8.16 is marked by a coincidence of one of the n mode excitation frequencies f_o (Equation 8.6, $C_1 = 0.25$,

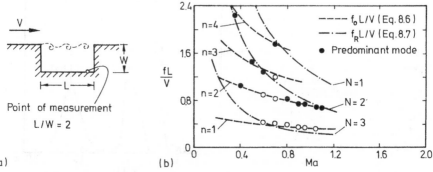

(a)   (b)

Figure 8.16. Response frequencies f of an acoustic wave in a shallow cavity to excitation by an impinging shear layer (after Rockwell & Naudascher, 1978; Data are taken from Rossiter, 1964).

$V_c/V = 0.66$) with one of the N mode standing wave frequencies f_R (Equation 8.7). In Figure 8.16b, Equations 8.6 and 8.7 are represented graphically along with the frequencies associated with the peaks in the pressure-fluctuation spectra measured by Rossiter (1964). In general, the frequency of the predominant pressure oscillation (indicated by dark circles) tends to coincide with that of the longitudinal cavity resonance. Indeed, as indicated by Rossiter's phase measurements along the cavity wall, standing wave patterns occurred mostly in the second and sometimes in the third mode ($N = 2, 3$) within the range of investigation. For the dark circle lying between the $N = 1$ and $N = 2$ curves, phase measurements showed that no standing wave existed.

More information on oscillations of flow past cavities is contained in Rockwell & Naudascher (1978, 1979), and further examples of fluid-oscillator response to IIE are given in Rockwell & Naudascher (1990).

8.4 MOVEMENT-INDUCED EXCITATION OF FLUID OSCILLATORS

The excitation of a fluid oscillator is called movement-induced if movements of the oscillator lead to a transfer of energy from the flow to the oscillator. The oscillations of such a self-excited fluid oscillator are often referred to as auto-oscillations.

Truly movement-induced oscillations of this type are not to be confused with oscillations extraneously induced by a self-excited body oscillator. In a typical *pumping system* as illustrated in Figures 8.17a and b, for example, a number of structural components exist that may undergo self-excited vibrations and thus lead to oscillations of a fluid oscillator. A case in point is a leaky valve in a closed position (Streeter & Wylie, 1967, p. 131, or Jaeger, 1963). If the valve's geometry and elastic characteristics are such that the leakage gap decreases with increasing pressure, the valve will experience self-excitation with periodic variations of flow rate (Figure 7.45a). Another possibility for self-excitation exists in the compressor of Figure 8.17a; under certain operating conditions, individual blades of this compressor are prone to stall flutter (Section 7.3.2). Naturally, both the fluctuations in the leakage flow rate and the vibrating compressor blades are potential sources for the excitation of a standing pressure wave in the duct or pipe and/or a surging in the plenum chamber or surge tank of the system.

The true movement-induced excitation (MIE) of a fluid oscillator does not require any of these MIE of structural components. Instead, this type of MIE involves an instability of the fluid-oscillator system, consisting, in the case considered, of duct and plenum including valves and pumping devices such as axial and centrifugal compressors and pumps. Such instability may arise if the system's valve is set at a nominally steady state (Figure 8.17d) and its pumping device is operated at stalled-flow condition (Figure 8.17e); it is an instability in the sense that small fluctuations in the mass flow rate $\dot{m} = \rho Q$ are amplified and

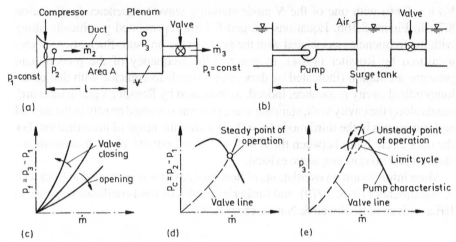

Figure 8.17. (a, b) Typical pumping systems (valve does not vibrate). (c) Valve characteristic; (d) Pump characteristic with point of operation determined with the aid of Figure 8.17c; (e) Limit cycle for operation under surge condition.

lead to a self-excited response of the fluid oscillator(s) in the system. This MIE is analogous to galloping in the sense that the oscillatory behavior can be determined from steady-state performance characteristics.

Autooscillations excited in this fashion may lead to impaired system performance such as decrease in efficiency and rise in noise as well as to catastrophic failures of the system. Applications involve both combustion and hydraulic systems, and the possibility of cavitation and two-phase flow occurring. According to the extensive review of the stability of pumping systems presented by Greitzer (1981), most instability problems arise if the systems are not run under optimal performance condition.

For quasi-steady conditions in the system, the time mean rate of flow is a constant, $\bar{m} = \bar{m}_2 = \bar{m}_3$. For small fluctuations $\tilde{m}$ of the mass rate of flow $\dot{m} = \bar{m} + \tilde{m}$, the equation for the onset of surging in the duct-plenum system of Figure 8.17a, according to Stenning (1980), is

$$\frac{l p_f' V}{A c_a^2} \ddot{z} + \left[\frac{l}{A} - \frac{p_c' p_f' V}{c_a^2} \right] \dot{z} + [p_f' - p_c'] z = 0 \tag{8.8}$$

Herein, c_a = speed of sound (Equation 4.5); $p_c = p_2 - p_1$ = pressure difference across the compressor; $p_f = p_3 - p_1$ = pressure difference across the valve; $p_f' = dp_f/d\dot{m}$; $p_c' = dp_c/d\dot{m}$; z can be $\tilde{m}$, $\tilde{p}_2$, or $\tilde{p}_3$; and l, A, and V are explained in Figure 8.17a. A comparison of the first and third terms with the equivalent terms in Equation 4.10 reveals that the oscillator described by this equation is a Helmholtz resonator with resonant frequency

$$f_R = \frac{c_a}{2\pi} \sqrt{\frac{A}{l \Psi}} \sqrt{\frac{p'_f - p'_c}{p'_f}} \qquad (8.9)$$

If p'_c is nearly zero, the second root approaches unity and this expression becomes identical with that of Equation 4.11.

Self-excitation (MIE) of this Helmholtz resonator sets in if the coefficient of $\dot{z}$ in Equation 8.8 becomes negative (negative damping). The system is then dynamically unstable and will develop surge oscillations from rest of increasing amplitude. Since p'_f is always positive, the slope of the curve of p_c versus $\dot{m}_2$ in Figure 8.17d (i.e., p'_c) must be positive for excitation to occur. Oscillations begin if

$$p'_c = \frac{dp_c}{d\dot{m}_2} \geq \frac{l c_a^2}{A \Psi p'_f} \qquad (8.10)$$

If p'_c is considerably larger than this value, the surge oscillation may build up into a large-amplitude limit cycle as illustrated in Figure 8.17e. According to Greitzer (1981), this sytem response depends, for a given compressibility-governed system, on the parameter

$$U_r \equiv \frac{U}{2\pi f_R l} \approx \frac{U}{c_a} \sqrt{\frac{\Psi}{Al}} \qquad (8.11)$$

where U = rotor speed. A critical value of U_r determines whether the mode of instability will be surge $[U_r > (U_r)_{cr}]$ or rotating stall $[U_r < (U_r)_{cr}]$ regardless of whether this value has been obtained using a large volume and a low speed or vice versa.

Some pumping systems become unstable even when the turbomachine is operating at its design flow rate. An example is a system with a *cavitating turbopump*. Again, the possible instabilities fall into two general categories: the rotating cavitation, characterized by an usteady cavitation pattern at the inducer inlet that rotates with respect to the inducer blades similar to rotating stall; and the auto-oscillation or surge, associated with mass-flow oscillations in the entire system similar to the surge oscillations discussed in the preceding paragraphs (Greitzer, 1981). Interestingly enough, these oscillations may occur even in the absence of any compliant volume external to the pump, because cavitation in the pump provides internal compliance in this case.

Figure 8.18 shows an example of a system in which the pressure-recovery characteristics of a diffuser rather than the pump characteristic determine the condition for dynamic instability. The system is typical of *gas turbine* annular combustors. Again, as in the case described in Figure 8.17e, instability demands a positive slope in the curve of presure-rise versus flow rate; this positive slope is brought about here by flow separation in the diffuser. If the system is operated at a rate of flow for which the static pressure rise in one of the diffuser branches increases with increasing flow fraction entering that branch (dark circle in Figure 8.18b), self-excited surge oscillations similar to the ones just described are likely

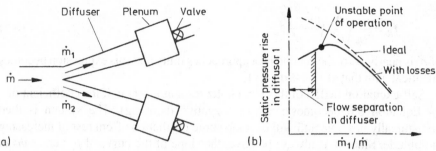

Figure 8.18. (a) Branched diffuser-plenum-valve system (part of a gas turbine annular combustor). (b) Static pressure recovery characteristic (after Ehrich, 1970).

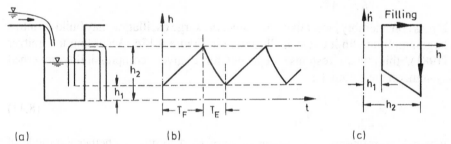

Figure 8.19. (a) Hydraulic system leading to relaxation vibration. (b) Histogram; (c) Phase diagram with limit cycle (after Magnus, 1965).

to occur. The pressure variations resemble a saw-tooth wave with twice the frequency of the flow-rate variation in that case (Ehrich, 1970).

A system with extreme nonlinearity leading to nonsinusoidal self-excited fluid oscillations is depicted in Figure 8.19. The water rises from h_1 to h_2 almost linearly during the fill period T_F. At $h = h_2$, the siphon starts emptying the vessel until, after a period T_E, it draws air at $h = h_1$ thus completing the cycle. Oscillations with characteristics like those in Figures 8.19b, c are called oscillations of the relaxation type. A self-excited oscillation of this type involving water hammer due to intermittent valve closures is being utilized in hydraulic rams for the purpose of delivering water to a height greater than that of the supply head (Gibson, 1952; Iversen, 1975; Krol, 1976).

A number of MIE of fluid oscillators are characterized by movements of a shock wave in the system. Perhaps the best-known among these is the *Hartmann oscillator* schematically shown in Figure 1.1f (Morch, 1964; Thompson, 1964). Figure 8.20 illustrates an example which is significant for hydraulic engineers because of the analogy between shock waves in supersonic internal flows (Ma > 1) and hydraulic jumps in supercritical free-surface flows (Fr > 1). Figure 8.20b depicts two typical instants during an oscillation cycle for a pressure ratio of

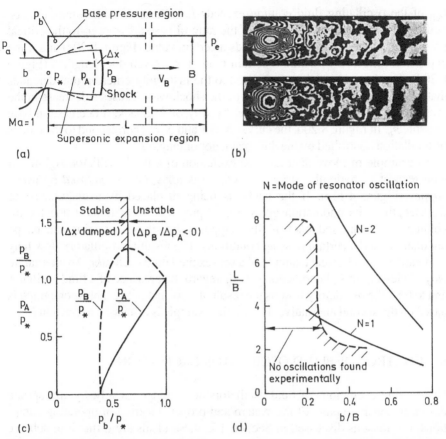

Figure 8.20. (a) Schematic of flow field in a rectangular duct with abrupt enlargement; (b) Interferograms of corresponding oscillations of shock-wave pattern; (c) Calculated pressure upstream and downstream of shock wave as a function of base pressure; (d) Relative length of resonator as a function of area ratio (after Anderson et al., 1977, 1978).

$p_a/p_b = 0.364$; in one, the shock front is further downstream than in the other. Oscillations between these two states are associated with large pressure fluctuations in the duct resonator at frequencies f close to

$$f_R = \frac{c_a}{\lambda}\left[1 - \left(\frac{V_B}{c_a}\right)^2\right] = \frac{(2N-1)c_a}{4L}(1 - \mathrm{Ma}_B^2), \qquad N = 1, 2, 3 \ldots$$

which is the resonant frequency predicted from Equations 4.33 and 4.19 without the effects of added mass. According to Anderson (1977), nearly all data for different expansion ratios ($6 < B/b < 14$) and length-to-width ratios ($2.4 < L/b < 5.3$) lie in the range $0.8 < f/f_R < 1.0$. The corresponding frequency reduction can be attributed, in great measure, to the fact that the effective length

L_{eff} of the oscillating fluid column exceeds L because of added-mass effects. Self-excited oscillations occur only if the ratio of base pressure p_b to the critical pressure p_* at the nozzle throat exceeds a certain value (Figure 8.20c) and if the expansion ratio b/B is within a certain range for a given duct length L (Figure 8.20d). The former condition is related to the instability criterion that demands that a decrease in pressure p_A in front of the shock wave causes a decrease in the pressure p_B past the shock and hence, by way of feedback, a decrease in base pressure p_b. In Figure 8.20d, the curves $N = 1$ and $N = 2$ correspond to the modes of oscillation controlled by the duct resonator of length L.

An example of movement-induced excitation of a fluid oscillator, significant with respect to hydraulic structures, concerns *air volumes enclosed by water nappes* (Figure 4.11c). If the nappe is being displaced by a disturbance of wavelength λ, its undulation gives rise to pressure fluctuations inside the air volume which are fed back to the nappe origin, thus sustaining the nappe undulations under certain phasing conditions. The resulting oscillatory flow may be regarded as a classical example of a self-excited fluid oscillator. An alternative way of viewing this phenomenon is to disregard the presence of a fluid oscillator and to treat the oscillatory flow as the result of flow instability. This monograph is based on the second alternative. Hence, the example is discussed in Section 6.5.

8.5 CONTROL OF FLUID-OSCILLATOR EXCITATION

Undesirable excitations of fluid oscillators are avoided, ideally, by appropriate design of sensitive parts of the system and proper selection of operating conditions. For reasons discussed in Section 1.2, these ideals are difficult to achieve with complex systems, so remedial measures are frequently necessary to suppress disturbing oscillations and to restore the proper functioning of completed installations. Since an understanding of the mechanisms of excitation is essential in either case, emphasis has been placed on their description in the previous sections.

In assessing a system with respect to fluid-oscillator excitations, one has to take the following steps:

1. Identify all fluid oscillators in the system and determine the resonant frequencies that may be excited;
2. Identify all sources of excitation and their nature;
3. Determine the dominant frequencies of the extraneously and instability-induced excitations (EIE), (IIE);
4. Compare the frequencies mentioned under (1) and (3) and investigate changes in design and/or operating conditions to avoid resonance;
5. Consider the criteria for potential movement-induced excitations (MIE) and investigate measures to avoid them;
6. Evaluate the necessity for inserting devices that eliminate or attenuate disturbing oscillations.

Since many of the sources of IIE are similar for fluid and body oscillators, it is possible to utilize the information presented in Chapters 6 and 9 in the determination of IIE frequencies (point 3). Moreover, one can attenuate the excitation (point 6) by adopting one or a combination of the devices discussed in Section 6.8.

8.5.1 *Means of avoiding resonance*

Regarding resonance in piping systems, a small *change in length* may substantially attenuate oscillations by removing key resonant frequencies f_R from the range of the excitation frequencies. Streeter & Wylie (1967, p. 227), for example, report a reduction of pressure-head oscillations in the suction line of a quintuplex pump from ± 30 m to ± 9 m by means of a decrease in suction-line length from 2.14 m to 1.92 m.

Similar results may be accomplished by changes in the speed of wave propagation c. As head fluctuations are proportional to c, moreover, a reduction in c reduces amplitudes as well. *Wavespeed reductions* in piping systems may be obtained by: (a) insertion of a section of pipeline made of rubber or other flexible material (Equation 4.4); (b) placing of a small flexible hose filled with air inside the metal pipeline (Remenieras, 1952); and (c) entraining air into the liquid flow (Equation 4.6).

Bleeding air into the system is widely used to eliminate or reduce the severity of cavitation. Usually, air-inlet valves are installed at low-pressure points upstream of where cavitation may occur. If air is entrained into a liquid flow, moreover, a variety of instability-induced excitations can be controlled and even eliminated. An example is the swirling instability in draft tubes discussed in Section 6.8.5.

The performance of reciprocating-pump systems can be improved by changes in arrangement, size, and length of piping components, and by changes in pump design. The transient response of piping systems can be controlled by *pressure-limiting devices* such as surge tanks or *auxiliary valving* which allows fluid to escape at predetermined instances (Streeter & Wylie, 1967; Lundgren, 1961). Changes in transient behavior are also attainable by the use of various combinations of capacitance, inertance, and resistance devices (Waller, 1959). In existing piping systems, harmful effects may be diminished by *stiffening and bracing* the system at the points of maximum pressure amplitude (Malamet, 1965). Finally, if feasible, changes in the pump speed can alter the frequency of the forcing function, thereby reducing the unsteadiness.

A measure against excessive standing waves (both acoustic and free-surface) and against sloshing is a *baffle plate* placed in the pressure-node plane where the wave-induced velocity fluctuations are maximum. Of course, such measures are feasible only if these plates do not unduly obstruct the flow.

For tube-bank casings, detuning baffles proved to be effective in attenuating resonant pressure oscillations excited by tube cross-flow instabilities (Chen,

1968). By installing such baffles parallel to the direction of flow so that the total width L of the enclosure is subdivided into different reaches L_i, the fundamental-mode oscillation is broken up in oscillations with different resonant frequencies, all of which can be above the lowest excitation frequency. In order to avoid amplification of one of the harmonics, L_i should differ from values equal to L/N ($N = 2, 3, ...$). Detuning baffles can be both solid and porous (Eisinger, 1980; Byrne, 1983). Their drawback is that they increase the resistance to flow through the casing.

Other techniques related to tube-bank casings involve removal of tubes, repositioning of tubes, and installation of fins and helical spacers (Chen, 1987, p. 340). Resonance can be completely eliminated by removing tubes located along the pressure node of the standing acoustic wave (Walker & Reising, 1968). Tests by Zdravkovich & Nuttall (1974) proved, moreover, that acoustic resonance can also be eliminated by means of unequal longitudinal pitches in successive rows. The attenuating effects of fin barriers and helical spacers were investigated by Eisinger (1980). The former are made of fins welded to the tubes such as to form 'fin walls' parallel to the flow which act somewhat like detuning baffles. Further information on the suppression of resonance in tube-bank casings is given by Blevins & Bressler (1987) and by Fitzpatrick (1986).

If fluid-oscillator resonance is brought about by instability-induced excitation, its cure can also be approached by either eliminating or attenuating the excitation at its source (Section 6.8) or by altering its frequency so that it falls outside the range of resonant frequencies. As a rule, changes in excitation frequency are more difficult to attain than changes in f_R.

8.5.2 *Use of filters*

For attenuating fluid oscillations in a given frequency range, a filter (or muffler) can be effective. One may use both reactive and absorptive filters. A reactive filter consists of a discontinuity in the flow passage which acts in such a way that only a fraction of the energy of an oncoming wave is transmitted while the rest is reflected. Wave transmission is thus hindered essentially without loss of energy. In an absorptive or dissipative filter, in contrast, the transmission of waves is impeded through damping by means of energy-absorbing and flow-resistive material. In the following, only reactive filters are described and those briefly. Filters of this type are available for both compressibility-governed and free-surface systems.

A filter is called a low- or high-pass filter depending on whether it transmits fluctuations below or above a certain cut-off frequency; a filter of the band-stop type transmits no fluctuations within a defined band of frequencies. Figure 8.21 shows examples of such filters for waves of low frequency f or large wavelength λ ($2\pi L/\lambda \ll 1$ in Figure 8.21a, $2\pi a/\lambda \ll 1$ in Figures 8.21b, c). The examples include: (a) an expansion chamber or basin; (b) a side branch; and (c) a Helmholtz

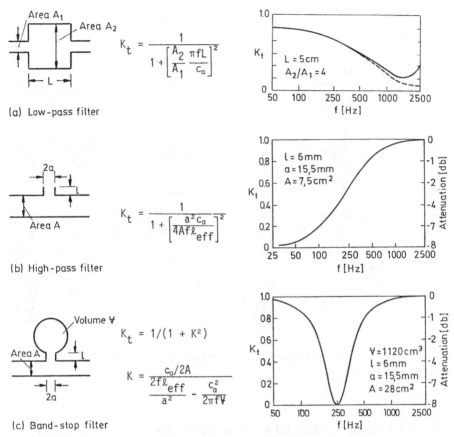

Figure 8.21. Filters for attenuating waves of low frequency (large wavelength) in ducts, pipes, or channels. Transmission coefficients K_t (after Kinsler & Frey, 1962).

resonator or resonator basin. The transmission coefficient K_t shown there, defined as the ratio of transmitted to incident wave power, was computed neglecting viscous effects. Actually, energy dissipation is small in all cases considered. By the action of the filter, a fraction of the energy is being absorbed during one part of the oscillation cycle and returned to the pipe or channel during another part of the cycle with such a change in phase that the result is a reflection of that fraction of energy back towards the source. Of course, all formulas and diagrams need to be slightly modified to fit real-flow situations.

For waves of high frequency or small wavelength, the relationships for K_t change completely. If the wavelength λ is reduced to four times the length L of the expansion chamber, for example, K_t becomes a minimum. A further decrease of λ (or increase in frequency $f = c/\lambda$) causes K_t to run through a series of maxima (at $\lambda = 2L, L, L/2 \ldots$) and minima (at $\lambda = 4L, 4L/3, 4L/5 \ldots$), until finally, for

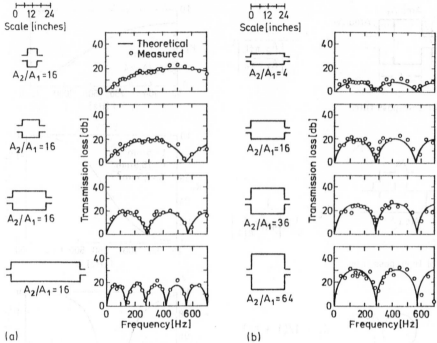

Figure 8.22. Characteristics of expansion-chamber filters for waves of high-frequency (small wavelength) (after Davis et al., 1954).

$\lambda/D \ll 1$, it remains at unity (D = pipe diameter). This final attainment of 100% transmission is characteristic of all filters depicted in Figure 8.21. In Figure 8.22, the filter characteristic for high-frequency waves is shown for an expansion chamber of circular cross-section. Instead of K_t, the transmission *loss* is shown in the diagrams, defined as the energy in the incident wave divided by the energy transmitted at the filter outlet into a non-reflecting termination.

By employing a series of filters of different types in combination with orifices, enlargements, and constrictions in the flow passage, one can construct *filter networks* with specified properties. The design and optimization of such networks is facilitated by utilizing the analogy between fluid-flow and electrical systems (Kinsler & Frey, 1962; Beranek, 1954), at least at a preliminary stage. In the final stage of the design of a filter or filter network, experimental noise or surface-wave measurements with the filters in place may be required (Davies, 1977). Since filter performance is difficult to predict for complex systems, this procedure allows for appropriate modifications on the basis of the tests. Filter networks for both gas and liquid systems are described by Franken (1960), Kinsler et al. (1982), and Waller (1959).

Figure 8.23 shows some applications of filters to free-surface systems. The filter in these examples consists of a lateral-basin and a surge-type resonator. Again, the principle on which these filters work is their excitation at the incident-wave frequency by means of which they return a portion of the incident wave and hinder its transmission past the filter. An important part in the design of a filter is its proper tuning. According to experiments of Valembois (1957), a range of frequencies from f to $2f$ can be covered by three to four resonators side by side with their resonant periods, $1/f_R$, increasing in geometric progression within that range (Figures 8.24d, 8.25b). Thus, solutions of this type will only work for

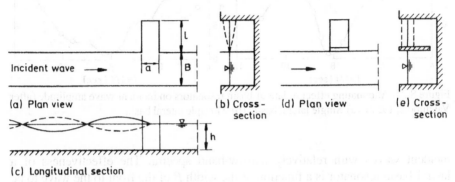

(a) Plan view   (b) Cross-section   (d) Plan view   (e) Cross-section

(c) Longitudinal section

Figure 8.23. (a-c) Schematic of an open channel with lateral-basin resonator; (d, e) Open channel with surge-type resonator.

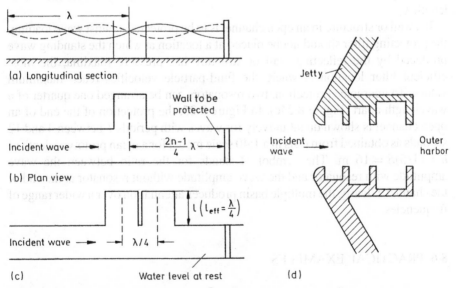

(a) Longitudinal section

(b) Plan view

(c)   Water level at rest   (d)

Figure 8.24. (a-c) Schematic of open channel with wall to be protected; (a, b) Side-branch resonator ineffective when located as indicated. (c) Effective protection with two side-branch filters. (d) Protection of an outer harbor (after Valembois, 1957).

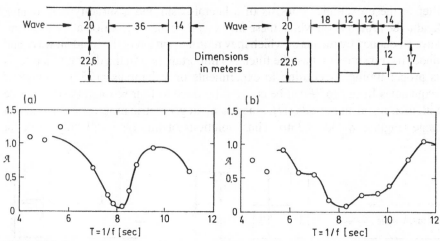

Figure 8.25. Attenuating effect of lateral-basin resonators on incident-wave amplitude (after Valembois, 1957). (a) Single lateral basin; (b) Multiple lateral basin.

incident waves with relatively narrow-band spectra. The effectiveness of a lateral-basin resonator is a function of the width B of the front to the wave to be reflected. Model tests by Valembois indicate that with resonators on both sides, wave fronts typical for harbor entrances of the order of 100 m can be acted upon effectively. Difficulties are encountered with wave fronts wider than the wavelength λ.

If a wall or structure in an open channel is to be protected against wave loading, the protecting filter should not be placed at a location at which the standing wave produced by the reflecting wall or structure has pressure maxima; the most efficient filter location is where the fluid-particle velocities reach maximum values. To guarantee protection, two resonators can be arranged one quarter of a wavelength apart (Figure 8.24c). In Figure 8.25, the protection of the end of an open channel is shown quantitatively for waves with periods T between 4 and 12 seconds as obtained from tests in a 1:40 scale model for a mean prototype depth of $\bar{h} = (4/5)B = 16$ m. The symbol $\mathcal{A}$ stands for the ratio between the wave amplitude with resonator and the wave amplitude without resonator at the end of the channel. Clearly, the multiple basin produces attenuation over a wider range of frequencies.

8.6 PRACTICAL EXAMPLES

8.6.1 *Compressibility-governed systems*

The following practical examples involving compressibility-governed or acoustic

fluid oscillators have been presented in the preceding sections:
1. Piping systems (Figures 8.3, 8.4, 8.5, 8.17, Section 8.5.2);
2. Air-filled room (Figure 8.6);
3. Cooling-water pool (Section 8.2);
4. Plates or cascades in ducts (Figures 4.15, 8.12);
5. Tube-bank casing (Figure 8.14, Section 8.5.1);
6. Gas turbine (Figure 8.18).

Because piping systems are important in hydraulic engineering, a few more examples are included here involving the excitation of the compressibility-governed or acoustic modes of penstocks, pipelines, and conduits.

A common source of extraneous excitation is the pressure fluctuation caused by flow separation in bends, joints, and other discontinuities in piping systems (Chen, 1980). In general, the spectra of these fluctuations show a marked peak in the range $0.1 < fD/V < 1.5$, where f = frequency, D = pipe diameter, and V = mean velocity. In one case of a Y-branch upstream of a butterfly valve (Figure 8.26a), a strong noise near the Y-branch, accompanied by vibrations of the valve, was produced at a particular discharge. The problem was cured by streamlining the Y-branch using a filler block and thus eliminating the separation that had excited the system.

A frequent instability-induced excitation (IIE) of fluid oscillations in penstocks is due to draft-tube surging. As in other examples of IIE, essential for the intensity of the response is an overlapping of the ranges of excitation and resonant frequencies. In a powerplant with two penstocks of 3.28 m diameter and 113 m/123 m lengths, severe pressure fluctuations prevented turbine operation beyond 78% of capacity (Falvey, 1980, Chart F). Field measurements revealed draft tube surges with frequencies between 1.0 and 1.9 Hz and an overlap with

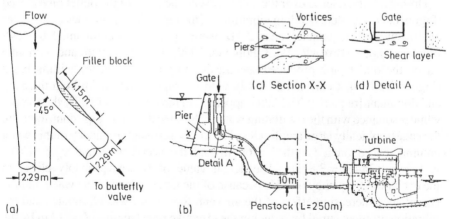

Figure 8.26. (a) Y-branch with improvement (after Falvey, 1980, Chart H). (b-d) Hydro-power system with gate vibrations associated with penstock resonance (after Hardwick et al., 1980).

resonant frequencies in the range between 1.4 and 1.9 Hz. Maximum pressure-head fluctuations in the penstock amounted to 16.9 m for a mean head of 84.4 m. Since the usual measures against draft-tube surging, i.e., air injection and fins on the draft-tube throat (see also Falvey, 1980, Chart A), had little effect, the turbine characteristics had to be changed by trimming the trailing edges of each turbine blade.

Other sources of instability-induced excitation that occur frequently are vortex shedding and impinging shear layers (Sections 6.3, 6.4). In a case of 'triple coincidence', neither the vortex shedding from the pier (Figure 8.26c) nor the impinging shear layer below the fully withdrawn gate (Figure 8.26d) would have been sufficient to create a gate-vibration problem without the matching of the dominant shedding frequency (between 1 and 2 Hz), the natural frequency of vertical gate vibration (about 1.8 Hz), and the resonant frequency of the penstock (about 1.9 Hz). The problem was cured by fixing damping pads on the gate which engaged near the highest region of gate travel and prevented the gate from vibrating.

If an outlet pipe with upstream control is located in a spillway face as shown in Figures 8.27 and 8.33, and if the spillway is operated with the outlet gates closed, pressure oscillations in the pipe can be excited in a number of ways. Above the outlet opening, the spillway flow produces a shear layer that impinges on the outlet invert, thus creating the IIE module treated in Section 6.4. If the water inside the pipe is free to move (e.g., on account of air vents downstream of the control gates), this shear layer undergoes large movements accompanied by what resembles more closely a movement-induced excitation (MIE) according to the description of Falvey (1980, p. 387). If the pipe outlet is located at the foot of a stilling basin, moreover, an additional source of excitation is EIE due to pulsations in the hydraulic jump.

During a flood release over the spillway with the gates of the outlet pipe closed (Figure 8.27a), violent bending vibrations of the pipe with amplitudes of the order of 10 cm and frequencies near 2 Hz were observed, accompanied by low-frequency noise (Božović, 1973; Popović, 1970). Field tests with and without a lid on the outlet pipe produced spectra due to pulsations in the stilling basin (Figure 8.27b) that correlated closely with the pipe vibrations in both intensity and dominant frequency. The latter appeared to have been excited by the former, while resonance with the oscillating water column in the outlet pipe contributed to the catastrophically large amplitudes. The lowest resonant frequency of this water column, based on a length $L \simeq 60$ m and the speed of sound $c_a \equiv \sqrt{E/\rho}$, is $f_R = c_a/(2\pi 4L) \simeq 3.8$ Hz. The actual value of f_R was probably closer to the excitation frequency of 2 Hz because of the air entrained in the water and the compliance-increasing effect of the air vent. Resonance was eliminated and the vibration problem cured by reducing the effective pipe length to $L = 17$ m by the insertion of gates near the pipe outlet. (A possible fluid-dynamic attenuation device is shown in Figure 8.32.)

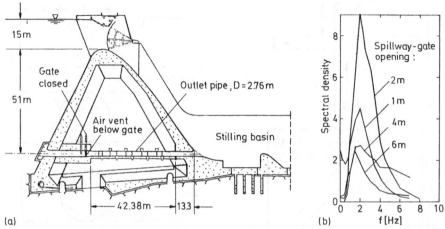

Figure 8.27. (a) Pulsations in a stilling basin during the spillway flow may excite resonant pressure oscillations in the outlet pipe if the latter is closed upstream. (b) Spectra of pressure fluctuations at the foot of the stilling basin (after Božović, 1973).

A frequent cause of resonant oscillations in piping systems is the unstable flow near a T-branch (Paidoussis, 1980, Chart XXVII; Chen, 1980, Chart K; Bakes et al., 1980). Due to fluctuations in the strength, size, and position of the vortices forming near the junction (Figure 8.28c), pronounced cyclic variations in pressure and in flow into the opposing laterals have been observed in several widely different pipe systems (Appel & Yu, 1966). In one case, this flow instability (IIE) led to intensive vibrations of a loading rig delivering gasoline to barges at a waterfront. Major fluctuations in flow rate and vibrations can take place, first, if the head loss at the junction is a significant part of the total loss in the system and, second, if the dominant frequency of excitation coincides with one of the resonant frequencies of the piping system. Figure 8.28b represents the best estimate of this frequency available at present, but departures of as much as 50%, or more, are possible. To prevent the excitation, one can place a dividing plate in the main pipe (Appel & Yu, 1966), or transform the T-branch into a Y-branch of reduced head loss (Bakes et al., 1980). If a substantial part of the flow continued past the laterals, the system was stable.

In the conduits of a navigation lock (Figures 8.29a, b), large pressure pulsations (Figure 8.29c) were observed in the initial phase of the emptying operation. At the beginning of this operation, the water level in the lock was 13.5 m above the tailwater level and the conduit discharge was 60 m³/s. Spectral analysis of the pressure histogram revealed peaks near 1.2 Hz, 3.6 Hz, and 6 Hz; these values are close to the resonant frequencies $f_R = c_a/\lambda = 1.2$ Hz, 3.0 Hz, and 5.7 Hz, computed with $c_a = 1000$ m/s (acoustic wave celerity for water in concrete enclosure) and $\lambda = 4L, 4L/3$, and $4L/5$, respectively (L = conduit length between

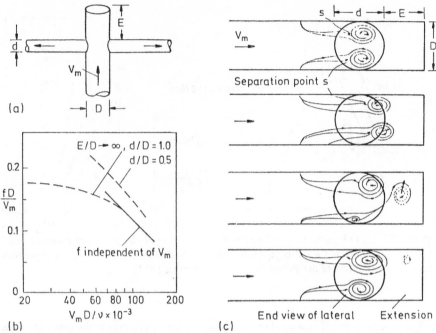

Figure 8.28. (a) T-Branch pipe system. (b) Estimate of average frequency *f* of excitation. (c) Cycle of vortex movement and formation for T-branch with $d/D = 1.0$ (after Appel & Yu, 1966).

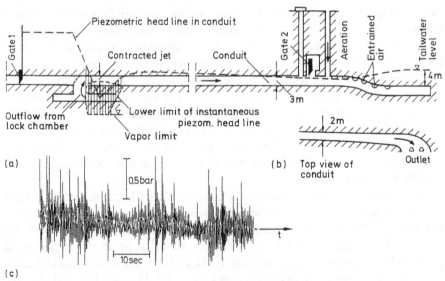

Figure 8.29. (a) Cross-section of the conduit of a navigation lock with piezometric head line during emptying operation; (b) Top view of conduit outlet. (c) Histogram of pressure pulsations recorded at gate 1.

gate 1 and outlet). The source of these standing pressure waves was first thought to be the blow-out of entrained air downstream of gate 2. Field tests and a computation of the piezometric head line showed, however, that the increase of instantaneous flow velocities in the contracted jet (Figure 8.29a) due to flow separation and turbulence lowered the local, instantaneous pressure to the vapor limit, thus giving rise to extraneously induced excitation due to cavitation (Naudascher & Kanne, 1990). The pressure pulsations were eliminated by opening gate 2 in two steps: first to 2.1 m until the head was reduced to a permissible level, and then to 3 m. In similar installations, flow constrictions like the one shown in Figure 8.29 should be avoided.

8.6.2 *Free-surface systems*

Several practical examples involving free-surface type fluid oscillators are contained in the foregoing sections and in the IAHR Monograph on Hydrodynamic Forces (Naudascher, 1991). They involve:

1. Harbors (Figures 4.4, 4.17, 8.7-8.10, 8.24);
2. Open channels (Figures 4.16, 8.13);
3. Navigation locks (Naudascher, 1991, Figures 2.57, 2.58).

In the following, additional examples of items 2 and 3 are presented and one, each, involving

4. Pipelines (Figures 8.33, 8.34); and
5. Closed-conduit spillways (Figure 8.35).

Figure 8.30a shows a bridge crossing over a canal at which vortex shedding from the bridge piers excited waves with amplitudes of 30 cm and a frequency of 0.38 Hz during flow at a mean velocity of $V = 1.1$ m/s. Wave trains progressed upstream with a beat frequency of 0.07 Hz. The most likely standing-wave pattern for this configuration, in light of Figure 4.13, is sketched in Figure 8.30b. The pattern has two possible interpretations: one based on mode (1.1) for one row of piers with $L/B = 10.87/12.39$, and one based on mode (2.1) for two rows of piers with $L/B = 21.06/12.39$. A rough estimate of the corresponding resonant frequencies can be obtained with the aid of Figure 4.15a and $c \simeq \sqrt{g4.8} = 6.87$ m/s (Equation 4.8b); the results for f_R are 0.37 Hz and 0.40 Hz, respectively. If these resonating waves were the source of the upstream wave trains, their superposition would produce a wave-train frequency of $(0.40 + 0.37)/2 = 0.385$ Hz and a beat frequency of $(0.40 - 0.37)/2 = 0.015$ Hz. The correspondance with the observed values of 0.38 and 0.07 Hz is surprisingly close, in view of the many effects that were neglected in the estimate (pier geometry, variable depth, effect of mean flow, etc.). The waves were probably excited by subharmonic resonance with vortex shedding from the piers (f_R corresponds to half the vortex-shedding frequency during lock-in, $f_o = $ Sh $V/d \simeq 0.26 \times 1.1/0.38 = 0.76$ Hz; subharmonic resonance of this type has also been mentioned by Toebes, 1965). Once the longitudinal spaces between the piers had been filled in, all but the vortex shedding

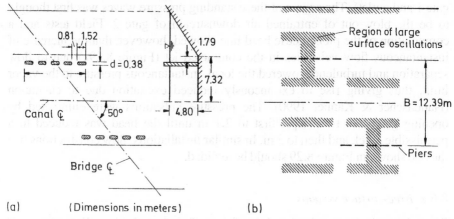

(a)   (Dimensions in meters)   (b)

Figure 8.30. (a) Bridge piers in trapezoidal canal gave rise to waves progressing upstream for about 1 km (after Falvey, 1980, Chart D). (b) Schematic of hypothesized standing wave patterns (in analogy with Figure 4.13).

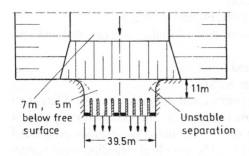

Figure 8.31. Top view of sluice basin. With discharge through 6 of 10 openings of 3.05 m as shown, a transverse wave was observed in the basin (Kolkman, 1984, p. 46).

furthest downstream was suppressed, and the waves were eliminated.

During the operation of a sluice with four of ten openings closed by gates as shown in Figure 8.31, transverse wave oscillations were observed with a node along the sluice centerline; their amplitudes were 1 m near the outer closed gates and their frequency was about 0.05 Hz. Unstable separation of the flow from the side walls played a dominant role in the excitation according to Kolkman (1984a). Most probably, however, there existed self-excitation (MIE) associated with transverse velocities set up by the waves, on the one hand, and cyclic discharge variations caused by the fluctuating head near the open gates, on the other. The transverse standing wave was successfully supressed by changing the condition of partial-gate operation.

A well-known phenomenon of self-excited longitudinal wave motion in open channels is that associated with sluice gates hardly touching the water surface (Binnie, 1979; Kolkman, 1980). The essential feature of this MIE is similar to the

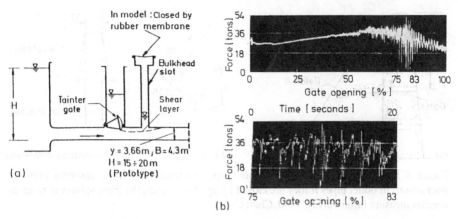

Figure 8.32. (a) Model of a reversed Tainter gate in a lock filling system. (b) Force on prototype gate strut (after Hecker & Dale, 1966).

one described in the second paragraph of Section 8.4. In both cases, the area of flow passage decreases with an increase in head at the control structure.

An example of a Tainter-gate vibration excited due to a resonating water-column surging in a bulkhead slot is shown in Figure 8.32. In the prototype, the top of the bulkhead slot was closed by a solid cover so that the enclosed air acted like a spring on the water column underneath. Thus, the bulkhead slot resembled the fluid oscillator sketched in Figure 4.3c. The force fluctuations on the Tainter-gate strut (Figure 8.32b) could be reproduced in the model only after the natural frequency of the fluid oscillator was properly simulated. This simulation, in turn, required a reduction in stiffness of the air spring, which could only be accomplished by enlarging the top of the bulkhead slot and covering it with a special membrane. Clearly, a model study without that precaution would never have identified even the possibility of vibration, because the excitation due to the shear layer impinging near the surging water column (Figure 8.32a) is extremely weak without a resonating fluid oscillator.

In the installation shown in Figure 8.33, the outlet pipes were only partially filled with water, thus providing the possibility for free-surface type fluid oscillators. With spillway discharges exceeding 16 m³/s per m width, surging of 0.025 to 0.052 Hz occurred with pressure-head amplitudes up to 25 m near the end of the pipe. Moreover, heavy audible pulsations occurred in the air vents downstream of the closed gates. The source of excitation was the same as in the case shown in Figure 8.27, namely shear-layer impingement on the outlet invert. Here, the excitation was eliminated by the installation of flow splitters (Figure 8.33b) that caused the spillway jet to impinge downstream of the outlet invert. The air entrained at the splitters also protected the concrete against cavitation damage.

Figure 8.34a illustrates a resonance problem observed in an irrigation-water

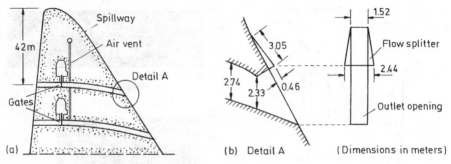

Figure 8.33. (a) Spillway flow over outlet-pipe openings may cause resonant pressure fluctuations in outlet pipes if they are closed by upstream gates; (b) Flow splitter or ramp as counter measure (after Falvey, 1980, Chart C).

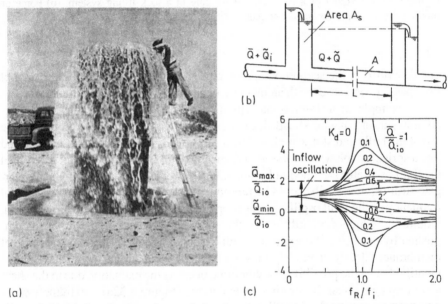

Figure 8.34. (a) Overflowing of a stand pipe due to surging in a pipeline water distribution system (from Simmons, 1965). (b) Sketch of pipeline system with stand pipes as control structures. (c) Amplification of discharge oscillations after a theory of Holley (1969).

distribution system consisting of reaches of pipeline connected by overflow stand pipes (Figure 8.34b). The major function of the stand pipes in such systems is to limit the static pressures in the line and to prevent excessive water-hammer effects. For discharges less than the design value, water spills into the downstream portion of the stand pipe and entrains air. With the accumulation and intermittent release of this air (blow out), surges of flow rate and piezometric head develop that, in their most violent form, may prevent the delivery of water. For a given amplitude $\tilde{Q}_{io}$ in the inflow-rate fluctuations, Holley (1969) computed the largest

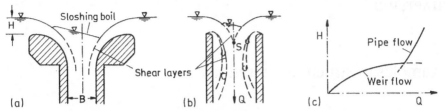

Figure 8.35. Surging in closed-conduit spillways. (a, b) Two-way drop inlets with (a) well-rounded crest, (b) thin crest. (c) Typical head-discharge curve (after Hebaus, 1983).

and smallest deviations, $\tilde{Q}_{max}$ and $\tilde{Q}_{min}$, from the mean flow rate $\overline{Q}$ as a function of the damping coefficient K_d of the reach, the relative flow rate $\overline{Q}/\tilde{Q}_{io}$, and the ratio f_R/f_i, where f_i is the frequency of the inflow oscillations and f_R is the resonant frequency according to Equation 4.13 (Figure 8.34c):

$$f_R = \frac{1}{2\pi}\sqrt{\frac{gA}{LA_s}}, \qquad K_d = \frac{C_L\tilde{Q}_{io}}{2AL}\sqrt{\frac{LA_s}{gA}}$$

(C_L is the friction-loss coefficient). If the reaches of the pipes are identical, their resonant frequencies f_R are the same ($f_i/f_R = 1$), and the system is essentially a surge amplifier. Measures of surge control include the prevention of air entrainment, the installation of surge tanks downstream of the stand pipes, and the design of reaches of different lengths L and/or different pipe-stand areas A_s. According to Figure 8.34c, making f_R/f_i smaller than unity is more effective than making it larger. Thus, the best surge control consists of designing consecutive pipe reaches successively longer or consecutive pipe stands successively wider.

An example of fluid oscillations involving the sloshing of a water mass is depicted in Figure 8.35. Closed-conduit spillways with rectangular cross-sections as shown have been used by the US Soil Conservation Service in many reservoirs. As they may exceed 20 m in height, flow-induced vibrations can endanger them. Hebaus (1983) found that a source of structural vibration develops within a narrow transition range between weir and pipe flow (Figure 8.35c). It involves a water mass ('boil') between the two nappes sloshing back and forth due to an excitation that is brought about by disturbances in the shear layers (Figures 8.35a, b). Even with a horizontal antivortex plate over the inlet, free-surface flow with a sloshing boil may exist below that plate. The resonant frequencies of sloshing resemble closely the f_R values obtained for water masses on boundaries that are shaped like the upper nappe profiles (Figure 4.19a). In cases of thin-crest structures, the sloshing is coupled with a nearly periodic formation of vortices along the shear layers past the crest as illustrated in Figure 8.35b. However, this vortex formation is not a prerequisite for sloshing excitation as sloshing also occurs for well-rounded crests. Sloshing was brought under control when a vertical dividing wall was inserted and extended at least to point S in Figure 8.35b (Hebaus, 1983).

CHAPTER 9

Examples of structural vibrations

9.1 PRISMATIC BODIES AND GRIDS OF PRISMS (TRASHRACKS)

9.1.1 Transverse and torsional vibrations of prisms

The preceding sections and the IAHR Monograph on Hydrodynamic Forces (Naudascher, 1991) describe a wide range of flow-induced excitations of prismatic bodies. The focus there is on the basic mechanisms, their detection and evaluation. In the following, the focus is on particular types of prismatic structures (e.g., a prism, a beam, a strut, a pylon, or a bridge deck) and the flow-induced vibrations they can undergo depending on their geometry and dynamic characteristics. What complicates the analysis of these vibrations is the wide variety of mechanisms that can excite vibrations either individually or in combinations.

Figure 9.1 depicts the regimes of three types of instability-induced (IIE) and two types of movement-induced (MIE) excitation for rectangular prisms with a *transverse* degree of freedom. If the slenderness ratio e/d is smaller than approximately 3 for smooth flow or 2 for turbulent flow (regime 1, cf. Figure 6.20a), the dominant IIE module is vortex shedding from the leading edge (LEVS). If e/d is greater than approximately 16 (regime 3), IIE is due to vortex shedding from the trailing edge (TEVS). Within the range of $2 < e/d < 16$ (regime 2), the dominant IIE module for an elastically supported prism is impinging leading-edge vortices (ILEV, cf. discussion of Figure 6.20a). The possible MIE modules in regime 1 are either galloping (Figure 7.23b) or wake breathing (Figures 7.37c, 9.13); their occurrence depends on whether the prism is free to vibrate in the transverse or streamwise direction. Vibrations excited by wake breathing or low-speed galloping (Nakamura & Hirata, 1991) are generally of limited intensity.

A typical response diagram for a prism in regime (1) is shown in Figure 9.2. Transverse vibration due to resonance with leading-edge vortex shedding (LEVS) occurs in the vicinity of $V_r = 1/\text{Sh}$. Important for the response behavior is the magnitude of the mass-damping parameter

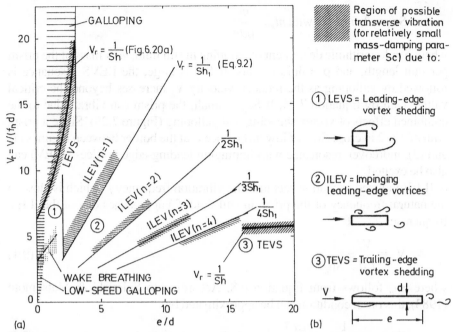

Figure 9.1. (a) Ranges of possible flow-induced vibration for lightly-damped rectangular prisms in low-turbulence cross-flow ($\alpha = 0°$). Cross-hatched and dotted regions represent prisms with transverse and streamwise degree of freedom, respectively. (b) Modes of vortex formation (after Naudascher & Wang, 1993).

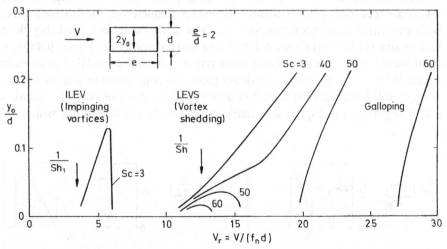

Figure 9.2. Typical response diagram for a rectangular prism with side ratio $e/d = 2$, free to vibrate in the transverse direction (after Washizu et al., 1978): Tu $\simeq 0.3\%$, $Vd/\nu = 0.2$ to 3.3×10^5.

$$Sc_\delta \equiv 2\,m_r\delta \qquad \text{with } m_r \equiv \frac{m}{\rho de}$$

where δ = logarithmic decrement of damping in still fluid, m = mass of the prism per unit length, and ρ = fluid density. If Sc_δ is large, the LEVS resonance is followed by galloping as the reduced velocity V_r increases beyond the critical value given by Equation 7.24. If Sc_δ is small, the prism can vibrate due to the combined effects of vortex shedding and galloping (Figure 7.25). Since a prism with $e/d = 2$ in cross-flow of low turbulence is at the border between regimes (1) and (2), moreover, resonance with impinging leading-edge vortices (ILEV) can also be excited.

ILEV-induced excitation sets in if the vibration frequency f (which is close to the natural frequency of the prism in still fluid, f_n) approaches one of the ILEV frequencies, i.e., if

$$V_r \equiv \frac{V}{f_n d} \simeq \frac{1}{Sh_n} \tag{9.1}$$

where Sh_n follows from Equation 6.8. According to Shiraishi & Matsumoto (1983), this onset condition can be approximated by

$$(V_r)_{cr} \simeq \frac{1}{n\,Sh_1} = \frac{e/d}{0.6}\frac{1}{n}, \quad n = 1, 2 \tag{9.2}$$

for the first- and second-harmonic ILEV resonances (Figure 9.3a).

Figure 9.4 illustrates the mechanism of ILEV excitation for an H-section prism in cross-flow. The shear layers separating from the leading edges roll up into vortices that impinge on the trailing edges of the prism before they reorganize into a vortex street further downstream. The vortex frequency f_o for this stationary-body experiment corresponds to $Sh \equiv f_o d/V = 0.108$, which is remarkably close to the value 0.115 from Figure 6.19b. Proof that the vortices in Figure 9.4a stem from shear-layer or vortex impingement rather than vortex shedding is given in Figure 9.4b: even if the vortex-shedding process is suppressed by a splitter plate, vortices still form with the same frequency f_o. The ranges of possible vibration within regime (2) in Figure 9.1a mark the values of e/d and V_r for which the

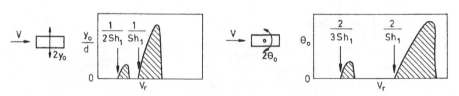

(a) Transverse vibration   (b) Torsional vibration

Figure 9.3. Schematics of response diagrams for rectangular prisms with side ratio e/d between 2 and 8 (after Shiraishi & Matsumoto, 1983).

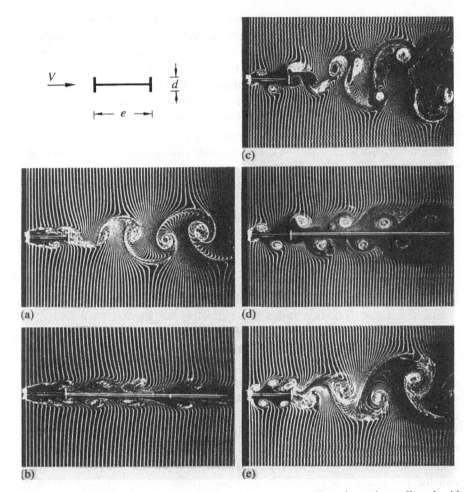

Figure 9.4. Visualization of low-turbulence flow past an H-section prism, aligned with the flow, with side ratio $e/d = 5$ and $eV/\nu \simeq 1200$ (from Nakamura & Nakashima, 1986). (a, b) Stationary prisms with and without splitter plate. (c-e) Prisms vibrating transversely with relative amplitude $y_o/d = 0.19$: (c) $V_r \equiv V/(fd) = 8.65$; (d) $V_r = 8.65$ with splitter plate; (e) $V_r = 4.3$.

vortex shedding is likely to lock-in to the dominant frequency of the impinging leading-edge vortices, or a rational fraction thereof.

The effect of transverse prism vibration on the vortex formation is shown in Figures 9.4c-e. In Figures 9.4c, d, the prism vibrates with a frequency that corresponds to the fundamental ILEV mode (i.e., $fd/V \simeq 0.115$); these figures therefore illustrate conditions of fundamental vortex resonance ($n = 1$). They both show strong vortices being generated just downstream of the leading edges in synchronism with the body movements, and flow patterns downstream that are completely locked-in to the prism vibration. Comparison of the photos in Figures 9.4c, d indicates, furthermore, that the characteristics of the movement-induced leading-edge vortices over the sides of the prism with and without the splitter plate are nearly the same. Figure 9.4e, finally, depicts a prism vibrating with a frequency that is double that of the fundamental mode (i.e., $fd/V \simeq 0.23$). This situation represents a condition of resonance with the second-harmonic vortex formation mode ($n = 2$); two leading-edge vortices instead of one now occur in the shear layers along the sides of the prism. The vortex street at about half the vibration frequency (i.e., $fd/V \simeq 0.115$) indicates that pairs of these vortices are coalescing past the trailing edge.

If a rectangular prism or an H- or T-section prism in regime (2) (Figure 6.19b) has a *torsional* rather than a transverse degree of freedom, it can be excited by impinging leading-edge vortices if the reduced velocity exceeds values according to

$$(V_r)_{cr} \simeq \frac{1}{Sh_1} \frac{2}{2n - 1} = \frac{e/d}{0.6} \frac{2}{2n - 1}, \quad n = 1, 2 \tag{9.3}$$

Here, $V_r \equiv V/(f_n d)$ and (f_n) is the natural frequency of torsional prism vibration. Again, this onset condition holds approximately for modes $n = 1$ and 2, as illustrated in Figure 9.3b. Using a lightly-damped, elastic H-section prism that was allowed to vibrate in its fundamental bending and torsional modes between sections rigidly clamped to force transducers, Schewe (1989) found superharmonic resonances up to the sixth order. Figure 9.5 depicts normalized root-mean-square values of the lift and moment fluctuations from this study. The corresponding bending and torsional vibrations for $n \geq 2$ set in at lower reduced velocities $V/(f_{nB}d)$ and $V/(f_{nT}d)$ than those marking the onset conditions given by Equations 9.1 to 9.3. Also, the higher the order of the vortex-formation mode, the larger is the shift. (Information on the superharmonic resonances, $n = 5/2$ and $n = 3/2$, and the possibility for a weak subharmonic resonance, $n = 1/2$, is given by Schewe, 1989.)

The difference in the onset condition for transverse and torsional prism vibrations (Figures 9.3, 9.5) can be explained with the aid of the vortex patterns and pressure distributions shown in Figure 9.6. Each vortex produces a reduction of pressure on the adjacent surface of the prism. In this way, the developed vortex D [Figure 9.6b (iv)] produces an upward force in combination with the mainly positive pressure on the opposite side. The corresponding transverse force $F_y(t)$ is thus synchronous with the upward prism velocity $\dot{y}$, and the work $F_y\dot{y}$ done on the prism per unit time is positive. The same reasoning applies to the situation in Figure 9.6b (ii), where $F_y\dot{y}$ is again positive. As a consequence, transverse prism vibrations are excited for the condition indicated, i.e., for V_r slightly greater than $1/Sh_1$.

If, with torsional prism vibrations, the up and down movements of the leading edges depicted in Figure 9.6a lead to patterns of leading-edge vortices similar to

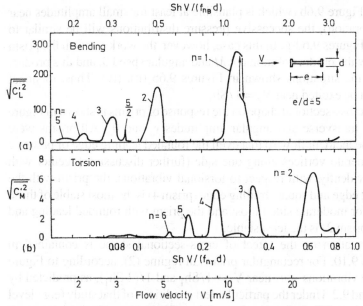

Figure 9.5. (a) Bending and (b) torsional responses of an elastic H beam in a low-turbulence air flow of zero incidence (after Schewe, 1989). $f_{nB}d^2/\nu \simeq 815$; $d = 11$ mm; Sh = 0.115; aspect ratio $L/e = 10.9$. (The H beam vibrated in the fundamental modes characterized by the natural frequencies and damping values $f_{nB} = 100.3$ Hz, $\delta_B = 0.0018$ and $f_{nT} = 401$ Hz, $\delta_T = 0.005$, for bending and torsion, respectively.)

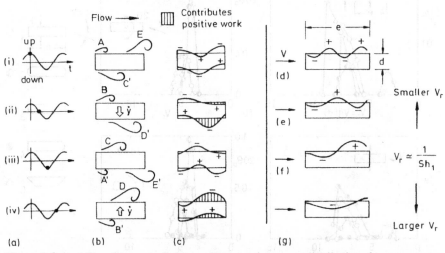

Figure 9.6. (a-c) Development of vortex patterns and pressure distribution along a rectangular prism in cross-flow vibrating transversely for V_r slightly greater than $1/\text{Sh}_1$. (d-g) Distribution of work done by pressure forces during one cycle of transverse prism vibration and its variation with V_r (Komatsu & Kobayashi, 1980).

those shown in Figure 9.6b (which is plausible at least for small amplitudes near the onset of vibration), the successive pressure distributions will be similar to those shown in Figures 9.6d-g. In this case, however, the work done on the prism is determined by the product of moment M and angular speed $\dot{\theta}$, and this product is negative for the situations shown in Figures 9.6b (ii), (iv). Thus, torsional vibration cannot be excited near $V_r = 1/Sh_1$.

The effect of cross-sectional shape on the response of prisms is shown in Figure 9.7 in terms of transverse and angular amplitudes y_o and θ_o. All prisms were inclined 7° to the incident flow so that even the rounded leading edge of prisms 2 and 4 could generate vortices along one side (further discussion together with Figure 9.9c). Evidently, with respect to torsional vibrations, the prism with the rounded leading edge and square trailing edge (prism 4) is the most stable of those tested, at least for moderate side ratios; and the prism with rounded leading and trailing edges (prism 2) is the least stable.

Further information on the effect of cross-sectional shape is contained in Figures 9.8 and 9.10. For rectangular prisms in regime (2), according to Figure 9.8a, transverse vibrations start near $V_r = 1/Sh_1$ and $1/(2\,Sh_1)$, as predicted by Equations 9.1 and 9.2. Under the particular test conditions of that study (e.g., level of approach-flow turbulence rising from 1% to 5% as the velocity increases), vibrations are excited only by specific vortex modes: by the fundamental mode

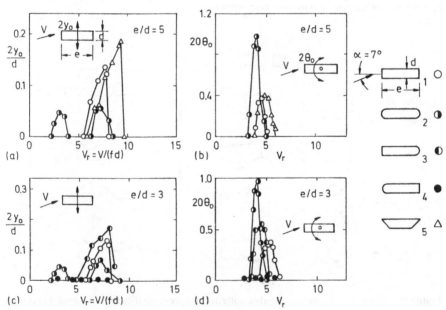

Figure 9.7. Effect of cross-sectional shape on the response of prisms in low-turbulence air flow at $\alpha = 7°$ incidence (after Shiraishi & Matsumoto, 1983). (a, c) Transverse vibration; (b, d) Torsional vibration.

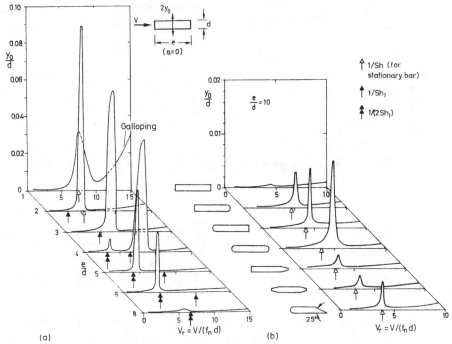

Figure 9.8. Response of prisms that are free to undergo transverse vibration in cross-flow, $\alpha = 0°$ (after Nguyen & Naudascher, 1991). (a) Rectangular prisms with different side ratio e/d; (b) Prisms of various shapes with $e/d = 10$, $\alpha = 0°$, $m_r = 170$. $[f_n d^2/\nu = 200$ where f_n = natural frequency in still water; $d = 10$ mm; Tu = 1 to 5%; aspect ratio $L/e \simeq 10$; δ = const = 0.041 in still water at amplitude $y_o = 0.1$ mm.]

($n = 1$) with $e/d = 2$ and 3, by the second harmonic ($n = 2$) with $e/d = 5, 6$, and 8, and by both with $e/d = 4$. The prism with $e/d = 1$ exhibits the behavior typical of regime (1).

All prisms represented in Figures 9.8b and 9.10 have a side ratio $e/d = 10$, one that is frequently used for trashrack bars. At zero flow incidence, they can be excited by either impinging leading-edge vortices or trailing-edge vortex shedding depending on whether the leading edge is square or rounded. Figure 9.8b shows that the excitation depends strongly on the shapes of *both* the leading and trailing edges: excitation is almost imperceptible for a rectangular prism and most intense for a prism with rounded edges. In contrast to a prism with $e/d = 3$ (Figure 9.7c), the prism with $e/d = 10$ is destabilized rather than stabilized if the leading edge is rounded and the trailing edge square.

The possibility of suppressing prism vibrations by means of side plates is illustrated in Figure 9.9. The principle of this fluid-dynamic vibration control is explained in conjunction with Figure 7.69. Evidently, side plates are effective

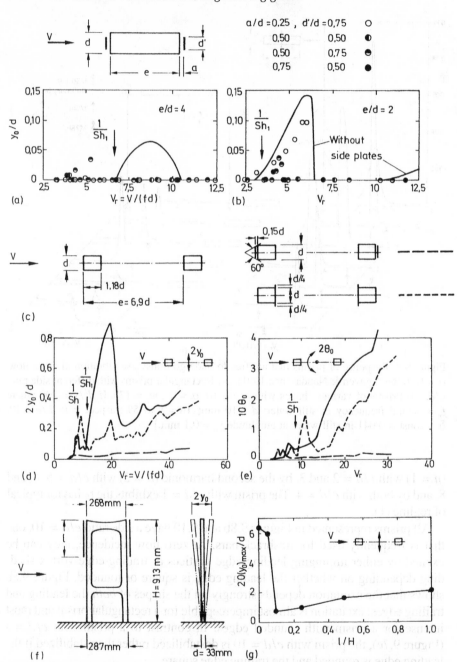

Figure 9.9. (a, b) Effect of side plates on the transvere response of rectangular prisms in low-turbulence air flow at $\alpha = 0°$ incidence (after Nakajima & Kobayashi, 1983); (c) Tandem prisms forming a rigid frame; (d, e) Effect of side plates on transverse and torsional response of these tandem prisms in low-turbulence air flow at $\alpha = 0°$ incidence; (f, g) Effect of side plates on maximum amplitude of suspension-bridge pylon (after Matsumoto, 1983).

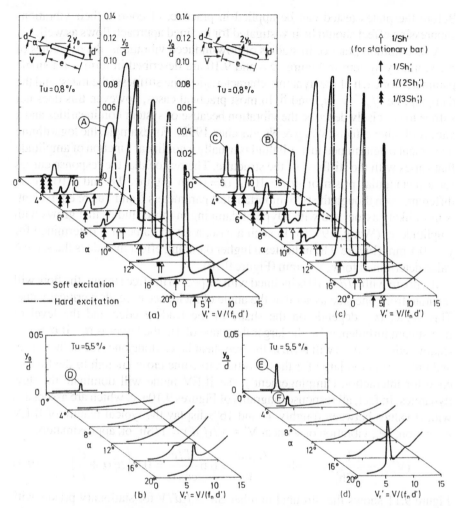

Figure 9.10. Response of rectangular prisms with side ratio $e/d = 10$ that are free to undergo transverse vibration, as a function of angle of incidence α and free-stream turbulence level Tu (after Naudascher & Wang, 1993). (a, b) Prism with square and (c, d) rounded edges. $[f_n d^2/\nu \simeq 35$ to 140, where f_n = natural frequency in still water; $d = 5$ mm; aspect ratio $L/e \simeq 10$; $m_r = 40$; δ = const = 0.041 in still water at amplitude $y_o = 1$ mm.]

control devices with respect to *both* self-excited and vortex-induced vibrations of prisms in cross-flow, as long as their geometry (a/d and d'/d) is properly selected to produce separating streamlines that attach tangentially to the sides of the prism. Even in the case of a suspension-bridge pylon, transverse as well as torsional vibrations, due to a combination of galloping and vortex excitation, were practically eliminated by side plates (Figures 9.9d, e); this was accomplished, surprisingly, with just the upper 10% of the pylon covered by such plates (Figure 9.9g).

Before the plates tested can be applied in practice, of course, their vibration-suppressing effect should be investigated for inclined approach flows as well.

As is common practice in studies of flow-induced vibrations, the dynamics of the vibratory system in Figures 9.8 and 9.10 are described in terms of a single parameter for each of the dynamic characteristics: the stiffness; the mass; and the damping; i.e., $f_n d^2/v$, m_r, and δ. In most practical cases, however, this does not suffice to concisely describe the vibration because of system nonlinearities and a variety of other effects (Wang & Naudascher, 1992). For example, the logarithmic decrement δ determined in a still fluid is usually a nonlinear function of amplitude that varies with the stiffness of the structure. The reason that the response curves for a rectangular prism of $e/d = 10$ at $\alpha = 0°$ in Figures 9.8 and 9.10 are so different, even though the two important test parameters δ and Tu are equivalent, is most likely associated with a type of damping in those studies that grows with amplitude y_o (Wang, 1992b). In such a case, a given value of δ determined for $y_o = 0.1$ mm (Figure 9.8) indicates a higher damping effect than does the same δ value determined for $y_o = 1$ mm (Figure 9.10).

If a prism with $e/d = 10$ is inclined with respect to the free stream, the flow will separate from one side as soon as the angle of incidence exceeds a critical value. This value, α_{cr}, depends on the shape of the leading edge and the level of free-stream turbulence Tu; the larger the value of Tu, the larger is α_{cr}. If $\alpha > \alpha_{cr}$, leading-edge vortices will form in the free shear layer along one side of the prism; and unless α is too large for these vortices to come close enough to the trailing edge for interaction ('impingement'), the ILEV mode will dominate the flow dynamics. In fact, all response diagrams of Figures 9.10a, c, which are associated with α values between roughly 4° and 16°, display the typical feature of ILEV excitation, i.e., vibration onset near $V'_r \equiv V/(f_n d') = 1/Sh'_n$ or, approximately,

$$(V'_r)_{cr} \simeq \frac{1}{n\,Sh'_1}, \qquad Sh'_1 \equiv \frac{(f_o)_1 d'}{V} \simeq 0.6\,\frac{d'/e}{\cos\alpha} = 0.6\left(tg\,\alpha + \frac{d}{e}\right) \qquad (9.4)$$

Figure 9.11 shows the Strouhal number $Sh' \equiv f_o d'/V$ for stationary prisms with $e/d = 10$, where f_o is the vortex frequency measured along one of the near-wake boundaries and d' is equal to $e\sin\alpha + d\cos\alpha$. Evidently, Sh' is rather sensitive to the particular test conditions (level and scale of turbulence, Reynolds number, aspect ratio L/e, etc.) and, to some extent, even to the data-reduction technique used; this is particularly so in an intermediate α range in which the kinematics of a weak ILEV mode competes with that of the far-wake vortex formation.

If the impinging leading-edge vortices are strengthened by transverse prism movements, they take over control and cause resonant prism vibrations near their dominant frequencies (Equation 6.8). This result occurs even for $\alpha < \alpha_{cr}$ (Figure 9.10a), and, surprisingly, also for a prism without 'leading edges' (Figure 9.10c), if disturbing effects are small enough. (When Tu changed only from 0.8% to 2.1% in the experiments of Wang, for example, the peaks A and B in Figures 9.10a, c disappeared. On the other hand, when the Reynolds number Vd/v was reduced to

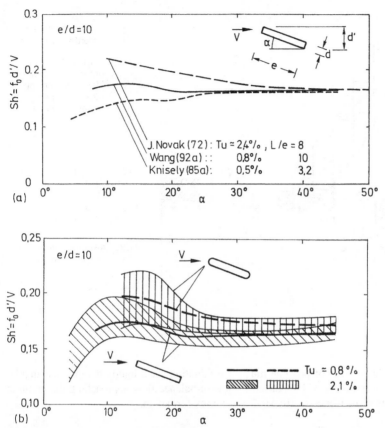

Figure 9.11. Strouhal number for stationary rectangular prisms with side ratio $e/d = 10$ as a function of incidence angle α. (a) Data for square-edged prism; (b) Data for prisms with square and rounded edges. [Re $\equiv Vd/\nu \simeq 1000$ to 2500, 1500, and 2000, respectively, in the experiments of J. Novak (1972), Wang (1992a), and Knisely (1985a).]

the extremely low value of 340, an additional peak, associated with the fundamental ILEV resonance near $V_r' = 1/Sh_1'$, showed up in the resonance curve for the prism with rounded edges at $\alpha = 4°$.) For a square-edged prism in smooth flow, the ILEV mode controls the excitation of transverse prism vibrations down to zero flow incidence.

With a turbulence level of Tu = 5.5% near the prisms, produced by a biplane grid of 1 cm × 1 cm bars with 4 cm × 4 cm mesh size, located 75 cm upstream of the test section, the prisms exhibited only the typical turbulence-buffeting response within the intermediate ranges of α. The remaining peaks in Figures 9.10b, d are due to either trailing-edge vortex shedding at small α values (TEVS, Figure 9.1b) or alternate-edge vortex shedding for $\alpha \le 20°$ (AEVS, Figure

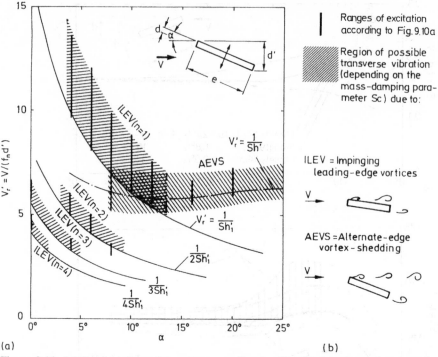

Figure 9.12. (a) Ranges of possible transverse vibration for a lightly-damped rectangular prism with side ratio $e/d = 10$ induced by low-turbulence flow approaching the prism at various angles of incidence α. (b) Modes of vortex formation.

9.12b). In contrast to the data related to ILEV and AEVS in Figure 9.10, the data related to TEVS (peaks C-F) cannot be generalized; the test Reynolds number was so small that the trailing-edge separation was laminar even with Tu = 5.5%. For a practical case with high Reynolds number, the separation will be turbulent, the Strouhal number will be large (e.g., Sh ≈ 0.265 instead of 0.178 for $\alpha = 0°$ according to Wang, 1992a), and excitation will be reduced.

Figure 9.12 depicts the ranges of possible flow-induced vibration for a rectangular prism with side ratio $e/d = 10$ as a function of incidence angle α in a way that is equivalent to Figure 9.1. The sources of vibration are, clearly, impinging leading-edge vortices for $\alpha \le 8°$ and alternate-edge vortex shedding for α values larger than about 14°. In the α range between roughly 8° and 14°, the mode of vortex excitation changes from the one to the other. Over part of this transition range, the velocity limit of vortex-induced vibration depends on whether vibration develops from rest or upon an initial push, as shown in Figure 9.10a. Similar soft and hard excitations prevail for a prism with rounded edges (Figure 9.10c).

9.1.2 *Streamwise and plunging vibrations of prisms*

Square and rectangular prisms can also be excited to vibrate in the streamwise direction. A typical example for this type of vibration is shown in Figure 9.13. The vortex shedding observed during streamwise vibration alternates between anti-symmetric and symmetric (Figures 9.13a, b). During the antisymmetric shedding mode, which was dominant at a ratio of approximately 3:1 for the case depicted, each side of the prism sheds one vortex every two vibration cycles. During symmetric shedding, it sheds one pair of vortices during each cycle. In both cases, the exciting force is due to the movement-induced mechanism of wake breathing (Section 7.3.5). From a plot of the phase-averaged distances $\hat{\Delta}$ between the free shear layer and the downstream corner of the prism, averaged over the two sides of the prism (see dash-dotted lines in Figures 9.13c-e), one finds that the wake

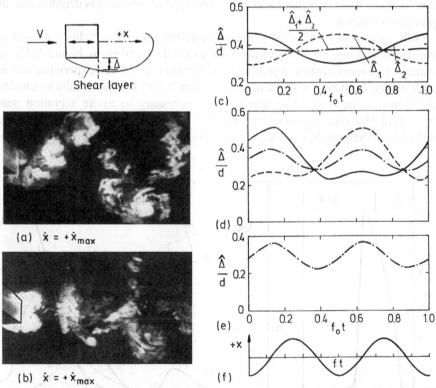

Figure 9.13. Streamwise vibration of a square prism in low-turbulence cross-flow ($\alpha = 0°$) according to Naudascher (1987b). (a) Antisymmetric and (b) symmetric vortex shedding for a freely-vibrating prism [$V/(fd) = 4$, $(x_o)_{rms} = 0.03d$, $Vd/v = 4.3 \times 10^3$]. (c-e) Phase-averaged distance $\hat{\Delta}$ (shown in sketch) during one period $T_o = 1/f_o$ of far-wake vortex formation for (c) stationary prism; (d, e) vibrating prism during (d) antisymmetric and (e) symmetric vortex shedding. (f) Prism displacements corresponding to (d, e).

width varies almost sinusoidally for both shedding modes. This breathing motion is phased with the prism motion (Figure 9.13f) in such a way that the near-wake reaches a maximum when the prism approaches its maximum velocity during the downstream stroke, and a minimum during the upstream one. In this way, an exciting force is generated that is much like the one described for the circular cylinder in Section 7.3.5.

Figure 9.14a presents the response diagram for a lightly-damped square prism that was excited by wake breathing in a water flume. The prism, which was free to vibrate in the direction of flow, reached maximum amplitudes near $V_r = 1/(2\,\text{Sh})$, corresponding to a flow velocity about half that at which transverse vibrations were excited by leading-edge vortex shedding (cf. Figure 9.1a). If the flow approaches the prism at larger incidence angles (Figure 9.14b), the peak response of vibration in the x direction shifts towards larger values of V_r, and the amplitudes increase substantially. For $\alpha = 45°$, for example, the maximum amplitude is about five times that for $\alpha = 0°$. This type of vibration is denoted here as plunging vibration.

Still greater amplitudes of plunging vibration occur for inclined prisms of elongated rectangular cross-section. An example is given in Figure 9.15. In contrast to a square prism, a prism with side ratio $e/d = 5$ can experience soft as well as hard excitation at incidence angles of $\alpha \geq 30°$. For the conditions cited in the figure, a push of finite amplitude is necessary to excite vibration near $V_r = 1/(2\,\text{Sh})$, whereas vibration near $V_r = 1/\text{Sh}$ starts from rest. For reduced velocities increasing beyond $V_r \simeq 30$, successively-larger values of initial ampli-

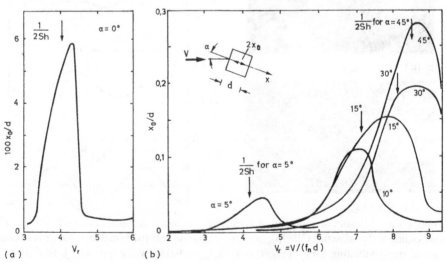

Figure 9.14. Response of a square prism in a free stream (after Callander, 1987). (a) Streamwise vibration; (b) Plunging vibration (in x-direction, as shown in sketch). [$f_n d^2/\nu \simeq 1250$; $d = 40$ mm; Tu $\simeq 1\%$; $m_r = 7.1$; $\delta = 0.054$ in still water.]

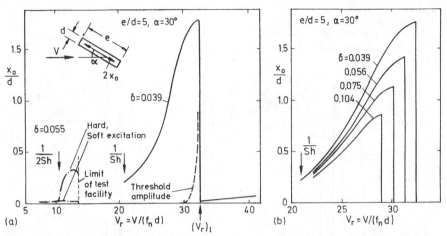

Figure 9.15. Response of a rectangular prism with side ratio $e/d = 5$, free to undergo plunging vibration in a free stream (after Callander, 1987). [——$f_n d^2/\nu \simeq 85$, $m_r = 21$; Tu = 1%; $-\cdot-\cdot- f_n d^2/\nu \simeq 650$, $m_r = 4.9$, Tu = 1%; δ = logarithmic decrement in still water.]

tude are necessary for vibration to occur; finally, beyond some limiting value $(V_r)_l$, even hard excitation is impossible. A further distinction with respect to plunging vibration of the square prism shown in Figure 9.14 is that $V_r = 1/\mathrm{Sh}$ marks the vibration onset rather than the peak response. Vortices are shed alternately from opposite front and rear corners in both vibration ranges (AEVS). However, one of the alternating vortices forms every two cycles of prism vibration in the first range and every cycle in the second. The former type of vortex excitation corresponds to a superharmonic resonance of second order (more details are presented by Callander, 1987). The effect of damping for the case of fundamental resonance is shown in Figure 9.15b: whereas $(V_r)_l$ diminishes with increasing damping, the onset of vibration occurs consistently near $V_r = 1/\mathrm{Sh}$.

Plunging-response diagrams for rectangular prisms with a side ratio $e/d = 10$ are shown in Figures 9.16 and 9.17a. In these diagrams, the reduced velocity and the Strouhal number are based on the effective width d' rather than the actual width d so that the value of Sh′ would be constant, at least in the range of α from 30° to 45° (Figure 9.11a). The vibration behavior is similar in part to that for $e/d = 5$: again, the prisms are excited from rest as the vortex-shedding frequency f_o for a stationary prism exceeds the still-water natural frequency f_n, or as V_r' exceeds $1/\mathrm{Sh}' \equiv V/(f_o d')$; an initial push of larger and larger amplitude is necessary to trigger prism vibration as V_r' is increased; and vibration ceases beyond some limiting reduced velocity. In contrast to the case with $e/d = 5$, however, it was impossible to stimulate superharmonic resonance near $V_r' = 1/(2\,\mathrm{Sh}')$ for any of

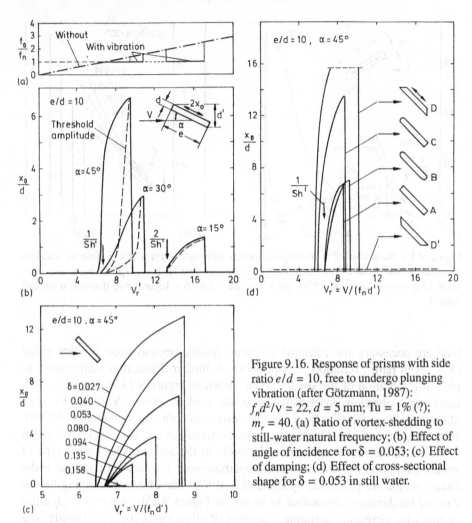

Figure 9.16. Response of prisms with side ratio $e/d = 10$, free to undergo plunging vibration (after Götzmann, 1987): $f_n d^2/\nu \simeq 22$, $d = 5$ mm; Tu = 1% (?); $m_r = 40$. (a) Ratio of vortex-shedding to still-water natural frequency; (b) Effect of angle of incidence for $\delta = 0.053$; (c) Effect of damping; (d) Effect of cross-sectional shape for $\delta = 0.053$ in still water.

the cases illustrated in Figures 9.16 and 9.17, probably because the mass damping in the latter cases was higher. With an incidence angle of 15°, no excitation occurred near $V_r' = 1/\mathrm{Sh}'$, and hard excitation commenced at about $V_r' = 2/\mathrm{Sh}'$. The last expression marks the subharmonic resonance condition for which $f_o = 2f_n$ (Figure 9.16a). The effect of damping on the plunging response was found to be similar to that for $e/d = 5$ in Figure 9.15b. As damping increased from $\delta = 0.027$ to 0.135 for $\alpha = 45°$, the maximum amplitude diminished by an order of magnitude, and for $\delta = 0.157$, the excitation ceased (Figure 9.16c). For all values of δ tested in conjunction with $\alpha = 45°$, vibration started almost exactly at $V_r' = 1/\mathrm{Sh}'$, and the vibration range decreased gradually with increasing δ.

Figure 9.17 shows the plunging response of prisms with side ratio $e/d = 10$ for a variety of flow-incidence angles. For the conditions tested, large-amplitude vibrations of prisms with $\alpha = 31°$, $21°$, $17°$, and $13°$ were obtained as V'_r increased beyond, approximately, values of $1/\text{Sh}'$, $1.5/\text{Sh}'$, $2/\text{Sh}'$, and $3/\text{Sh}'$. From the corresponding onset condition $V'_r \simeq 1/(n\,\text{Sh}')$ with $n = 1, 2/3, 1/2, 1/3$, one can conclude that plunging vibrations of prisms with elongated sections are mainly induced by fundamental or subharmonic resonance with alternate-edge

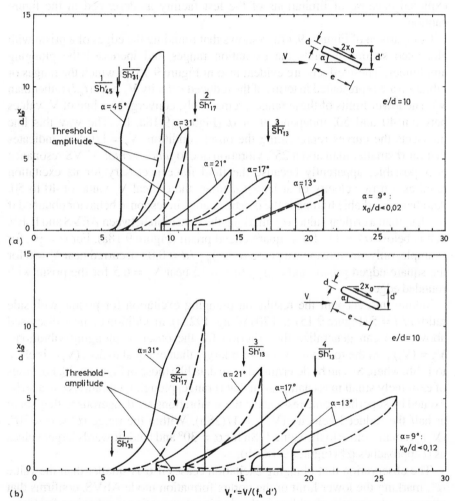

Figure 9.17. Response of rectangular prisms with side ratio $e/d = 10$, free to undergo plunging vibration (after Naudascher & Wang, 1993). (a) Rectangular prism, $\delta = 0.020$; (b) Prism with rounded edges, $\delta = 0.018$. [$f_n d^2/\nu \simeq 22$; $d = 5$ mm; Tu = 0.8%; aspect ratio $L/e \simeq 10$; $m_r = 40$; δ = logarithmic decrement in still water at amplitude $x_o = 20$ mm. Because of limitations of the test facility, the V'_r range could be explored only up to: 11.5 for $\alpha = 45°$; 13.7 for 31°; 27.7 for 13°; and 34.2 for 9°.]

vortex shedding. For incidence angles smaller than about 22°, only hard excitation occurred. In the V_r' range from 16 to 19, the square-edged prism at $\alpha = 13°$ was quite sensitive to small amplitude modulations in the sense that vibrations subsided when the amplitude was slightly reduced (Figure 9.17a). Two ranges of excitation occurred for the prism with rounded edges at $\alpha = 13°$ (Figure 9.17b), one commencing near $V_r' = 3/Sh'$ and one near $V_r' = 4/Sh'$. Whether subharmonic resonances of higher order exist at other incidence angles was not fully explored because of limitations of the test facility as described in the figure caption.

Comparison of Figures 9.17a, b shows that rounding the edges of a prism with elongated section extends the excitation ranges and increases the plunging amplitudes. These effects are evident also in Figure 9.18, in which the ranges of vibration are represented in terms of the reduced velocity $V_r \equiv V/(f_n d)$ rather than V_r'. The lower limits of these ranges, remarkably, converge to a line of V_r values between 40 and 50, independent of α (Figures 9.18a, b). The way this line intersects the curves representing the onset condition, $V_r = 1/(n Sh)$, indicates that for α smaller than about 25°, vibration due to *fundamental* AEVS resonance is impossible, apparently because the fluid force necessary for its excitation requires a flow velocity of at least $f_n d$ times the critical V_r value of 40 to 50. Another remarkable finding is the drastic change in response behavior obtained if α is less than a critical value, α^*, marking the transition between AEVS and ILEV, that is below $\alpha^* \simeq 11°$ for a square-edged prism (Figure 9.18a). For $\alpha = 9°$, for example, only small response peaks of $(x_o)_{max}/d \simeq 0.02$ occurred near $V_r = 6$ for the square-edged prism, and $(x_o)_{max}/d \simeq 0.12$ near $V_r = 6.5$ for the prism with rounded edges.

Taking into account the results on plunging excitation for prisms with side ratios $e/d = 5$ (Figure 9.15) and 20 (Wang, 1922a), in addition to those discussed above, one can generalize the criterion for the onset of plunging vibrations, $V_r = (V_r)_{cr}$, in the following way. For α larger than or equal to 30°, $(V_r)_{cr}$ is close to $1/Sh$, where Sh can be determined from data like those in Figure 6.20. For cases of extremely small mass damping in that α range, plunging vibrations can also be excited by superharmonic resonance; these vibrations are dangerous as they set in at half the reduced velocity $(V_r)_{cr} \simeq 1/(2Sh)$. Within the range $\alpha^* < \alpha < 30°$, $(V_r)_{cr}$ remains close to the value $1/Sh$ for $\alpha = 30°$ and tends towards larger values as α approaches α^* (Figures 9.18a, b).

The observation that plunging excitation ceases if α is smaller than the value α^*, marking the lower limit of the vortex formation mode AEVS, confirms that alternate-edge vortex shedding is the dominant excitation mechanism for large-amplitude plunging vibrations. However, as flow visualization indicates (Figure 9.19), a role in the excitation is also played by a movement-induced variation of the instantaneous angle of incidence α_i during the vibration cycle, much as for stall flutter. Depending on the shapes of the leading and trailing edges, the fluid forces induced by these two mechanisms can vary drastically, and with them the

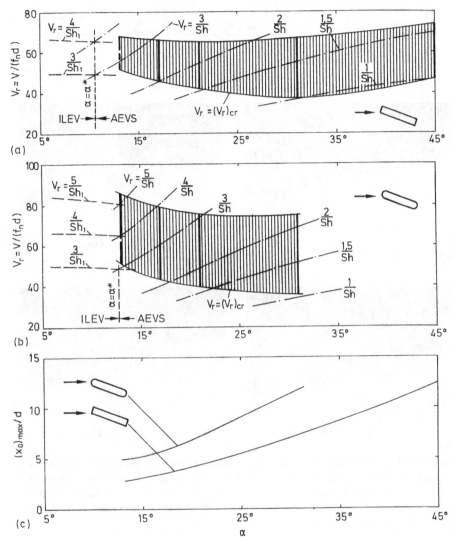

Figure 9.18. (a, b) Ranges of plunging vibration of rectangular prisms ($e/d = 10$) with mass-damping parameter $2m_r\delta$ between 1.44 and 1.60, induced by low-turbulence flow approaching the prism at various angles of incidence α. (c) Corresponding maxima of relative vibration amplitude.

body response (Figure 9.16d). For a prism with a cross-sectional shape marked D, the exposure of the front and rear faces of the prism to the wake region of large pressure variation is a maximum and leads to the largest amplitudes. For the same prism turned around (shape D'), this exposure is zero and no excitation occurs.

The effect of turbulence on the AEVS-induced plunging vibration appears to

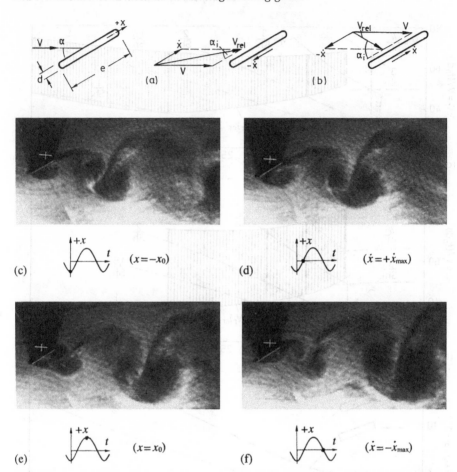

Figure 9.19. (a, b) Schematic of movement-induced changes in the relative approach-flow velocity V_{rel}. (c-f) Consecutive photos of hydrogen-bubble flow patterns obtained with one-sided illumination during a cycle of plunging vibration of a rectangular prism with rounded edges ($e/d = 10$) in a free stream at $\alpha = 31°$ incidence (after Naudascher & Wang, 1993). [$Vd/v \simeq 970$, $V_r' \simeq 7.6$, $x_o/d \simeq 1.3$.]

be similar to that on the well-known transverse vibration induced by leading-edge vortex shedding: turbulence essentially weakens the excitation but does not suppress it. This finding is not surprising in view of the fact that there is a gradual transition from the first to the second type of vibration as the angle of incidence is increased beyond $\alpha = 45°$.

Figure 9.20 depicts the wake-flow structure for an elongated prism with rounded edges at incidence angles of $31°$ and $17°$. The pairs of photographs a, b, and c, d, were each taken for identical flow conditions except that the prism was restrained in the left pair of photos and free to vibrate in the right one. Compari-

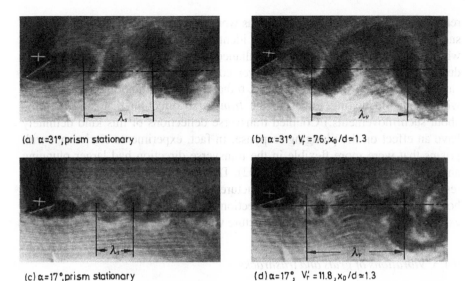

(a) α=31°, prism stationary

(b) α=31°, V'_r = 7.6, x_0/d ≈ 1.3

(c) α=17°, prism stationary

(d) α=17°, V'_r = 11.8, x_0/d ≈ 1.3

Figure 9.20. Hydrogen-bubble pictures of wake-flow structure obtained with one-sided illumination for an elongated prism (e/d = 10) with rounded edges (after Naudascher & Wang, 1993). (a, b) α = 31°, Vd/v ≃ 970; (c, d) α = 17°, Vd/v ≃ 970.

sons of the left and right pictures show clearly that the vortices in the wakes are stronger and more regular for the vibrating prism. The fact that every single pair of alternating vortices is amplified during a cycle of prism vibration for α = 31° and only every second pair for α = 17° is in perfect agreement with the corresponding onset conditions V'_r ≃ 1/Sh′ and V'_r ≃ 2/Sh′, respectively (cf. Figure 9.17b). These first and one-half harmonic resonance conditions change the vortex patterns drastically: whereas the ratio of wavelengths λ_v/λ_s for a vibrating (λ_v) and a stationary prism (λ_s) is close to unity for α = 31° (Figures 9.20a, b), it is close to two for α = 17°. (More precisely, λ_v/λ_s is equal to about Sh′V'_r due to the lock-in of the vortex frequency to f_n.) Compared to the vortex street past the *stationary* prism at incidence angle 17°, in other words, the vortex pattern past the vibrating prism contains only every second vortex pair (Figure 9.20d). It appears quite plausible that this strong deformation of the normal vortex street (Figure 9.20c) requires hard excitation with ever larger push amplitudes as V'_r increases beyond the onset value.

For all tests that led to the results depicted in Figures 9.14 to 9.20, the Reynolds number Vd/v, or the parameter $f_n d^2/v$, had no effect on the response behavior. The tests were performed in a flume 1 m wide and with water depths of 0.5 m to 0.6 m. The effect of flow confinement exerted by the flume walls was negligible for the cases reported. When the flow was more closely confined, this effect gave rise to substantial movement-induced excitation involving fluid-flow pulsations at relatively low values of V_r. The prisms were fixed to an oscillating platform in such a way that they would undergo identical translatory vibration over their whole length when excited. Small simultaneous flexural movements of the prisms during their plunging vibration, however, could not be avoided completely. (One can easily deduce from Figures 9.19a, b that plunging vibrations are associated with strong fluctuating forces in the *transverse* direction that tend to bring about such movements.) Coupled transverse deflections of this kind definitely have an effect on the plunging response. In fact, experiments carried out with prisms that were more flexible in the transverse direction had larger plunging amplitudes (Wang & Naudascher, 1992). The observation has significance with respect to trashrack vibrations: any structure composed of bars that can vibrate in both the streamwise *and* transverse directions is more susceptible to flow-induced excitation than is a corresponding structure with only one degree of freedom.

9.1.3 *Vibrations of grids and trashracks*

Grids of adjacent prisms or bars are used in a variety of structures exposed to gas or liquid flow. They are utilized, for example, as headlight screens to separate traffic on dual carriage motorways or as trashracks to prevent trash from entering a conduit or canal used for water supply or a duct leading to a turbine or pump. Such grids consist typically of panels composed of a series of vertical prisms or bars and several horizontal members provided to stiffen the individual bars and to maintain equal distances between them. In most cases, the ratio of the bar's longitudinal to its transverse dimension is large. The individual bars are hence more susceptible to transverse than to plunging excitation. Contrary to general opinion, however, the grid panels as a whole can well be excited to vibrate in the plunging mode because the panel stiffness in the streamwise direction is usually small.

In pumped-storage systems, trashracks are often used at draft-tube outlets where they are particularly vulnerable to vibration failure. In fact, most failures due to rack vibration reported in recent years occurred at pumped-storage facilities (cf. Rao, 1989 and Section 9.1.3c). Trashracks in such systems are subject to two-way flows and to velocities that can readily reach values two or three times the standard design value of about 1 m/s; local values can be even ten times greater (Hamilton, 1982).

Among the causes of grid or trashrack vibrations are:
- buffeting due to approach-flow turbulence or other extraneous sources of excitation (such as bar interference at large local flow incidence angles, turbine-load changes, draft-tube surges),
- resonance with vortex shedding or impinging vortices, and
- galloping, wake breathing, and the plunging-vibration mechanism as described in the preceding section.

(a) *Transverse vibration.* Transverse vibrations of bars in a grid have been shown to be almost identical to those of isolated bars, no matter whether the bars vibrate by themselves or in unison (Nguyen & Naudascher, 1991). Therefore one can assess a trashrack on the basis of the detailed information given in Figures 6.20, 9.1, 9.2, and 9.7 to 9.12. To do this, one needs primarily to modify the vibration-onset conditions.

The onset conditions for excitation by impinging leading-edge vortices (ILEV), which, for the isolated prism or bar, was presented in terms of the Strouhal number Sh'_1 in Equation 9.4, can be evaluated for a grid composed of a row of such bars by replacing the approach-flow velocity V by the effective velocity V_{eff} between the bars. For adjacent bars with spacing s and angle of incidence α, the ratio V/V_{eff} can be estimated as

$$\frac{V}{V_{eff}} \simeq 1 - \frac{d'}{s \cos \alpha}, \qquad d' = e \sin \alpha + d \cos \alpha \qquad (9.5)$$

so long as the confinement by bar spacing and inclination is not too large. Assuming V_c/V_{eff} to remain approximately equal to the value 0.6 observed for the isolated bar, one obtains the grid Strouhal number for the fundamental ILEV mode ($n = 1$) as

$$(Sh'_1)_{grid} \equiv \frac{(f_o)_1 d'}{V} \simeq \frac{0.6 \, d'}{e(\cos \alpha - d'/s)} \qquad (9.6)$$

The effect of relative bar spacing s/d on the general Strouhal number $Sh \equiv f_o d/V$ for a grid composed of bars with rounded edges and side ratio $e/d = 3$ is depicted in Figure 9.21. This effect is similar to the flow-confinement effect obtained for a bar in a channel of finite width s, as a comparison of the two curves in the figure shows.

(b) *Plunging and streamwise vibration.* Plunging vibrations of grids or trashracks composed of bars of elongated cross-section are possible wherever the flow approaches the bars at some incidence angle. Small angles of flow incidence are almost impossible to avoid in practice, and large angles can occur locally near side piers or accumulated trash (Figure 9.22), or in cases of intake racks to a row of only partially operating turbines. If part of the rack is clogged with trash or obstructed by a tree trunk, for example, the flow is not only deflected but its velocity is larger in the unclogged parts; both effects increase the likelihood of excitation. Moreover, evidence of fatigue damage due to bending stresses in rectangular trashrack bars is no reason to exclude plunging vibration as the cause; the latter generates strong cyclic forces on the individual bars that act transverse to the vibratory motion and cause periodic bending stresses.

The plunging response of a grid or trashrack composed of rectangular bars is quite similar to that of an isolated bar with comparable geometric and dynamic characteristics. A comparison of Figures 9.16b and 9.23 shows that the isolated

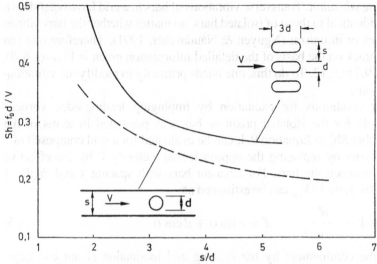

Figure 9.21. Effect of flow confinement on Strouhal number. ——— Grid or trashrack composed of rectangular bars with rounded edges (after Crausse; see Levin, 1957): $e/d = 3$, $\alpha = 0$; – – – – Circular cylinder placed in rectangular duct (cf. Figure 6.18b).

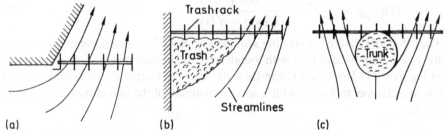

Figure 9.22. Situations leading to locally-inclined approach flows incident upon trashracks.

bar and the grid with the bars spaced at $s/d = 20$ both start to vibrate as V_r' exceeds a critical value, and both cease to vibrate abruptly at a still larger value of V_r'. If the reduced velocity is determined in terms of the velocity of the flow passing the adjacent bars, that is, approximately by V_{eff} given by Equation 9.5, then even the onset conditions $V_r' = 1/(n\text{Sh}')$ (with $n = 1$ for $\alpha = 30°$ and $45°$) are nearly fulfilled. The vibration frequency of the grid was about 96% to 99% of the natural frequency f_n, measured in still water, throughout the ranges of resonance (Wang, 1992a). For $\alpha = 15°$ and $s/d = 10$ and 20, only low-level buffeting occurred; it was similar to that for random turbulence.

Increased interference due to a decrease in the relative bar spacing s/d, in general, reduces the plunging response amplitude and increases turbulence buffeting. At sufficiently large bar spacings ($s/d > 10$), the plunging excitation

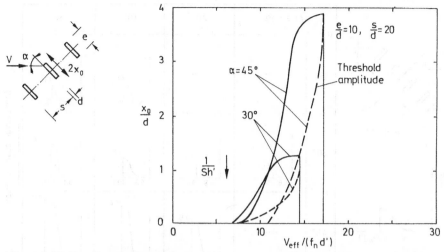

Figure 9.23. Plunging response of inclined grid or trashrack composed of rectangular bars of side ratio $e/d = 10$ in a free stream (after Rao et al., 1987). [$f_n d^2/\nu \simeq 25; d = 5$ mm; Tu = 3% (?); $m_r = 40; \delta = 0.069$ in still water.]

becomes stronger with increasing incidence angle. For relatively small values of s/d, however, this effect is counteracted by the interference of the individual near-wakes. A summary of these effects is shown in Figure 9.24. With the aid of this figure, it is possible to estimate the danger of plunging grid vibration on the basis of data for an isolated bar of comparable geometric and dynamic characteristics (Figures 9.14-9.18).

Figure 9.25 shows the effect of partial obstruction of a grid or trashrack that was placed at right angles in a laboratory flume of 1 m width. The effective angle of flow incidence is a maximum near the obstructing plate (near bar 1 in Figure 9.25a) and diminishes towards zero away from the obstruction. Consequently, the wake-flow structure differs widely for bars in various locations. As shown in Figure 9.25b, the plunging response in this case depends on the damping, δ, and on the number of bars that are so interconnected as to move in unison. Even though δ increased as more bars were interconnected, plunging vibrations were found to set in at about $\overline{V}/(f_n d) \simeq 20$ in all cases tested, where $\overline{V}$ is the average mean velocity over the unobstructed part of the grid. In terms of the average velocity V_a approaching the grid, in other words, the reduced velocity at vibration onset, $(V_r)_{cr}$, is roughly 10, compared to the values between 40 and 50 for an unclogged grid (Figure 9.18). The results illustrate the increased susceptibility of trashracks to plunging excitation due to trash accumulation even in cases in which the grid or trashrack is placed at right angles to the flow. In any case, it is essential to use the effective velocity near the trashrack bars when one determines the onset value $(V_r)_{cr}$ of the reduced velocity.

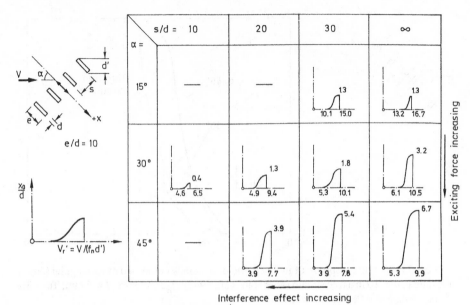

Figure 9.24. Effect of relative bar spacing s/d and angle of incidence α on the plunging response of a grid or trashrack composed of rectangular bars of side ratio $e/d = 10$. [$f_n d^2/\nu \simeq 25$; $d = 5$ mm; Tu = 3% (?); $m_r = 40$; $0.04 < \delta < 0.07$ in still water.]

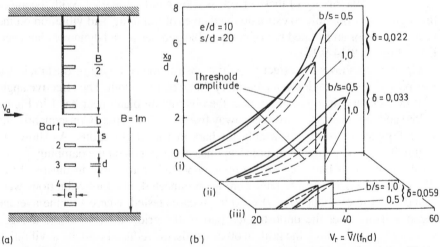

Figure 9.25. Excitation of bars of a grid that is partially obstructed by a plate (after König, 1986) [$f_n d^2/\nu \simeq 25$; $d = 5$ mm; Tu = 3% (?); $\overline{V} \simeq 2V_a$]. (a) Sketch of test arrangement; (b) Plunging response if (i) bar 1, (ii) bars 1 and 2 interconnected, (iii) bars 1 through 4 interconnected.

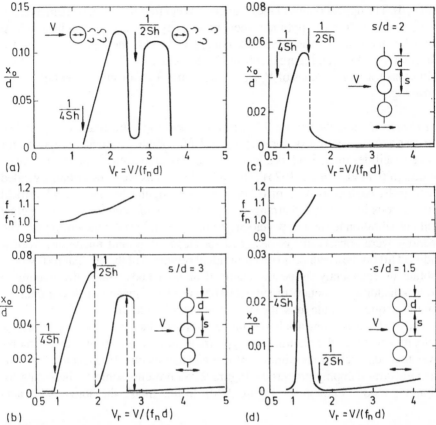

Figure 9.26. Streamwise response of circular rods in low-turbulence cross-flow. (a) Single rod (after King, 1974) [$f_n d^2/v \simeq 2000$; Tu = 1%; $Sc_\delta = 0.32$ in still water]. (b-d) Grids of different rod spacings (after Callander, 1988, and Naudascher, 1987b) [$f_n d^2/v \simeq 3600$; Tu $\simeq$ 1%; Sc = 0.22].

Even though *streamwise vibration* of bars excited by wake breathing (Figure 9.13) has rarely been observed in grid or trashrack installations, it can occur, and it can even cause structural damage in cases in which the stiffness and damping of the grid are low. Wake-breathing excitation can reach a dangerously high intensity for rods with circular cross-section. According to Figure 9.26a, streamwise vibration of an isolated rod is possible in two ranges, one starting near $V_r = 1/(4Sh)$ and involving symmetric vortex shedding, and the other starting near $V_r = 1/(2Sh)$ and involving antisymmetric vortex shedding (Naudascher, 1987b). For a *row* of circular cylinders in cross-flow, the streamwise response is quite similar to that for an isolated rod if the relative spacing, s/d, is larger than about 2.5 for the conditions tested (Figure 9.26b). For smaller relative spacings, vibrations are less intense in the first range and do not occur in the second (Figures

9.26c, d). The upper bounds of these ranges, except for $s/d = 1.5$, were marked by either a sudden drop in amplitude when the velocity was increased or by a sudden rise when the velocity was decreased. For $s/d = 3$ and 4, these jumps occurred at different V_r values (cf. Figure 9.26b and Naudascher, 1987b). The low-amplitude vibrations for larger values of V_r shown in Figure 9.26 are caused by turbulence buffeting.

(c) *Practical example*. Figure 9.27 shows a model of a trashrack composed of circular rods for a pumped-storage project. Unfavorable experiences reported for similar facilities made it advisable to test this model thoroughly at an early stage of design (Crandall et al., 1975). The individual racks were, essentially, welded steel grills, each consisting of 6 horizontal rectangular beams and 47 vertical circular rods in a staggered arrangement. The trashrack sections were to operate without vibration for any flow velocity from 1.2 to 4.9 m/s in the prototype. The sections were conservatively designed for steady drag and fluctuating vortex-induced lift on the assumption that individual segments of the circular rods would vibrate independently as beams clamped at both ends. Since the maximum vortex-shedding frequency turned out to be an order of magnitude lower than the natural frequency of the rods, the system was considered safe.

The model tests, using a half-size model, were conducted for the prototype flow velocity and twice the natural prototype frequency. An effort was made, moreover, to model the various vibration modes and their natural frequencies as well as the structural damping properties. Figure 9.28a presents some of these data for plunging-mode vibration (cf. Figure 2.20). In that mode, the horizontal beams

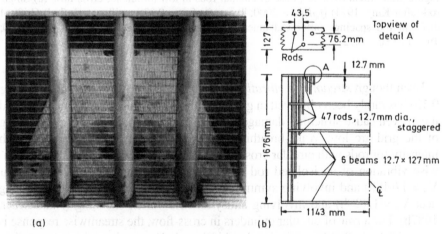

(a)   (b)

Figure 9.27. (a) Front view of model trashrack of Racoon Mountain Project made up of 20 individual racks supported by concrete piers; (b) Front view of individual rack section (half-size model) (after Crandall et al., 1975).

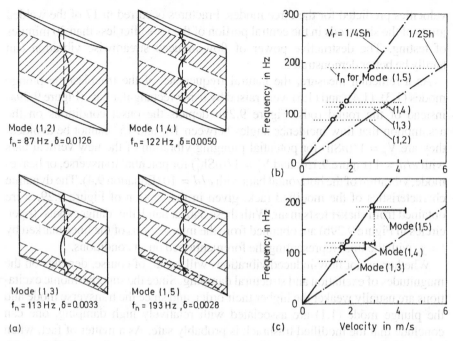

Figure 9.28. Dynamic and response characteristics of model trashrack shown in Figure 9.27 (after March & Vigander, 1980). (a) Sketches of mode shapes for plunging vibration of a trashrack panel; (b) Natural frequencies compared to four-fold and two-fold Strouhal frequencies; (c) Vibration frequency and ⊢——⊣ ranges of vibration. (All frequencies cited pertain to the half-scale model.)

were bent out of the trashrack plane, and they deflected the vertical rods in a way that is depicted for the center rod.

Had the data shown in Figure 9.26 been known at the time, the susceptibility of the system to vibration could have been predicted. Figure 9.26b is relevant to the case considered since $s/d = 43.5/12.7 = 3.4$. It shows that plunging vibrations are possible provided that the structural damping is low enough; these vibrations start near $V_r \equiv V/(f_n d) = 1/(4Sh)$ for the first vibration range and near $V_r = 1/(2Sh)$ for the second. Using a value of $Sh \simeq 0.25$, obtained from Figure 9.21, one can construct a diagram of these onset conditions and indicate the points of vibration onset, $f_n = f_{\text{onset}} = 4ShV/d$, that belong to the first resonance range (open circles in Figure 9.28b). Even without an estimate of the effect of structural damping, one can deduce from this information that the plunge modes (1,3), (1,4), and (1,5), which are associated with extremely low damping (Figure 9.28a), are more likely to be excited than the plunge mode (1,2). The test results depicted in Figure 9.28c bear this out: the trashrack model did in fact vibrate in these three modes as the velocity V increased beyond 1.7 m/s, i.e., the lowest of the onset velocities predicted for the three modes. Fractures occurred in 17 of the welded joints of the short rods in the central portion of the rack after less than 30 minutes of testing. The destructive power of flow-induced streamwise vibration can hardly be better demonstrated.

As a remedial measure, the natural frequencies of the troublesome plunge modes (1,3), (1,4), and (1,5) were raised by using rectangular bars (Figure 9.29a) instead of the round rods. Figure 9.29b depicts the onset conditions on the assumption that flow incidence angles between 10° and 15° cannot be avoided; they are $V_r = 1/(n\mathrm{Sh})$ for potential plunging vibration of the new vertical bars with $e/d = 4$ (Figure 9.15); and $V_r' = 1/(n\mathrm{Sh}_1')$ for potential transverse, or heave-mode, vibration of the horizontal bars with $e/d = 10$ (Equation 9.4). The dynamic characteristics of the modified rack, given in the caption of Figure 9.29, were obtained from shaker tests in air. With these data at hand, the 'danger spots' (open circles in Figure 9.29b) are obtained from the intersections of the lines marked by $f = f_n$ with the lines representing the forementioned onset conditions.

Whether or not flow-induced vibrations will occur, of course, depends on the magnitudes of excitation and structural damping. Since the superharmonic excitations are usually weaker the higher their order, and since the transverse mode and the plunge mode (1,1) are associated with relatively high damping, one can conclude that the modified trashrack is probably safe. As a matter of fact, when

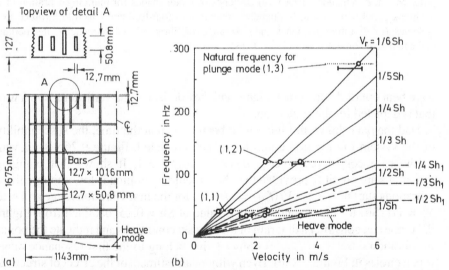

Figure 9.29. (a) Front view of modified trashrack for Racoon Mountain Project (after Crandall et al., 1975). (b) Natural frequencies (dotted lines) compared to onset conditions $V_r = 1/(n\mathrm{Sh})$ for plunge modes (full lines) and $V_r = 1/(n\mathrm{Sh}_1)$ for heave modes (dashed lines); $\vdash\!\!\dashv$ ranges of vibration. [Plunge modes of 12.7 × 50.8 mm bars: mode (1,1) $f_n = 41$ Hz, $\delta = 0.1652$; mode (1,2) $f_n = 119$ Hz, $\delta = 0.0075$; mode (1,3) $f_n = 275$ Hz, $\delta = 0.0060$. Heave mode of 12.7×127 mm bar ($n = 1$): $f_n = 33$ Hz, $\delta = 0.2419$.]

the rack was tested in cross-flow, it responded with extremely small amplitudes in one heave- and two plunge-vibration ranges, as indicated in Figure 9.29b. The corresponding maximum strain in the bars amounted to only 1% of the yield strain. The placement of neoprene-rubber pads between the top of the test rack and the bottom of the adjacent dummy rack completely suppressed these vibrations over the range of investigation, $V \leq 5.5$ m/s. A similar suppression of vibrations was obtained by mounting a baffle plate 152×305 mm at the center of the rack.

9.1.4 *Conclusions and design recommendations*

Contrary to common opinion, the dangerous vortex formation modes for bars in a grid or trashrack are impinging leading-edge vortices (ILEV) and alternate-edge vortex shedding (AEVS) rather than leading-edge or trailing-edge vortex shedding (LEVS or TEVS). ILEV is the main source of transverse bar vibration, and AEVS may excite both transverse and plunging vibrations.

If the bars have square leading edges, ILEV occur independent of flow incidence angle α, but they excite the most violent *transverse* vibrations of the bars at moderate α values (e.g., $5° < \alpha < 13°$ for bars with side ratio $e/d = 10$). With rounded leading edges, surprisingly, leading-edge vortices are still possible, and they can excite even larger vibration amplitudes than occur in corresponding cases involving square-edged bars. The onset of ILEV-induced transverse bar vibrations is characterized by reduced velocities equal to, approximately, $1/(n\mathrm{Sh}_1)$, where Sh_1 denotes the Strouhal number of the fundamental ILEV mode ($n = 1$) and $n = 2, 3,...$ denote the higher harmonics. For small mass-damping parameters Sc_δ, transverse bar vibrations can be excited by superharmonic resonance of relatively high order (e.g., up to $n = 4$ for $e/d = 10$ and $\mathrm{Sc}_\delta \simeq 3.3$). With approach-flow turbulence, excitation by ILEV is drastically reduced and replaced by turbulence buffeting.

The principal source for the excitation of *plunging* bar vibrations is fundamental or subharmonic resonance with AEVS, that is, vortex shedding from a leading and a trailing edge on opposite sides of an inclined bar. Plunging vibrations can only occur, therefore, if the flow incidence angle exceeds a critical value, α^*, marking the transition from ILEV to AEVS (e.g., $\alpha^* \simeq 11°$ to $13°$ for bars with side ratio $e/d = 10$). The amplitudes of plunging vibration exceed those of the transverse vibration by, typically, two orders of magnitude. Bars with rounded edges develop larger plunging amplitudes than those with square edges. Whereas transverse vibrations are excited from rest, plunging vibrations, in most cases, occur after an initial push, and the magnitude of this push must be the greater the larger the reduced velocity V_r is. The onset of plunging bar vibrations is associated with a more or less constant value of V_r, independent of the mass damping [e.g., $(V_r)_{\mathrm{cr}} \simeq 20$ for $e/d = 5$ and between 40 and 50 for $e/d = 10$]. In cases of grids or trashracks, the effective velocity between the bars should be used in the expression for V_r. For extremely small mass-damping parameters Sc_δ, plunging bar vibrations can also be excited by superharmonic resonance close to one half of the onset velocity defined above [e.g., near $(V_r)_{cr} \simeq 10$ for $e/d = 5$ and $Sc_\delta \simeq 0.55$]. If the value of Sc_δ is increased, the amplitudes and the ranges of V_r for plunging vibration are reduced, until the excitation finally ceases altogether (e.g., as Sc_δ exceeds 12.5 for $e/d = 10$ and $\alpha = 45°$). Approach-flow turbulence has less effect on the excitation than in the case of transverse vibrations, except near the critical angles of incidence α^*.

The following detailed conclusions and design recommendations are presented, first with respect to *transverse* vibrations of bars in a grid or a trashrack:

  – *For angles of flow incidence near zero*, square-edged bars with a side ratio e/d less than approximately 3 may be excited into transverse vibrations by either leading-edge vortex shedding or galloping. With e/d between about 2 and 16 for smooth flow and 2 and 8 for turbulent flow, transverse vibrations can be excited by impinging leading-edge vortices. This excitation is strongest in the range $3 < e/d < 7$. Trailing-edge vortex shedding is a relatively weak source of excitation for rectangular bars with rounded leading edges.

  – With the flow approaching a grid composed of rectangular bars *at an angle of incidence* α, impinging leading-edge vortices may excite transverse bar vibrations of large amplitude for side ratios even larger than 7, provided the bar spacing relative to the bar thickness exceeds the value of about $s/d \simeq 10$. For $e/d = 10$, typically, the greatest vibration amplitudes occur at incidence angles between 6° and 12° for square-edged bars and between 8° and 16° for bars with rounded edges. For α larger than about 20°, the only source of excitation for rectangular bars is turbulence buffeting.

  – The smaller the relative spacing of the bars in a grid, the more intense is the buffeting as a result of wake turbulence and wake interference. As is typical of buffeting response, the vibration amplitudes increase progressively with increasing velocity.

  – From the standpoint of minimizing the danger of transverse vibrations, the adjacent bars in a grid should preferably have side ratios larger than about 8 with the flow approaching them parallel to their long side. Sharp edges are preferable to rounded edges for rectangular profiles.

  – Serious transverse vibrations of grid bars can be avoided if the grid has sufficient stiffness and damping. The most effective way to control vibration is to use bars whose unsupported length is short. In this way, the natural frequency of the bars can be raised enough that the maximum reduced velocity stays below the critical values marking the onsets for all possible sources of excitation.

With respect to minimizing the danger of *plunging* or *streamwise* vibrations of grids or trashracks composed of bars, the recommendations are:

– Bars with circular profiles should not be used. If rectangular bars with large side ratio are used, the angles of flow incidence should be less than 10°.

– Trash accumulation should be kept small.

– The stiffness of the grid panels should be large enough so that the maximum reduced velocity remains below the critical values marking the onset of plunging excitation. The panel stiffness can be increased either by selecting stronger bars or by reducing the free distances between supports.

– If, for some reason, the stiffness of the grid panel cannot be increased sufficiently, one may be able to increase the mass-damping parameter beyond the value at which plunging excitation is suppressed by mounting special damping devices (Blevins, 1977).

Fluid-dynamic attenuation of bar or grid vibrations by means of a modification of the bar profile is difficult to attain for the usual case that more than one mechanism can excite vibrations. Although rounded leading edges and sharp trailing edges help suppress transverse and torsional vibrations for bars with side ratios $2 < e/d < 6$, for example, this modification has little effect on transverse vibration with side ratios $e/d > 7$ or on plunging vibration in inclined flow. A good overall performance is attained with the simplest cross-sectional shape: the square-edged rectangular bar with a moderately large side ratio.

9.2 GATES AND GATE COMPONENTS

9.2.1 *Overview*

Flow conditions leading to vibrations of gates and gate seals are treated in several sections of this volume and in the IAHR Monograph on Hydrodynamic Forces (Naudascher 1991). Specifically, the following cases of excitation are discussed there as follows:

(a) Predominantly extraneously induced excitation (EIE):
    1. Skinplate of sluice gate (Naudascher 1991, Figure 2.24);
    2. Tandem gate during emergency closure (Naudascher 1991, Figure 2.25);
    3. Tainter gate with two-phase flow (Naudascher 1991, Figures 2.26, 2.27);

(b) Predominantly instability-induced excitation (IIE):
    4. High-head leaf gate with unstable jet flow (Naudascher 1991, Figures 3.52-3.54);
    5. Cylinder gate with bistable underflow (Naudascher 1991, Figures 3.57, 3.58);
    6. Double-leaf gate with bistable underflow and with simultaneous over- and underflow involving vortex shedding (Naudascher 1991, Figures 4.64-4.69);

7. Multiple-leaf gate with simultaneous over- and underflow involving vortex shedding (Figure 6.49);
8. Stoplog with unstable separation and vortex shedding (Figure 6.63);
9. Leaf gate with overflow involving impinging shear layer (Naudascher 1991, Figure 4.57);
10. Various leaf gates with underflow involving impinging shear layers (Figures 6.28-6.30);
11. Tainter gates with underflow involving impinging shear layers (Naudascher 1991, Figures 4.70, 4.71);
12. Flap gate with oscillating nappe (Figures 6.34, 6.68, 6.69).

(c) Predominantly movement-induced excitation (MIE):

13. Leaf gates excited by modified type of galloping (Figures 7.26-7.28);
14. Various gates and gate seals with pulsating flow rate (Figures 7.48j-s, 7.71);
15. Tainter gate with pulsating flow rate (Figure 7.49).

(d) Excitation mainly due to a resonating fluid oscillator in the system:

16. High-head leaf gate withdrawn into gate chamber (Figure 8.26);
17. Various high-head gates in conduits in closed position (Figures 8.27, 8.29, 8.33);
18. Tainter gate in conduit (Figure 8.32);
19. Miter gate during flood release (Naudascher 1991, Figures 2.57, 2.58).

Moreover, practical experiences with flow-induced vibrations of gates and gate seals and information on their attenuation have been collected by Naudascher & Rockwell (1980) and Kolkman (1984b). Suffice it, therefore, to demonstrate in this section that, despite great differences in appearance, vibrations of gates and gate components can be traced back, in general, to mechanisms discussed in the foregoing chapters.

9.2.2 Vertical vibrations of gates with underflow

Figure 9.30 shows a variety of high-head and free-surface gates that are raised to allow flow underneath them. During such underflow, the gate bottom of thickness e (Figures 9.30a, e), as well as the undersides of the skinplate and the bottom seal of total thickness e' (Figure 9.30c) are exposed to hydrodynamic forces that may excite vibrations in the vertical or horizontal direction. As shown below, the mechanism of these excitations has much in common with that of transverse and streamwise excitations of prismatic bodies, presented in Section 9.1. The treatment of gate excitation is complicated by the great variety of boundary shapes encountered.

A few common shapes of gate bottoms are depicted in Figures 9.31 and 9.32. According to the graphs in these figures, boundary geometry greatly affects the vibration responses of gates and the ranges of relative gate opening s/e within which excitation occurs. Clearly, the excitation is induced by the instability of an

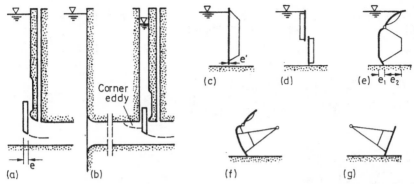

Figure 9.30. Some types of (a, b) high-head gates and (c-g) free-surface gates. (a) Intake-type leaf gate; (b) Tunnel-type leaf gate; (c, d) Single- and double-leaf gates; (e) Long-span gate with attached flap; (f, g) Tainter gates.

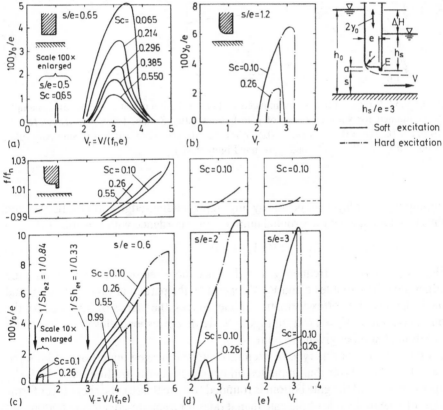

Figure 9.31. Response diagrams for gates, elastically suspended in the vertical direction (after Kanne, 1989). (a) $a/e = r/e = 0$, $\Delta H/e \simeq 0.065$ V_r^2; (b) $a/e = 0$, $r/e = 0.2$, $\Delta H/e \simeq 0.13$ V_r^2; (c-e) $a/e = r/e = 0.2$, $\Delta H/e = 0.13$ V_r^2. [$f_n e^2/\nu = 6250$ to 13 000; $e = 25$ and 50 mm; Tu = turbulence level $\simeq 5$ to 8%; ζ = damping ratio in still water at amplitude $y_0 = 1$ mm; Sc $\equiv 2m_r\zeta$ = mass-damping parameter.]

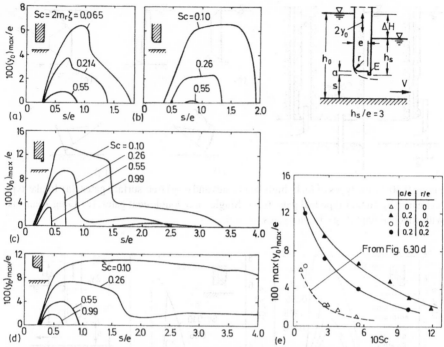

Figure 9.32. Ranges of instability-induced vertical excitation of gates and maximum amplitudes attained for various shapes of gate bottom (after Kanne, 1989). (a) $a/e = r/e = 0$; (b) $a/e = 0$, $r/e = 0.2$; (c) $a/e = 0.2$, $r/e = 0$; (d) $a/e = r/e = 0.2$; (e) Summary plot. [Values of $f_n e^2/v$, Tu, $\Delta H/e$, and ζ given in Figure 9.31.]

impinging shear layer (or better: by impinging leading-edge vortices, ILEV), and thus it is associated with Strouhal numbers in accordance with Equation 6.1,

$$\text{Sh}_e \equiv f_o e/V = (n + \varepsilon) V_c/V, \qquad n = 1, 2, ... \qquad (9.7)$$

Herein, f_o = vortex frequency, e = distance between leading and impinging edges, and the variables ε and V_c/V depend on the particular flow conditions and the order n of the vortex-formation mode. In fact, the values of Sh_e cited for the onset conditions $V_r = 1/\text{Sh}_e$ in Figure 9.31c do correspond closely to the values of Strouhal number given in Figure 6.28.

Prerequisites for this type of excitation are submerged underflow and a separated free shear layer that impinges on or passes near a 'trailing edge' E. If the upstream edge of the gate bottom is rounded (Figure 9.32b), the shear layer forms closer to the underside of the gate than it does for a square-edged gate bottom. As a consequence, the maximum amplitudes occur for larger gate openings. If the skinplate is extended a small distance a (Figures 9.32c, d), the feedback process inherent to ILEV is strengthened and remains functional over a larger range of

s/e. If the extension a is larger than some limiting value, however, the shear layer remains reattached to the skinplate in a stable fashion, and excitation is impossible as discussed in connection with Figure 9.33.

The extreme values of the maximum amplitude $(y_o)_{max}$ in Figure 9.32 for gates with rounded upstream edges and gates with extended skinplates correspond to hard excitation, a fact that can be verified from the response diagrams in Figure 9.31. Hard excitation occurs, for example, if the velocity V increases slowly beyond the range of soft excitation, and it does not occur if V decreases after the vibration has subsided. The reason for this behavior is that the more the reduced velocity V_r exceeds the point of resonance, $V_r = 1/Sh_e$, the larger is the up-and-down movement of the gate bottom that is required to maintain amplified feedback by keeping the frequency of the leading-edge vortices locked-on to the frequency of gate vibration f. The range over which this type of movement-induced amplification is effective depends mainly on the opening ratio s/e and, of course, the mass-damping parameter $Sc \equiv 2m_r\zeta$. The larger the gate opening s, the greater is the distance between the shear layer and the 'impingment edge' E. Beyond a certain value of s/e, therefore, movement-induced amplification ceases to be effective (Figures 9.32b-d).

Evidently, the susceptibility of a gate with a given bottom configuration to excitation by impinging shear layers depends strongly on the distance between the mean shear-layer position and the impingement edge E. This distance, d^*, in turn, depends on the position of the separation streamline below the gate as follows (Figure 9.33a):

$$d^* \simeq a_s - a^* \qquad (9.8)$$

Here, a^* is the vertical distance between the point of flow separation S and point E, and a_s is the vertical distance from S to the point at which the separation streamline intersects the vertical through E. For practical purposes, a^* can be expressed by $a + r$ for the bottom shapes depicted in Figure 9.32. As shown in Figure 9.33a, moreover, a_s can be estimated on the basis of information concerning either the shape of the separation streamlines (Betz & Petersohn, 1931) or the contraction coefficient C_c. The latter varies appreciably with the ratio of effective gate opening to upstream depth of flow, s^*/h_o (Figure 9.33b), and with approach-flow conditions (Figures 9.32c-e). [For the data reported in Figures 9.31, 9.32, as well as in Figures 6.29, 6.30, the upstream depth h_o was a variable as indicated by the values of $\Delta H/e$ given in the corresponding figure captions.]

In order to protect a gate against flow-induced vibration in the vertical direction, including the self-excited type shown in Figure 7.28, one must either provide sufficient damping or select a shape of gate bottom from which the flow remains unseparated or stably *reattached* to the gate bottom for all operating conditions. In the latter case, a^*/e must exceed the largest possible value of a_s/e, which can be estimated with the aid of Figure 9.33. This recommendation for the situation sketched in Figure 9.33a is more easily fulfilled for a tunnel gate with

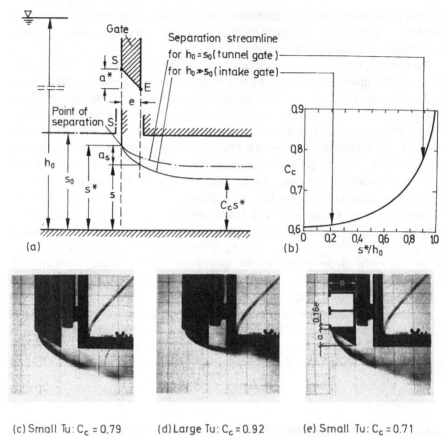

(c) Small Tu: $C_c = 0.79$ (d) Large Tu: $C_c = 0.92$ (e) Small Tu: $C_c = 0.71$

Figure 9.33. (a) Schemes for assessing the susceptibility of a leaf gate to flow-induced vertical vibration. (b) Contraction coefficient C_c for submerged, turbulence-free flow. (For tunnel gates, $h_o = s_o$ if effect of corner eddy shown in Figure 9.30b is disregarded.) (c-e) Effect of approach-flow conditions on position of free shear layer and on C_c for an intake gate with $a/e = 0.6$ at $s/e = 6$ (after Nguyen & Naudascher, 1983). (c, d) Smooth upstream face; (e) Open girders. [Tu = turbulence level.]

limited upstream depth of flow, s_o, than it is for an intake gate of identical cross section but large upstream depth, h_o. Similarly, a tunnel gate with a given ratio $a*/e$ becomes safer from flow-induced vibration as the ratio of tunnel height s_o to gate thickness e becomes smaller. Figure 9.34b shows this to be true. The ranges of 'unstable reattachment' depicted in this figure were estimated from the degree of unsteadiness of pressure readings taken along the underside of stationary gate models (Naudascher, 1964). In order to exclude excitation of vertical gate vibration on the basis of Figure 9.34, therefore, one should select a value of $a*/e$ well above the upper range-boundaries.

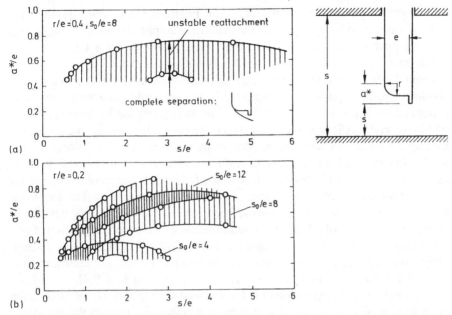

Figure 9.34. Conditions of unstable reattachment of flow onto the extended skinplate of a tunnel gate as a function of a^*/e. (a) $r/e = 0.4$; (b) $r/e = 0.2$.

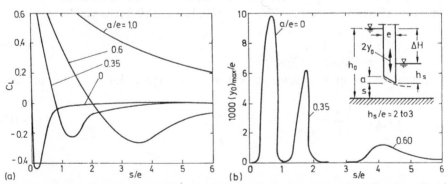

Figure 9.35. Application of instability indicator to ILEV-induced vertical vibration of leaf gates with inclined underside (after Nguyen, 1990): (a) Mean-lift coefficient; (b) Summary plot of maxima of response diagrams presented in Figure 6.29. [Sc $\equiv 2m_r\zeta \simeq 0.5$; $f_n e^2/\nu = 51\,000$; Tu $\simeq 1\%$; $\Delta H/e = 1.06\,V_r^2$.]

The importance of fluid-elastic or movement-induced amplification for gate vibrations generated by ILEV is illustrated in Figure 9.35. In accordance with Figure 7.27, one of the susceptibilities to movement-induced excitation is indicated by a positive slope in the plot of the mean-lift coefficient C_L versus the relative gate opening s/e. The same indicator, clearly, applies in the present case:

The ranges of positive slope $dC_L/d(s/e)$ in Figure 9.35a do correspond to the ranges of vibration in Figure 9.35b; the maximum amplitudes do occur where that slope is greatest; and the one gate that is characterized by a negative slope throughout (i.e., the gate with $a/e = 1.0$) did not vibrate at all.

From the fact that gate responses are different with and without a free surface downstream, as reported in Figure 6.29, one can infer that gate vibrations generated by ILEV are sensitive to amplification produced by fluid resonators within the system (Section 6.2.3). After having passed the gate lip, the impinging leading-edge vortices form larger vortices by successive vortex pairing (Figure 9.36a). In cases with a free tailwater surface, these vortices excite gravity waves associated with pressure oscillations that are fed back to the leading edge of the gate bottom and amplify the vortex generation (Figure 9.36b). The larger the distance d^* (Equation 9.8) between the free shear layer and the trailing edge E, the stronger is the effect of this fluid-resonant feedback on the (soft) excitation of gate vibrations. For a typical case of *hard* excitation depicted in Figure 9.31c, Kanne (1989) found the effect of h_s/e on a_{max}/e to be small [$100 a_{max}/e$ increased from 4.9 to 5.6 as h_s/e was raised from 3 to 12, at $s/e = 0.6$ and Sc = 0.4].

The mechanism of excitation of ILEV-induced vertical gate vibration is analogous to that of ILEV-induced vibration of a prismatic body in the transverse mode. Figures 9.37b, c illustrate this analogy by showing corresponding phases of the vortex formation and the accompanying distributions of pressure acting on gate and prism (cf. Figure 9.6). As a consequence, the conditions for vibration onset are also alike; in terms of $Sh_e \equiv f_o e/V$, Equation 9.2 now yields

$$(V_r)_{cr} \equiv \left(\frac{V}{f_n e}\right)_{cr} = \frac{1}{Sh_e} \tag{9.9}$$

Substituting the approximations $\varepsilon \approx 0$ and $V_c V \approx 0.5$ in Equation 9.7, one obtains

$$(V_r)_{cr} = \frac{1}{n}\frac{V}{V_c} \approx \frac{1}{0{,}5n}, \qquad n = 1, 2, \dots \tag{9.10}$$

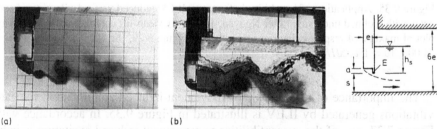

(a)   (b)

Figure 9.36. Flow patterns downstream of a stationary gate with extended skinplate ($a/e = 0.35$) without and with free tailwater surface (after Nguyen, 1990). (a) Tunnel flowing full, $s/e = 2.5$; (b) Free surface in tunnel, $h_s/e = 2$ to 3, $s/e = 2.0$.

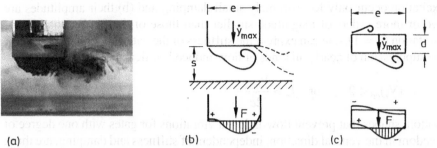

Figure 9.37. (a) Photo of vibrating gate with dye injected into shear layer (after Kanne, 1989): $s/e = 1.2$, $V_r = 3.4$. (b, c) Instantaneous vortex patterns and corresponding pressure distributions along gate bottom and side of rectangular prism during vibratory motion ($y = y_{max}$) for V_r slightly greater than onset value for vortex mode $n = 1$.

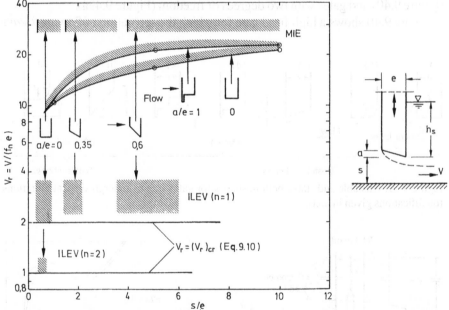

Figure 9.38. Ranges of possible flow-induced vertical vibration for lightly-damped gates with various bottom shapes, Tu $\simeq$ 5 to 8% (summary of Figures 6.29, 6.30, 7.26-7.28).

The summarizing Figure 9.38 and the data contained in Figures 9.31 and 9.32 show that this simplified formula describes the onset conditions for ILEV-induced vertical vibrations surprisingly well, quite independently of the shape of the gate bottom. Since MIE-induced vertical gate vibrations are confined to regions of much larger reduced velocities V_r (Figure 9.38), Equation 9.10 yields a significant design criterion. Because (a) the higher harmonics ($n > 1$) of ILEV excitation occur only for extremely small damping and (b) their amplitudes are two or more orders of magnitude smaller than those of the fundamental ILEV excitation ($n = 1$), one can express the stiffness of the gate support, necessary for the suppression of nearly all kinds of flow-induced vertical gate vibration, as

$$(V_r)_{cr} < 2 \quad \text{or} \quad f_n > \frac{V_{max}}{2e} \tag{9.11}$$

Bottom shapes that prevent flow-induced vibrations for gates with one degree of freedom in the vertical direction, independent of stiffness and damping, are those from which the flow never separates, or to which it remains stably reattached under all operating conditions (Figures 9.39f, g). The a^*/e value that separates stable and unstable gate-bottom shapes can be deduced with the aid of Figure 9.33: a^* should exceed a_s for all operating conditions by some 10% to 20%. There are exceptions to this rule involving, e.g., gates with flow separation upstream (Figure 9.40) and gates with two degrees of freedom (Figure 9.49a).

Figure 9.40 shows a high-head gate for which the separation of flow *upstream*

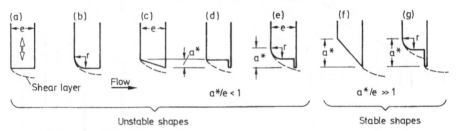

Figure 9.39. Unstable and stable bottom shapes for gates free to undergo vertical vibrations (qualifications given in text).

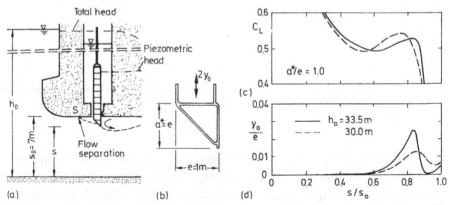

Figure 9.40. Application of instability indicator to extraordinary case of leaf gate with destabilizing upstream flow separation (after Nguyen, 1990). (a, b) Gate arrangement and detail of gate bottom; (c) Corresponding mean-lift coefficient, and (d) response diagram for estimated values $V_r \simeq 5$ and $Sc \equiv 2m_r\zeta \simeq 1$.

of the gate led to vertical gate vibrations despite the fact that $a^* > a_s$. Destabilizing flow separation from the tunnel ceiling can also occur at tunnel gates due to corner eddies upstream (Figure 9.30b); such eddies are absent only if 'boundary-layer suction', generated by flow through the gate chamber, averts their formation (Naudascher, 1991). In the case shown in Figures 9.40a and b, field tests (US Army Corps of Engineers, 1956) revealed gate vibrations in the range of roughly $0.7 < s/s_o < 0.9$ (Figure 9.40d). The reduced velocity and the mass-damping parameter were estimated to be $V_r \simeq 5$ and $Sc \equiv 2m_r\delta \simeq 1$ (Nguyen, 1990). As shown in Figures 9.40c, d, the range of vibration coincides more or less with the range of positive slope $dC_L/d(s/e)$. The instability indicator (Figures 7.27, 9.35) suggested by Nguyen et al. (1986b) is thus verified by prototype data even for a complex case. What makes the gate in Figure 9.40a particularly vulnerable to vibrations is the reduced horizontal pressure force and the correspondingly small frictional damping.

Further examples of possible ILEV-induced gate vibrations are depicted in Figure 9.41. If a tunnel gate is withdrawn into the gate chamber with its lower edge close to the tunnel ceiling, it can still be excited to vibrate by impingements of the shear layer as shown in Figure 9.41a. Frictional damping is ineffective in this case because of no horizontal load; as a result, the gate can develop into a 'bang-bang gate' performing more or less random movements in the gate chamber (Chang & Hampton, 1980). A comparable source of excitation is pressure fluctuation within a gate slot (Figures 9.41b, 8.26d), unless the slot's downstream edge is set back sufficiently far (Figure 6.66). The dominant frequency of excitation for these two cases can be obtained from Figure 6.25. The possibility of fluid-resonant amplification is illustrated with the aid of Figures 8.26b-d.

Pressure fluctuations of extremely large intensity occur for a tandem gate arrangement (Figure 9.41c) if one of the gates becomes stuck and the other undergoes an emergency closure. Along with the fluctuating displacements of the shear layer originating at the bottom of the upstream gate, the high-velocity flow alternately hits and misses the skinplate of the downstream gate for a certain critical range of gate openings; this generates peak-to-peak pressure amplitudes

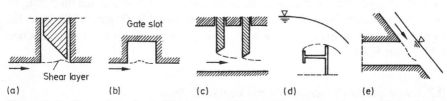

Figure 9.41. Examples of possible sources of gate vibration associated with impinging shear-layer instability. (a) Tunnel gate withdrawn in gate chamber; (b) Gate slot; (c) Emergency closure in case of a stuck service gate; (d) Leaf gate with overflow; (e) Leaf gate (not shown) closing a bottom outlet submerged by spillway flow.

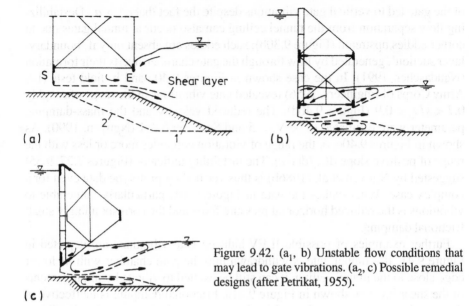

Figure 9.42. (a_1, b) Unstable flow conditions that may lead to gate vibrations. (a_2, c) Possible remedial designs (after Petrikat, 1955).

with an intensity as high as the free-stream dynamic pressure $\rho V^2/2$ and, in many instances, cavitation (Naudascher 1991, Figure 2.25). The provision necessary in these cases is an instruction for the operator not to interrupt an emergency closure near a critical gate position. A case of ILEV-induced excitation of the double-leaf gate shown in Figure 9.30d is sketched in Figure 9.41d (cf. Naudascher, 1991, Figures 4.38, 4.49, 4.57); and the excitation of a bottom-outlet gate in a closed position, referred to in Figure 9.41e, is described in conjunction with Figure 8.33.

Impinging shear-layer instability is not restricted to leaf gates. If its bottom girder is close to the free shear layer as sketched in Figure 9.42a_1, a long-span gate may be excited to vibrate in ways that are not much different from those depicted in Figure 9.31a. Possible remedies are to slope the apron shortly downstream of the skinplate S (Figure 9.42a_2) or to incline the bottom girder (Figure 9.42c). In the former case, the apron slope should not exceed 1:1.8 so that the emerging jet does not separate from the floor and become bistable (Figure 6.38, gate II). Further sources of excitation for long-span gates are discussed in conjunction with Figure 9.49.

9.2.3 *Horizontal vibrations of gates with underflow*

The analogies between the instability-induced excitation mechanisms for prismatic bodies and gates extends beyond the characteristics illustrated in Figure 9.37. For both prisms and gates, excitation occurs due to fundamental or superharmonic resonance with impinging leading-edge vortices (ILEV) and can lead to vibrations in both the horizontal and vertical directions.

Figure 9.43 illustrates the latter point with information deduced from photos taken during *controlled* horizontal gate vibrations at particular values of reduced velocity $V_r \equiv V/(fe)$. Since regions adjacent to the vortices in these photos are associated with zones of low pressure, the instantaneous pressure distributions along the bottom of the gate are approximately as sketched in Figures 9.43d, e. The corresponding values of V_r were deduced on the assumption that the vortex-convection velocity V_c is about 0.5V irrespective of the ILEV mode (Equation 9.10). Thus, V_r is related to the wavelength $\lambda = V_c/f$ of the pressure distribution as

$$V_r \equiv \frac{V}{fe} \simeq \frac{2\lambda}{e} \quad \text{for} \quad V_c = 0.5\,V \tag{9.12}$$

Using this relationship, one can identify the sketches in Figures 9.43d (ii) and (iv) with the onset condition for *vertical* gate vibration according to Equation 9.10 ($n = 2$ and 1), and the sketches in Figures 9.43d (iii) and (v) with the end of the

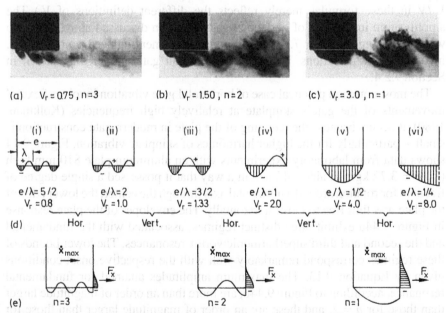

Figure 9.43. (a-c) Instantaneous flow patterns at various reduced velocities V_r during controlled horizontal gate vibrations showing modes $n = 3$, 2, and 1 of impinging leading-edge vortices (ILEV) (after Nguyen, 1990). (d, e) Sketches of instantaneous pressure distributions (for $\dot{x} = \dot{x}_{max}$) characterizing (d) onsets of horizontal and vertical gate vibrations and (e) typical resonance conditions of a gate vibrating horizontally with ILEV modes $n = 3$, 2, and 1.

ranges of vertical gate vibration according to Figure 9.31a. Typical flow patterns within these two ranges are depicted in Figures 9.43b, c; they show, respectively, two vortices and one vortex along the bottom of the gate.

The ranges of *horizontal* gate vibration are adjacent to the ranges of vertical vibration since their excitation requires different pressure distributions. Essential for an amplified excitation in this instance is a positive fluid-dynamic force acting horizontally on the gate during periods of positive gate velocities (Figure 9.43e), so that the work done on the gate during a cycle of vibration is positive. The onset condition for horizontal gate vibration is therefore

$$(V_r)_{cr} = \frac{2}{2n-1} \frac{V}{V_c} \simeq \frac{4}{2n-1}, \qquad n = 1, 2, \dots \qquad (9.13)$$

[Figures 9.43d (i), (iii), (v)]. Within the ranges of vertical gate vibration mentioned above, the force F_x is out-of-phase with the gate velocity, and horizontal excitation is impossible.

The remarkable analogy between the onset conditions for vertical and horizontal gate vibrations (Equations 9.10, 9.13), on the one hand, and transverse and torsional vibrations of prisms (Equations 9.2, 9.3), on the other, supports the hypothesis of similar excitation mechanisms. (The difference in the values of V_c/V in these formulas merely reflects the different definitions of V.) The amplification in the case of horizontal gate vibration discussed above is due to wake breathing (Section 7.3.5). Additional movement-induced amplification, involving flow pulsations, is possible with small gate openings as shown in Section 9.2.4.

The most common practical case of horizontal gate vibration involves flexural movements of the gate's skinplate at relatively high frequencies (Kolkman, 1980); it occurs because the damping of the plate in modern gate construction is small – particularly for the higher harmonics of skinplate vibration. Figure 9.44 shows data from laboratory experiments with an aluminum plate 810 mm high ($I_\Theta/B = 3.72$ kgm in air), stiffened in a way that it possessed a single degree of freedom for rotation around a horizontal axis through the center; the lower edge of the plate was thus free to move horizontally. The envelopes of the plate response in Figure 9.44a exhibit three distinct regimes, associated with the fundamental and the second and third superharmonic vortex resonances. The lower bounds of these regimes correspond remarkably well with the respective onset conditions given by Equation 9.13. The maximum amplitudes attained for fundamental resonance, according to Figure 9.44b, are more than an order of magnitude larger than those for $n = 2$, and these are an order of magnitude larger than those for $n = 3$. In fact, third-harmonic resonance occurred only with extremely small depths of submergence h_s, i.e., in cases of minimum fluid damping and maximum fluid-resonant amplification (cf. discussion of Figure 9.36). Since this resonance is restricted to a very narrow range of relative gate openings s/e, moreover, it is not significant in practice unless the gate is meant to regulate the discharge within that range.

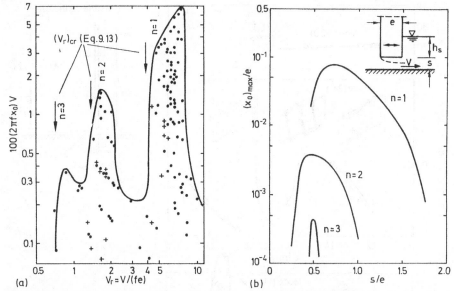

Figure 9.44. Characteristics of instability-induced excitation of lightly-damped gate plates (after Jongeling, 1988). (a) Envelope of skinplate responses in horizontal direction (x_o = tip amplitude); (b) Ranges of excitation and maximum amplitudes attained for resonance with ILEV modes n = 1, 2, and 3. [e = 20 mm, h_s/e = 4 to 25, $\Delta H = V^2/2g$, $\zeta \simeq 0.0025$ in air.]

A typical flexural response of a vertical plate with underflow, obtained for decreasing values of the velocity V, is shown in Figure 9.45. The jumps from one mode of plate vibration to another (Figure 9.45b) occurred during finite times of transition, during which the spectrum of plate deflection revealed the coexistence of several vibration modes. The plate, made of brass 4 mm thick, was supported over its entire height by two vertical bars 10 mm thick, fixed on the side walls of a flume 1 meter wide; a styrofoam strip 70 mm high and 8 mm thick was glued onto its downstream side to increase the lip thickness without changing the plate rigidity. The sketches in Figure 9.45b depict three of the observed modes of vibration (cf. Figure 2.20). The corresponding natural frequencies, measured in still water 100 mm deep, are f_n = 23.9 Hz for mode (1,1), 20.0 Hz for mode (2,0), 35.2 Hz for mode (3,0), and 215 Hz for mode (3,1). A plot of the response curves in terms of reduced velocity in Figure 9.45c shows that all curves fall within the V_r limits of about 4 to 8, which is in agreement with the range of fundamental vortex resonance marked by Figures 9.43d (v), (vi). Due to its smaller damping, a steel plate of 8 mm thickness was excited also within the second-harmonic resonance range, marked by Figures 9.43d (iii), (iv), though at substantially reduced amplitudes.

The stable lip shape shown in Figure 9.46d is difficult to realize in practice. However, since ILEV-induced excitations of gate plates are limited to a narrow

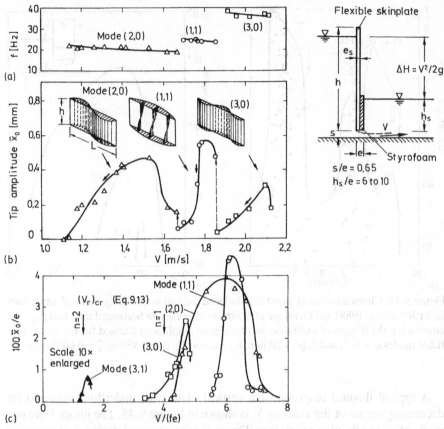

Figure 9.45. Examples of instability-induced excitation of gate plates, simply supported on two sides. (a, b) Frequency and amplitude responses for decreasing discharge at a gap ratio $s/e = 0.65$ (after Nguyen et al., 1987); (c) Dimensionless version of Figure 9.45b. [$h/L = 280/1000$ mm, $e = 3e_s = 12$ mm, $\zeta = 0.031$ for mode (1,0) in still water for $h_s = 100$ mm. ▲ $e = e_s = 8$ mm.]

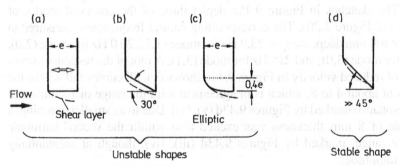

Figure 9.46. Unstable and stable bottom shapes for gates free to undergo horizontal vibrations.

range of gate openings ($s < 2e$, Figure 9.44b), one can design the lip of the plate with as small a thickness, e, as feasible and make sure that the gate passes quickly through the endangered gate-opening range. Naturally, any stiffening of the lip, which is the only part of the skinplate that is exposed to the exciting forces (Figure 9.43e), and any augmentation of structural damping also helps to attenuate the vibration. The greatest danger for destructive vibration arises from simultaneous action of the movement-induced type of excitation discussed in the next section.

9.2.4 *Vibrations of gates and gate seals involving flow-rate pulsations*

Among the most dangerous mechanisms for movement-induced excitation of gates and gate seals are those involving coupling with fluid-flow pulsations (Section 7.4). For a vibrating skinplate, for example, such pulsations occur if the plate is inclined with respect to the apron in such a way that it becomes a press-shut device (Figure 7.48p). In contrast to the instability-induced plate vibrations discussed in the preceding section, the movement-induced or self-excited vibrations arise even if the underflow is not submerged, merely from the coupling of fluid-inertia effects with plate vibrations via flow-rate pulsations.

For the system shown schematically in Figures 9.47a, b, Kanne et al. (1991) derived a stability criterion by applying the simplified theory of Kolkman (1976). Treating the system like a body oscillator with a single torsional degree of freedom and linear damping, $I_\Theta\ddot{\Theta} + B_\Theta\dot{\Theta} + C_\Theta\Theta = M(t)$ (Equation 2.18), they obtained the stability criterion in the form:

$$\frac{C_{\Theta cr}}{\rho V^2 l^2} = \frac{l\sin\alpha}{2s}\left[1 + \frac{4I_\Theta/L_{eff}}{\rho l^3 + 8\sqrt{C_{\Theta cr}I_\Theta}\,\zeta_\Theta C_c s/(lV)}\right]\left[1 - \frac{4\sqrt{C_{\Theta cr}I_\Theta}\zeta_\Theta}{\rho l^3 V C_c L_{eff}}\right] \quad (9.14)$$

Here, $C_{\Theta cr}$ is the value of the spring constant per unit width, $C_\Theta \equiv Cl_1^2$, that must be exceeded to secure stability; $V = \sqrt{2g\Delta H}$ = velocity in the vena contracta of height $C_c s$; I_Θ = moment of inertia of the gate mass per unit width; $\zeta_\Theta = B_\Theta/(2\sqrt{I_\Theta C_\Theta})$ = damping ratio; $\rho h L_{eff}$ = added mass per unit width; and l, h, s, ΔH, and α are as defined in Figure 9.47a. For zero damping, Equation 9.14 reduces to

$$\frac{C_{\Theta cr}}{\rho V^2 l^2} = \frac{l\sin\alpha}{2s}\left[1 + \frac{4I_\Theta}{\rho l^3 L_{eff}}\right] \quad (9.15)$$

According to this equation, the gate becomes less stable if either its stiffness or its opening s are reduced, or if the velocity V or the angle of inclination α are increased.

Figures 9.47c, d show the effective length L_{eff} of the added mass obtained from measured vibration frequencies f on the basis of

$$2\pi f = \sqrt{C_\Theta/(I_\Theta + \rho h L_{eff}l^2/4)} \quad (9.16)$$

For tunnel gates, L_{eff}/h is usually so large that $I_\Theta/(\rho l^3 L_{eff}) \ll 1$ and the effect of

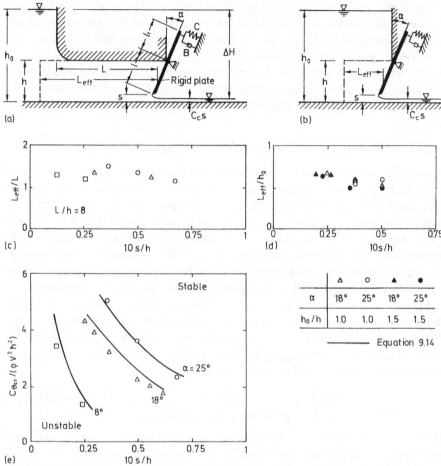

Figure 9.47. (a, b) Systems of gates susceptible to movement-induced excitation involving variations of gate opening during vibration. (c, d) Relative effective length of the added mass for systems 'a' and 'b', respectively. (e) Comparison of computed and measured critical stiffness, $C_{\Theta cr}$, for $I_{\Theta}/(\rho h^4) = 0.195$, $L/h = 8$, $h_o/h = 1.775$, and $\zeta_{\Theta} = 0.08$ to 0.2 in still water (Kanne et al., 1991).

L_{eff} on $C_{\Theta cr}$ can be neglected (Equation 9.15). As shown in Figure 9.47e, the prediction according to Equation 9.14 agrees well with measured values of $C_{\Theta cr}$ in such cases. For a free-surface gate as depicted in Figure 9.47b, on the other hand, the agreement is not good; here, both L_{eff} and ζ_{Θ} affect the results substantially, and the stability criterion has to be derived from a more elaborate theory (e.g., Ishii, 1992).

Further examples of gate and gate-seal vibrations, self-excited by mechanisms involving flow-rate pulsations, are referred to in Figures 7.48j-o and q-s, and

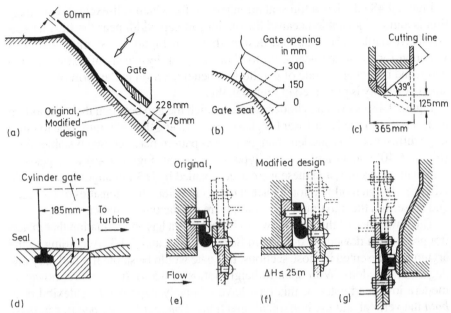

Figure 9.48. Examples of vibration problems and vibration control involving various gates and gate seals. (a) Bottom seal of submersible Tainter gate (after Brown, 1961). (b, c) Bottom of cylinder gate in intake tower (after Martin & Ball, 1955): (b) before and (c) after modification. (d) Bottom of cylinder gate for turbine entrance. (e-g) Top seals of tunnel gates (after Petrikat, 1980).

discussed in Section 7.4.3. They show that self-excited gate or gate seal vibrations can occur if a seal arrangement gives rise to flow through a leak channel with an upstream constriction. An obvious countermeasure, therefore, is to eliminate such a constriction or move it towards the downstream end of the flow passage. Figure 9.48a depicts such a case. The originally diverging flow passage caused not only cavitation but also serious skinplate vibrations in a number of installations (Brown, 1961; Neilson & Pickett, 1980) due to flow pulsations generated by a fluctuating gap width. With the modified design shown in the same figure, these vibrations were eliminated.

Violent vertical gate vibration of a similar origin, accompanied by cavitation and extreme downpull, occurred when the cylinder gate referred to in Figure 9.48d reached an opening of 1.6 mm. This vibration, as well as the cavitation and excessive downpull, could have been avoided, had the 'leak channel' at small gate openings not been divergent. Since a modification of gate-bottom or seat geometry in the completed scheme turned out to be too costly, additional damping over the first 10 mm of gate opening was recommended·(Naudascher et al., 1988). A damping device capable of suppressing self-excited gate vibration at small gate openings is shown in Figure 7.71b.

Figure 9.48e depicts a top seal arrangement for which self-excited seal vibration is almost inevitable because fluctuations in gap width near the flexible seal are unavoidable. The modified design shown in Figure 9.48f mitigated such vibration for moderate heads, up to about 25 mm, by means of fixing the gap width and reducing the rate of gap flow. A better solution in the form of a seal of small flexibility is presented in Figure 9.48g.

Figure 9.48b shows a cylinder gate for which the gap between the gate bottom and gate seat formed a diffuser for gate openings s less than 300 mm. Due to fairly large diffuser angles, sudden changes in flow pattern between two bistable states (Figure 6.40) and/or a plug-valve type of excitation (Figure 7.46) were possible for small values of s, and these were accompanied by jerky changes in downpull. After the design modification depicted in Figure 9.48c, the intensities of vibration and noise, and the maximum downpull, were substantially smaller.

The lower face of the gate shown in Figure 9.49a has an inclination that makes it a press-open device, as defined in Sections 7.4.1 and 7.4.3. According to the arguments presented in those sections, this gate would be stable if it had only one degree of freedom. With span-to-height ratios of 15 to 20 or more, however, modern low-head gates of this type have relatively large bending flexibility in *both* the vertical and the horizontal directions. *For small gate openings* in these cases, coupling of the vertical and horizontal vibration modes can lead to movement-induced excitation in the following way. When the gate undergoes horizontal deflections, the corresponding fluctuation of the hydrodynamic force

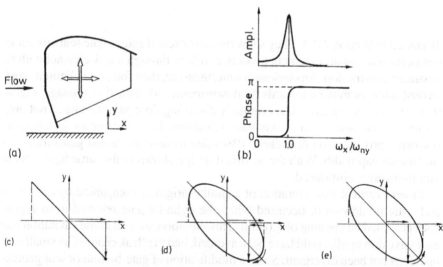

Figure 9.49. (a) Shell-type long-span gate with degrees of freedom in vertical and horizontal directions. (b) Amplitude and phase response of a classical oscillator. (c-e) Traces of gate amplitudes for (c) $\omega_x = \omega_{ny}$, (d) ω_x slightly greater than ω_{ny}, and (e) ω_x slightly less than ω_{ny} (after Ishii et al., 1993).

acting on the inclined lower face produces vertical gate deflections that, in turn, cause flow-rate pulsations; these pulsations generate additional force fluctuations that reinforce the initial horizontal deflections under certain conditions. Ishii & Knisely (1992) have analyzed these conditions and obtained the stability criterion

$$\frac{\omega_x}{\omega_{ny}} < 1 \quad \text{or} \quad \frac{\omega_x}{\omega_{ay}} > \left(\frac{\omega_x}{\omega_{ay}}\right)_{\lim} \tag{9.17}$$

For dynamic stability, in other words, the ratio of the in-water horizontal-vibration frequency ω_x to the natural in-water vertical-vibration frequency ω_{ny} of the gate should be smaller than unity; or the ratio of ω_x to the in-air vertical-vibration frequency ω_{ay} of the gate should be larger than a limiting value $[(\omega_x/\omega_{ay})_{\lim} \approx 1.5$ in Ishii's study]. If either of these criteria is violated, self-excited gate vibrations will set in, no matter whether the underflow is submerged or not, unless the mass-damping of the gate is large enough.

For typical shell-type long-span gates, the natural frequencies of vibration in the first bending modes in air are almost identical for the horizontal and vertical directions. In water with only underflow, the frequency of horizontal vibration is significantly smaller than that of vertical vibration due to added-mass effects. In accordance with Equation 9.17a, therefore, self-excited vibrations during underflow can occur only if these gates are suspended on wire-rope cables because the frequency of the gate vibrating vertically in the *rigid-body mode* is then smaller than the frequency of the horizontal bending mode. However, if the gate has substantial overflow simultaneously, its added mass for vertical motion will be much larger, and the stability criterion of Equation 9.17a may be violated also for vertical bending-mode vibrations. In fact, Ishii observed severe self-excited vibration of a full-scale shell-type gate during simultaneous over- and underflow, even though the nappe was broken by spoilers and the space underneath was fully ventilated; no vibration was observed in that installation when there was only underflow.

9.2.5 *Vibrations of gates with simultaneous over- and underflow*

Shell-type long-span gates (Figure 9.49a) with relatively large overflow and simultaneous underflow at small gate openings can develop *self-excited vibrations* even in the absence of an inclined upstream face. The mechanism of excitation is similar to the one described in conjunction with Equation 9.17. In the case of a gate with a vertical upstream face, however, the essential element in bringing about the mode coupling is the hydrodynamic force acting on the gate crest. Any streamwise gate vibration initiates a substantial added-mass flow (Figure 7.37a) that, in turn, produces pulsations in the rate of overflow. With these pulsations, the hydrodynamic force on the gate crest will fluctuate and 'force' the gate to vibrate vertically at the in-water frequency ω_x of horizontal gate vibration.

As a result, the rate of underflow will pulsate, thus generating horizontal force fluctuations that amplify the initial streamwise vibration of the gate.

If one approximates the vertical gate response to the fluctuating forces on the gate crest as a forced vibration of a classical body oscillator (Figures 2.3, 2.4), one obtains response diagrams as sketched in Figure 9.49b, where ω_{ny} is the natural in-water frequency of vertical gate vibration. In other words, one can expect a change in phase of 180° as the forcing frequency of ω_x progresses from values below resonance to those above. As a consequence, the vertical gate motion should exhibit a comparable phase reversal relative to the horizontal motion as shown in Figures 9.49d, e. It can be readily verified (Ishii et al., 1993) that the condition of Figure 9.49d, i.e., ω_x/ω_{ny} slightly greater than one, corresponds to positive work done on the vibrating gate and, hence, to self-excited gate vibration – quite in agreement with the criteria in Equation 9.17.

Gates with simultaneous over- and underflow can also be excited by *alternate shedding of vortices* much as any prismatic structure in a cross-flow (Section 6.3). Figure 9.50 shows an example of this instability-induced excitation: a multiple-leaf gate, used for emergency closure of a turbine intake for the case of a runaway

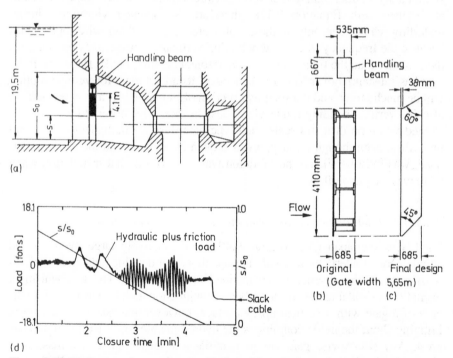

Figure 9.50. (a) Section of power house with first of three leaf gates in half-lowered position; (b, c) Cross-section of leaf gates tested in a model. (d) Combined hydrodynamic and friction load acting on original design of leaf gate and handling beam during closure of the first leaf (Elder & Garrison, 1964).

turbine. The combined hydrodynamic and friction load acting vertically on the first of three leaves during closure with a runaway discharge of 162 m³/s per intake (Figure 9.50d) indicates nearly periodic load fluctuations that start from the centered position of the leaf. The maximum force amplitude of 6.35 tons was almost sufficient to cause the lifting cables to become slack. Only when the lower edge of the first leaf gate was 0.6 to 0.9 m from the floor did the force fluctuation subside. When the second leaf gate was lowered to a position about 2 m short of

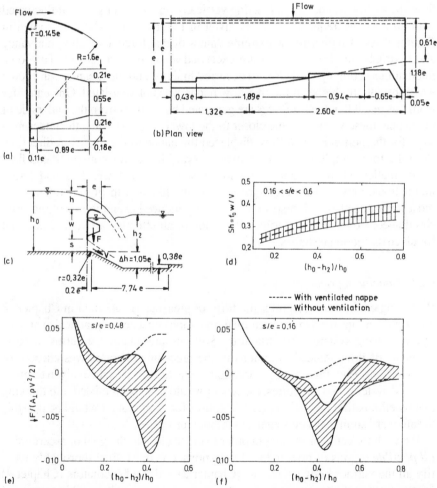

Figure 9.51. (a, b) Hook-type double-leaf gate. (c) Sketch of gate and stilling-basin arrangement. (d) Strouhal number for stationary gate with ventilated nappe. (e, f) Normalized hydrodynamic force F acting vertically on lower gate leaf (after Naudascher, 1959). $[h_o/e = 2.95; h/e = 0.64; e = 125$ mm; Re $= e\sqrt{2gh}/v = 1.4\times10^5; A_\perp = 53380$ mm² = area of loading.]

closure, it experienced a hydraulic uplift equal to its weight, and the downward movement stopped. From tests with several cross-sectional shapes of gate leaf, the one shown in Figure 9.50c did not cause slack crane cables and closed safely. Vortex excitation, however, persisted with a maximum force amplitude of about 5.5 tons. The variation of the excitation frequency and the dependence of the force amplitude on both the size of the handling beam and its distance from the leaf gate are presented in Figure 6.49.

A double-leaf gate of the free-surface type, designed for both overflow and underflow, is sketched in Figures 9.51a-c. During simultaneous over- and under-flow, the hydrodynamic force F acting vertically on the lower gate leaf fluctuated with an average frequency of f_o as given in Figure 9.51d. The data shown in Figures 9.49e, f represent the extreme values of that force, excluding buoyancy and hydroelastic effects, which are exceeded about once a second. The force fluctuations acting upward ($F < 0$) occurred in jerks, and the mean upward force was substantially higher if the overflow nappe was not ventilated. One can infer from this result that the lift forces are generated by vortices from the underside of the nappe; these vortices come closer to the lower gate leaf when the underpressure, for the non-ventilated flow, displaced the nappe in the upstream direction. Both the force amplitudes and the mean upward forces for non-ventilated flow were smaller when the upper and lower sides of the lower leaf had larger inclinations. Obviously, a cross-sectional shape like that in Figure 9.42c is the most favorable also with respect to instability-induced excitation resulting from simultaneous over- and underflow. Provision of an effective air vent is essential for all cross-sectional shapes.

9.2.6 *Concluding remarks*

Practicing engineers may prefer the form of presentation adopted in Chapters 8 and 9 and perhaps wish that this entire monograph consisted of practical examples of engineering systems and structures. Still, the presentation in terms of basic elements of flow-induced vibrations in the preceding chapters was chosen for good reasons. Had flow-induced vibrations been discussed exclusively in terms of affected systems and structures, the reader would have been misled into thinking that by eliminating the various *specific* causes for excitation, it would be possible to safeguard similar systems against vibration problems.

We wish to reemphasize that in order to come close to the goal of ascertaining *all possible* origins of flow-induced vibrations, one must, first, thoroughly iden-tify all the various body oscillators (Chapter 2) and fluid oscillators (Chapter 4) for the system of interest; and, second, one must discover all possible sources of parametric excitation (Section 2.5) as well as extraneously, instability-, and movement-induced excitations (Chapters 5-7) that might affect these body and fluid oscillators, either individually or in combination, for the given operating conditions. From one application to another, hardly any type of structure, say a grid or a gate in a water-power scheme, is identical in geometry, dynamic characteristics, integration into the overall system, and operating conditions. Therefore, a given type of structure may be susceptible to different types of excitation, depending upon the specific circumstances. Moreover, the possible types of excitation can take a multitude of forms and, as a consequence, may be difficult to detect in a given installation.

Engineers are advised, therefore, to use the examples in Chapters 8 and 9 merely as a means of training themselves in recognizing the basic elements of flow-induced vibrations treated in the preceding chapters. Simpler roads or short cuts to satisfactory engineering solutions do not seem to exist in this complex field of fluid-structure interaction.

REFERENCES

Abramowitz, M. & I.A.Stegun (eds) 1970. *Handbook of Mathematical Functions*. Dover, p. 411 and 468.

Abramson, H.N. 1966. Dynamic behavior of liquid in moving container. *Applied Mechanics Review*, Vol. 16, p. 501.

Abramson, H.N. 1969. Hydroelasticity: A review of hydrofoil flutter. *Applied Mechanics Review*, Vol. 22, No. 2, p. 115.

Achenbach, E. & E. Heinecke 1981. On vortex shedding from smooth and rough cylinders in the range of Reynolds numbers 6×10^3 to 5×10^6. *Journal Fluid Mechanics*, Vol. 109, p. 239.

Ackermann, N.L. & A. Arbhabhirama 1964. Viscous and boundary effects on virtual mass. *ASCE Journal Engineering Mechanics Division*, Vol. 90, No. EM4, p. 123.

Aguirre, J.E. 1977. Flow-induced, in-line vibrations of a circular cylinder. Dissertation, Imperial College of Science and Technology, London. (Discussed in: Naudascher, 1987b.)

Alster, M. 1972. Improved calculation of resonant frequencies of Helmholtz resonators. *Journal Sound and Vibration*, Vol. 24, p. 64.

Anderson, J.S. 1977. The effect of an air flow on a single side-branch Helmholtz resonator in a circular duct. *Journal Sound and Vibration*, Vol. 52, No. 3, p. 423.

Anderson, J.S., W.M. Jungowski, W.J. Hiller & G.E.A. Meier 1977. Flow oscillations in a duct with rectangular cross-section. *Journal Fluid Mechanics*, Vol. 79, Pt. 4, p. 769.

Anderson, J.S., G. Grabitz, G.E.A. Meier, W.M. Jungowski & K.J. Witczak 1978. Base pressure oscillations in a rectangular duct with an abrupt enlargement. *Archives of Mechanics*, Vol. 87, No. HY6.

Appel, D.W. & Y.S. Yu 1966. Pressure pulsations in flow through branched pipes. *ASCE Journal Hydraulics Division*, Vol. 92, No. HY6, p. 179.

Appelt, C.J., G.S. West & A.A. Szewczyk 1973, 1975. The effects of wake splitter plates on the flow past a circular cylinder in the range $10^4 < Re < 5 \times 10^5$. *Journal Fluid Mechanics*, Vol. 61, p. 187 and Vol. 71, p. 145.

Arndt, R.E.A. 1981. Cavitation in fluid machinery and hydraulic structures. *Annual Review of Fluid Mechanics*, Vol. 13, p. 273.

Aschrafi, M. & G.M. Hirsch 1983. Control of wind-induced vibrations of cable-stayed bridges. *Journal Wind Engineering and Industrial Aerodynamics*, Vol. 14, p. 235.

Au-Yang, M.K. 1986. Turbulent buffeting of a multispan tube bundle. *ASME Journal Vibration, Stress, and Reliability in Design*, Vol. 108, p. 150.

367

Baird, R.C. 1955. Wind-induced vibration of a pipeline suspension bridge and its cure. *Transactions ASME*, p. 797.

Bajaj, A. K. & V. Garg 1977. Linear stability of jet flows. *ASME Journal Applied Mechanics*, p. 378.

Bakes, F. & B. P. McGoodwin 1980. Flow-induced vibration in circulating water system. *ASCE Journal Hydraulics Division*, Vol. 106, No. HY8, p. 1394.

Ball, J.W. 1959. Hydraulic characteristics of gate slots. *ASCE Journal Hydraulics Division*, Vol. 85, No. HY10, p. 81.

Bardowicks, H. 1976. Effects of cross-sectional shape and amplitude on aeroelastic vibrations of sharp-edged prismatic bodies (in German). Doctoral Dissertation, Techn. Universität Hannover, Germany.

Bauer, H. F. & A. Siekmann 1971. Dynamical interaction of a liquid with the elastic structure of a circular container. *Ingenieur Archiv*, Vol. 40, p. 266.

Bearman, P.W. 1967. On vortex street wakes. *Journal Fluid Mechanics*, Vol. 28, Pt. 4, p. 625.

Bearman, P.W. 1967b. The effect of base bleed on the flow behind a two-dimensional model with a blunt trailing edge. *Aeronautical Quarterly*, Vol. 18, p. 207.

Bearman, P.W. 1969. On vortex shedding from a circular cylinder in the critical Reynolds number regime. *Journal Fluid Mechanics*, Vol. 37, Pt. 3, p. 577.

Bearman, P.W. 1972. Some measurements of the distortion of turbulence approaching a two-dimensional bluff body. *Journal Fluid Mechanics*, Vol. 53, Pt. 3, p. 451.

Bearman, P.W. 1984. Vortex shedding from oscillating bluff bodies. *Annual Review Fluid Mechanics*, Vol. 16, p. 195.

Bearman, P.W. & M.E. Davies 1977. The flow about oscillating bluff structures. *Intern. conference wind effects on buildings and structures* (K.J. Eaton, ed.), Cambridge University Press, p. 285.

Bearman, P.W. & E.D. Obasaju 1982. An experimental study of pressure fluctuations on fixed and oscillating square-section cylinders. *Journal Fluid Mechanics*, Vol. 119, p. 297.

Bearman, P.W. & D.M. Trueman 1972. An investigation of the flow around rectangular cylinders. *Aeronautical Quarterly*, Vol. 23, p. 229.

Bearman, P.W. & A.J. Wadcock 1973. The interaction between a pair of circular cylinders normal to a stream. *Journal Fluid Mechanics*, Vol. 61, Pt. 4, p. 499.

Bearman, P.W., I.S. Gartshore, D.J. Maull & G.V. Parkinson 1987. Experiments on flow-induced vibration of a square-section cylinder. *Journal Fluids and Structures*, Vol. 1, p. 19.

Benjamin, T.B. 1963. The threefold classification of unstable disturbances in flexible surfaces bounding inviscid flows. *Journal Fluid Mechanics*, Vol. 15, No. 3.

Beranek, L.L. 1954. *Acoustics*, McGraw-Hill.

Berger, E. 1978a. Some new aspects in fluid oscillator model theory. *Colloquium industrial aerodynamics*, Fachhochschule Aachen, Germany.

Berger, E. 1978b. On some progress in Fluid-Oscillator-Model Theory. Mitteilung Curt-Risch-Institut, Technische Universität Hannover, Germany, Vol. 1.

Betz, A. & E. Petersohn 1931. Application of the theory of free jets (in German). *Ingenieur-Archiv*, Vol. 2, p. 190.

Binnie, A.M. 1972. The stability of a falling sheet of water. *Proceedings Royal Society of London*, Series A, Vol. 326.

Binnie, A.M. 1979. Unstable flow under a sluice gate. *Proceedings Royal Society of London*, Series A, Vol. 367, p. 311.

Birkhoff, G. & E.H. Zarantanello 1957. *Jets, wakes and cavities.* Academic Press, p. 291/2.

Bishop, R.E.D. & A.Y. Hassan 1964. The lift and drag forces on a circular cylinder oscillating in a flowing fluid. *Proceedings Royal Society of London*, Series A, Vol. 277, p. 51.

Bisplinghoff, R.L., H. Ashley & R.L. Halfmann 1955. *Aeroelasticity.* Addison Wesley.

Blake, W.K. 1986. *Mechanics of fluid-induced sound and vibration.* Vol. 17-I and 17-II in Applied Mathematics and Mechanics, Intern. Monograph Series, Academic Press.

Blenk, H.F., D. Fuchs & F. Liebers 1935. On measurements of vortex frequencies (in German). *Luftfahrtforschung*, Vol. 12, p. 38.

Blevins, R.D. 1975. Vibrations of a loosely held tube. *Journal Eng. Ind.*, Vol. 97, p. 1301.

Blevins, R.D. 1977. *Flow-induced vibrations.* Van Nostrand Reinhold. (Updated edition 1990).

Blevins, R.D. 1984. Review of sound induced by vortex shedding from cylinders. *Journal Sound and Vibration*, Vol. 92, No. 4, p. 455.

Blevins, R.D. 1986. *Formulas for natural frequency and mode shape.* (Repr. Ed.). Robert E. Krieger, Florida.

Blevins, R.D. 1986a. Acoustic modes of heat exchanger tube bundles. *Journal Sound and Vibration*, Vol. 109, p. 19.

Blevins, R.D. & M.M. Bressler 1987. Acoustic resonance in heat-exchanger tube bundles. (1. Physical nature of the phenomenon; 2. Prediction and suppression of resonance.) *ASME Journal Pressure Vessel Technology*, Vol. 109, No. 3, p. 275/283.

Blevins, R.D., R.J. Gilbert & B. Villard 1981. Experiment on vibration of heat exchanger tube arrays in cross flow. *6th Intern. conference structural mechanics in reactor technology*, Paris, Paper B6/9.

Bokaian, A.R. & F. Geoola 1982a. Hydrodynamic galloping of rectangular cylinders. London Centre for Marine Technology, Dept. Civil Engineering, University of London, UK, April.

Bokaian, A.R. & F. Geoola 1982b. On the cross-flow response of cylindrical structures. London Centre for Marine Technology, Dept. Civil Engineering, University of London, UK, August.

Bokaian, A.R. & F. Geoola 1983. On the cross-flow response of cylindrical structures. *Proceedings Institution Civil Engineers*, London, Vol. 75, p. 397.

Bokaian, A.R. & F. Geoola 1984. Hydroelastic instabilities of square cylinders. *Journal Sound and Vibration*, Vol. 92, p. 117.

Bokaian, A.R. & F. Geoola 1984a. Wake-induced galloping of two interfering circular cylinders. *Journal Fluid Mechanics*, Vol. 146, p. 383.

Bokaian, A.R. & F. Geoola 1984b. Proximity-induced galloping of two interfering circular cylinders. *Journal Fluid Mechanics*, Vol. 146, p. 417.

Bourque, C. & B.G. Newman 1960. Reattachment of a two-dimensional, incompressible jet to an adjacent flat plate. *The Aeronautical Quarterly*, Vol. XI, August.

Božović, A. 1973. Vibration of the bottom outlet on the Bayina Bashta Dam. *11th Intern. congress on large dams*, Madrid.

Brackenridge, J.B. & W.L. Nyborg 1957. Acoustical characteristics of oscillating jet-edge systems in water. *Journal Acoustical Society of America*, Vol. 29, No. 4.

Brooks, N.P.H. 1960. Experimental investigation of the aeroelastic instability of bluff two-dimensional cylinders. M.A.Sc. Thesis, University British Columbia, Vancouver, Canada.

Brown, F.R. 1961. Fluctuation of control gates. *9th IAHR congress*, Dubrovnik, Yugoslavia.

Burton, T. E. 1980. Sound speed in a heat exchanger tube bank. *Journal Sound and Vibration*, Vol. 71(1), p. 157.

Byrne, K. P. 1983. The use of porous baffles to control acoustic vibrations in crossflow tubular heat exchangers. *Journal Heat Transfer*, Vol. 105, p. 751.

Callander, S.J. 1987. Flow-induced vibrations of rectangular bars (in German). Doctoral dissertation Universität Karlsruhe. (Also: Sonderforschungsbereich 210, Universität Karlsruhe, Germany, Report SFB 210/ET/37.)

Callander, S.J. 1988. Streamwise oscillations of circular-sectioned trashrack bars. Sonderforschungsbereich 210, Universität Karlsruhe, Germany, Report SFB 210/E/43.

Camichel, C., P. Dupin & M. Tessier-Soller 1927. Sur l'application de la loi de similitude aux périodes des formations des tourbilllons alternés de Bénard-Karman. *Comptes Rendus*, Vol. 185, p. 1556.

Camichel, E., D. Eydoux & M. Gariel 1919. *Etude theoretique et experimentale des coups de Bèlier*. Dunod.

Carlucci, L.N. 1983. Damping and hydrodynamic mass of a cylinder in simulated two-phase flow. *ASME Journal Vibration, Acoustics, Stress, and Reliability in Design*, Vol. 105, p. 83.

Carlucci, L.N. & J.D. Brown 1982. Experimental studies of damping and hydrodynamic mass of a cylinder in confined two-phase flow. *ASME Journal Mechanical Design*, Vol. 104, ASME Paper 81-DET-27.

Carman, A. 1960. Report on vibrating nappe. Dept. Civil Engineering, University Witwatersrand, Johannesburg, South Africa.

Carr, L.W. 1985. Dynamic stall progress in analysis and prediction. *AIAA atmospheric flight mechanics conference*, Snowmass, Colorado, Paper No. 85-1769-CP.

Carr, L.W., K.W. McAlister & W.J. McCroskey 1977. Analysis of the development of dynamic stall based on oscillating airfoil experiments. NASA TN D-8382.

Carta, F.O. 1978. Aeroelastic and unsteady aerodynamics. In: *The Aerothermodynamics of aircraft gas turbine engines*, G.C. Oates (ed.), AF APL-TR78-52, AD-AO59 784. Available from Nat. Techn. Inform. Service, US Dept. Commerce, Springfield, Virginia, 22161.

Carta, F.O. & A.O. St. Hilaire 1978. Experimentally determined stability parameters of a subsonic cascade oscillating near stall. *ASME Journal Engineering for Power*, Vol. 100, p. 111.

Casparini, D.A., L.W. Curry & A. Deb Chaudhury 1980. Passive viscoelastic systems for increasing the damping of buildings. *4th Colloquium industrial aerodynamics*, Fachhochschule Aachen, Germany, Vol. 2, p. 257.

Cassidy, J.J. & H.T. Falvey 1970. Observations of unsteady flow arising after vortex breakdown. *Journal Fluid Mechanics*, Vol. 41, p. 727.

Chan, Y.Y. 1974. Spatial waves in turbulent jets. *Physics of Fluids*, Vol. 17, No. 1, p. 46.

Chanaud, R.C. 1963. Experiments concerning the vortex whistle. *Journal Acoustical Society of America*, Vol. 35, p. 953.

Chanaud, R.C. 1965. Observations on the oscillatory motion in certain swirling flows. *Journal Fluid Mechanics*, Vol. 21, p. 111.

Chang, H.T. & I.G. Hampton 1980. Experiences in flow-induced gate vibration. *Intern. conference water resources development*. Taipei, Taiwan, p. 945.

Chen, S.S. 1975. Vibration of nuclear fuel bundles. *Nuclear Engineering and Design*, Vol. 35, p. 399.

Chen, S.S. 1977. Flow-induced vibrations of fluid-conveying cylindrical shells. *Journal Engineering for Industry*, Vol. 96, p. 420.

Chen, S.S. 1983. Instability mechanisms and stability criteria of a group of circular cylinders subjected to cross flow. Part 1: Theory. *ASME Journal Vibration, Acoustics, Stress, and Reliability in Design*, Vol. 105, p. 51.

Chen, S.S. 1984. Guidelines for the instability flow velocity of tube arrays in axial and cross flow. *Journal Sound and Vibration*, Vol. 93, p. 439.

Chen, S.S. 1987. *Flow-induced vibration of circular cylindrical structures*. Hemisphere Publ. Co./Springer.

Chen, S.S. & H. Chung 1976. Design guide for calculating hydrodynamic mass. ANL-CT-76-45, Argonne National Laboratory, Argonne, Illinois.

Chen, S.S. & J.A. Jendrzejczyk 1983. Stability of tube arrays in crossflow. *Nuclear Engineering and Design*, Vol. 75, p. 351.

Chen, S.S. & J.A. Jendrzejczyk 1987. Fluid excitation forces acting on a tube array. *ASME Journal Fluids Engineering*, Vol. 109, p. 415.

Chen, S.S. & M.W. Wambsganns 1972. Parallel flow induced vibration of fuel rods. *Nuclear Engineering Design*, Vol. 18, p. 253.

Chen, S.S., M.W. Wambsganns & J.A. Jendrzejczyk 1976. Added-mass and damping of a vibrating rod in confined viscous fluids. *ASME Journal Applied Mechanics*, Vol. 43, p. 325.

Chen, Y.N. 1968. Flow-induced vibration and noise in tube-bank heat exchangers due to von Karman Vortex Streets. *ASME Journal Engineering for Industry*, Feb., p. 134.

Chen, Y.N. 1977b. The sensitive tube spacing region of tube bank heat exchangers for fluid-elastic coupling in cross-flow. In: *Fluid-structure interaction phenomena in pressure vessel and piping systems, ASME*; M.K. Au-Yang & S.J. Brown (eds.), p. 1.

Chen, Y.N. 1978. Damping of vibration induced by Karman vortices in axial flows (in German). Sulzer-Forschungsheft.

Chen, Y.N. 1980. Experiences with flow-induced vibrations at Sulzer. In: *Practical experiences with flow-induced vibrations*, E. Naudascher & D. Rockwell (eds.). Springer.

Cheung, J.C.K. & W.H. Melbourne 1983. Turbulence effects on some aerodynamic parameters of a circular cylinder at supercritical Reynolds numbers. *Journal Wind Engineering and Industrial Aerodynamics*, Vol. 14, p. 399.

Chow, W.Y. & A.B. Rudavsky 1980. Vibrations of a cantilevered pump column. In: *Practical experiences with flow-induced vibrations*, E. Naudascher & D. Rockwell (eds.). Springer.

Chung, H. & S.S. Chen 1977. Vibration of a group of circular cylinders in a confined fluid. *ASME Journal Applied Mechanics*, Vol. 44, p. 213.

Clays, D. & G. Tison 1968. Vortex-induced oscillations at low-head weirs. Discussion of paper by Z.G. Hanko. *ASCE Journal Hydraulics Division*, Vol. 94, No. HY4, p. 1160.

Cooper, K.R. 1974. Wake galloping on aeroelastic instability. In: *Flow-induced structural vibrations*, Naudascher (ed.). Springer, p. 762.

Cooper, K.R. 1988. The use of a forebody plate to reduce the drag and to improve the aerodynamic stability of a cylinder of a square cross-section. *Journal Wind Engineering and Industrial Aerodynamics*, Vol. 28, p. 271.

Corless, R.M. & G.V. Parkinson 1988. A model of the combined effects of vortex-induced oscillation and galloping. *Journal Fluids and Structures*, Vol. 2, p. 203.

Courchesne, J. & A. Laneville 1982. An experimental evaluation of drag coefficient for rectangular cylinders exposed to grid turbulence. *ASME Journal Fluids Engineering*, Vol. 104, p. 523.

Cowdrey, C.F. & J.A. Lawes 1959. Drag measurements at high Reynolds numbers of a circular cylinder fitted with three helical strakes. Nat. Phys. Lab., UK, Aero Rep. No. 384.

Crandall, S.H., S. Vigander & P.A. March 1975. Destructive vibration of trashracks due to fluid-structure interaction. *ASME Journal Engineering for Industry*, Nov., p. 1359.

Crausse, E. 1939. Sur un phènomene d'oscillation du plan d'eau provoqué par l'écoulement autour d'obstacles en forme de piles de pont. *Comptes rendus de séances de l'Academie des Sciences*, p. 209.

Crow, D.A., R. King & M.J. Prosser 1978. Hydraulic model studies of the rising sector gate: Hydrodynamic loads and vibrations studies. Thames Barrier Design, Institution of Civil Engineers, London, p. 125.

Davenport, A.G. 1961. Application of statistical concepts to the wind loading of structures. *Proceedings Institution of Civil Engineers*, London, Vol. 19, p. 449.

Davenport, A.G. 1962. Buffeting of a suspension bridge by storm winds. *ASCE Journal Structure Division*, Vol. 88, p. 233.

Davenport, A.G. 1963. The buffeting of structures by gusts. *Intern. symposium wind effects on buildings and structures*, Nat. Phys. Lab., London.

Davies, D.D., G.M. Stokes, D. Moore & G.L. Stevens 1954. NACA Report 1192.

Davies, P.O.A.L. 1977. Bench test procedures and exhaust system performance. *Surface transportation exhaust noise symposium*, Environmental Protection Agency Report 550/9-78-206, p. 5.

Defant, A. 1961. *Physical Oceanography*, Vol. II. Pergamon Press.

Delany, N.K. & N.E. Sorensen 1953. Low-speed drag of cylinders of various shapes. NACA, TN 3028.

Den Hartog, J.P. 1985. *Mechanical vibrations*, Dover, 4th ed.

Derby, T.F. & J.E. Ruzicka 1969. Loss factor and resonant frequency of viscoelastic shear damped structural composites. Nat. Aeron. and Space Admin. Report NASA CR-1269.

Dickens, W.R. 1979. The self-induced vibration of cylindrical structures in fluid flow. *Proceedings Institution of Civil Engineers*, London, Vol. 67, p. 13.

Dickey, W.L. & G.B. Woodruff 1956. The vibration of steel stacks. *Transactions ASCE*, Vol. 121, p. 1054.

Dimotakis, P.E., R.C. Lye & D.Z. Papantoniou 1981. *15th Intern. symposium fluid dynamics*, Jachranka, Poland.

D'Netto, W. & D.S. Weaver 1987. Divergence and limit cycle oscillations in valves operating at small openings. *Journal Fluids and Structures*, Vol. 1, p. 3.

Dockstader, E.A., W.F. Swiger & E. Ireland 1956. Resonant vibration of steel stacks. *Transactions ASCE*, Vol. 121, p. 1088.

Donaldson, R.M. 1956. Hydraulic turbine runner vibration. *ASME Journal Engineering Power*, Vol. 78, p. 1141.

Dörfler, P. 1980. Mathematical model of the pulsations of Francis turbines caused by the vortex core at partial load. *Escher Wyss News* 1/2, p. 101.

Dougherty, N.S. & C.F. Anderson 1976. An experimental study on suppression of edgetones from perforated wind tunnel walls. AIAA Paper 76-50, 14th AIAA Aerospace Sciences Meeting, Washington, D.C.

Dowell, E.H. 1966. Flutter of infinitely long plates and shells, Part 1: Plate. *AIAA Journal*, Vol. 4, p. 1370.

Dowell, E.H. 1970. Panel flutter: A review of the aeroelastic stability of plates and shells. *AIAA Journal*, Vol. 8, p. 385.

Dowell, E.H. 1974. *Aeroelasticity of plates and shells*. Noordhoff Intern. Publishing.

Dowell, E.H., H.C. Curtiss, R.H. Scanlan & F. Sisto 1989. *A modern course in aeroelasticity*. Kluwer Academic Publishers, 2nd ed.

Drescher, H. 1956. Measurement of unsteady pressure on cylinder in cross-flow (in German). *Zeitschrift Flugwissenschaften*, Vol. 14, No. 112, p. 17.

Durgin, W.W., P.A. March & P.J. Lefebvre 1980. Low-mode response of circular cylinders in cross-flow. *ASME Journal Fluids Engineering*, Vol. 102, p. 183.

Eaton, J.K. & J.P. Johnston 1980. A review of research on subsonic turbulent flow reattachment. *AIAA 13th plasma and fluid dynamics conference*, Snowmass, Colorado, Paper No. 80-1438.

Eaton, J.K. & J.P. Johnston 1981. Low-frequency unsteadiness of a reattaching turbulent shear layer. *3rd Symposium turbulent shear flows*, University of California at Davis.

Ehrich, F.E. 1970. Aerodynamic stability of branched diffusers. *ASME Journal Engineering for Power*, Vol. 92, p. 330.

Eisinger, F.L. 1980. Prevention and cure of flow-induced vibration problems in tubular heat exchangers. *ASME Journal Pressure Vessel Technology*, Vol. 102, p. 138.

Elder, R.A. & J.M. Garrison 1964. Form-induced hydraulic forces on three-leaf intake gates. *ASCE Journal Hydraulics Division*, Vol. 90, No. HY3, p. 215.

Emmerling, R. 1973. Instantaneous structure of wall pressure in turbulent boundary-layer flow (in German). Mitteilungen Max-Planck-Institut für Strömungsforschung, Göttingen, Germany, No. 56.

Ericsson, L.E. 1980. Karman vortex shedding and the effect of body motion. *AIAA Journal*, Vol. 18, No. 8.

Ericsson, L.E. & J.P. Reding 1988. Fluid mechanics of dynamic stall, Part I: Unsteady-flow concepts; Part II: Prediction of full-scale characteristics. *Journal Fluids and Structures*, Vol. 2, p. 1 and 113.

Escudier, M.P. 1980. Swirling-flow induced vibrations in turbomachine exit chambers. In: *Practical experiences with flow-induced vibrations*, E. Naudascher & D. Rockwell (eds.). Springer.

Escudier, M.P. 1987. Confined vortices in flow machinery. *Annual Review Fluid Mechanics*, Vol. 19, p. 27.

Escudier, M.P. 1988. Vortex breakdown: Observations and explanations. *Progress in Aerospace Sciences*, Vol. 25, p. 189.

Escudier, M.P. & J.J. Keller 1983. Vortex breakdown: A two stage transition. AGARD, CP No. 342, Paper 25.

Escudier, M.P. & P. Merkli, 1979. Observations of the oscillatory behavior of a confined ring vortex. *Am. Inst. Aeronautics and Astronautics Journal*, Vol. 17, No. 3, p. 253.

Ethembabaoglu, S. 1973. On the fluctuating flow characteristics in the vicinity of gate slots. Division Hydraulic Engineering, Norwegian Institute of Technology, Trondheim.

Ethembabaoglu, S. 1978. Some characteristics of unstable flow past slots. *ASCE Journal Hydraulics Division*, Vol. 104, No. HY5, p. 649.

Etzold, F. & H. Fiedler 1976. Near-wake structure of a cantilevered cylinder in a cross-flow. *Zeitschrift Flugwissenschaften*, Vol. 24, No. 2, p. 77.

Every, M.J., R. King & O.M. Griffin 1982a. Hydrodynamic load on flexible marine structures due to vortex shedding. *ASME Journal Engineering Resources Technology*, Vol. 104, p. 330.

Every, M.J., R. King & D.S. Weaver 1982b. Vortex-excited vibrations of cylinders and cables and their suppression. *Ocean Engineering*, Vol. 9, No. 2, p. 135.

Falvey, H.T. 1971. Draft tube surges. A review of present knowledge and an annotated bibliography. US Bureau of Reclamation, Rep. REC-ERC-71-42.

Falvey, H.T. 1980. Bureau of Reclamation experience with flow-induced vibrations. In:

Practical experiences with flow-induced vibrations, E. Naudascher & D. Rockwell (eds.). Springer.

Falvey, H.T. & J.J. Cassidy 1970. Frequency and amplitude of pressure surges generated by swirling flow. *IAHR Hydrdaulic machinery symposium*, Stockholm.

Feld, L.S. 1971. Superstructure for World Trade Center. *Civil Engineering, ASCE*, p. 66.

Feng, C.C. 1968. The measurement of vortex induced effects in flow past stationary and oscillating circular and D-section cylinders. M.A.Sc. Thesis, University of British Columbia, Canada.

Fischer, F.J., W.T. Jones & R. King 1980. Current-induced oscillations of cognac piles during installation – Prediction and measurement. In: *Practical experiences with flow-induced vibrations*, E. Naudascher & D. Rockwell (eds.). Springer.

Fitzhugh, J.S. 1973. Flow-induced vibration in heat exchangers. In: *UKAEA/NPL Intern. symposium on vibration problems in industry*, Keswick, UK, Paper 427.

Fitzpatrick, J.A. 1986. A design guide proposal for avoidance of acoustic resonance in in-line heat exchangers. *ASME Journal Vibration, Acoustics, Stress, and Reliability in Design*, Vol. 108, No. 3, p. 296.

Fleeter, S. 1977. Aeroelasticity research for turbomachine applications. *AIAA dynamics specialist conference*, San Diego, California (March 24-25), Paper 77-437.

Fox, R.W. & S.J. Kline 1960. Flow regime data and design methods for curved subsonic diffusors. Report PD-6, Dept. Mechanical Engineering, Stanford University, California.

Franke, M.E. & D.L. Carr 1975. Effects of geometry on open cavity flow-induced pressure oscillations. AIAA Paper 75-492, *AIAA Aero-Acoustics Conference*, Hampton, Virginia.

Franken, P.A. 1960. Reactive mufflers. In: *Noise reduction*, L.L. Beranek (ed.). McGraw Hill, p. 414.

Fukushima, J. & L.U. Dadone 1977. Comparison of dynamic stall phenomena for pitching and vertical translation motions. NASA CR-2793.

Furness, R.A. & S.P. Hutton 1975. Experimental and theoretical studies of two-dimensional flixed-type cavities. *ASME Journal Fluids Engineering*, Vol. 97, p. 515.

Gal, A. 1971. Determination of added mass for flaps with overflow (in German). Institut für Wasserbau, Universität Stuttgart, Germany, Report No. 18.

Garrick, I.E. 1976. Aeroelasticity, frontiers and beyond. *AIAA Journal of Aircraft*, Vol. 13, No. 9, p. 641.

Garrick, I.E. & W.H. Reed 1981. Historical development of aircraft flutter. *AIAA Journal of Aircraft*, Vol. 18, No. 11, p. 897.

Gartshore, I.S. 1984. Some effects of upstream turbulence on the unsteady lift forces imposed on prismatic two-dimensional bodies. *ASME Journal Fluids Engineering*, Vol. 106, p. 418.

Gartshore, I.S., J. Khanna & S. Laccinole 1979. The effectiveness of vortex spoilers on a circular cylinder in smooth and turbulent flow. *5th Intern. conference wind engineering*, Fort Collins, Colorado.

Gaster, M. 1969. Vortex shedding from slender cones at low Reynolds numbers. *Journal Fluid Mechanics*, Vol. 38, Pt. 3, p. 565.

Gaster, M. 1971. Some observations on vortex shedding and acoustic resonances. Aeronautical Research Council, Current Paper No. 1141.

Gerrard, J.H. 1966. The mechanics of the formation region of vortices behind bluff bodies. *Journal Fluid Mechanics*, Vol. 25, p. 401.

Gibson, H.H. 1952. *Hydraulics and its applications*. 5th ed., London.

Götzmann, K.-H. 1987. Parameter study on the vibration excitation of trashrack bars in inclined flow (in German). Diploma Thesis, Institut für Hydromechanik, Universität Karlsruhe, Germany, No. 535.

Goldschmied, F. R. & D. N. Wormley 1977. Frequency response of blower/duct/plenum fluid system. *Journal Hydronautics*, Vol. 11, No. 1, p. 18.

Gorman, D. J. 1980. The effect of artificially induced upstream turbulence on the liquid cross-flow induced vibration of tube bundles. In: *Flow-induced vibration in power plant components*, M. K. Au-Yang (ed.). PVP-Vol. 41, *ASME*, New York. (Also: Gorman 1982, *Intern. conference vibration in nuclear plants*, Keswick, UK, Paper 1.6.)

Gowda, B. H. L. 1975. Some measurements on the phenomenon of vortex shedding and induced vibrations of circular cylinders. Institut für Thermo- und Fluiddynamik, Techn. Universität Berlin, Rep. DLR-FB 75 -01.

Graham, J. M. R. & D. J. Maull 1971. The effect of oscillating flap and an acoustic resonance on vortex shedding. *Journal Sound and Vibration*, Vol. 18, No. 3, p. 371.

Greenspan, J. E. 1961. Vibrations of cross-stiffened and sandwich plates with application to underwater sound radiators. *Journal Acoustic Society of America*, Vol. 33, p. 1485.

Greenway, M. E. & C. J. Wood 1973. The effect of a bevelled trailing edge on vortex shedding and vibration. *Journal Fluid Mechanics*, Vol. 61, p. 323.

Gregory, R. W. & M. P. Paidoussis 1966. Unstable oscillation of tubular cantilevers conveying fluid. *Proceedings Royal Society of London*, Series A, Vol. 293, p. 513 & 528.

Grein, H. 1980. Vibration phenomena in Francis turbines: Their causes and prevention. *IAHR hydraulic machinery symposium*, Tokyo.

Greitzer, E. M. 1981. The stability of pumping systems. *ASME Journal Fluids Engineering*, Vol. 103, p. 193.

Griffin, O. M. 1980. Vortex-excited cross flow vibrations of a single cylindrical tube. In: *Flow-induced vibrations, ASME Congress,* San Francisco, California, S. S. Chen & M. D. Bernstein (eds). p. 1.

Griffin, O. M. 1984. Vibrations and flow-induced forces caused by vortex shedding. In: *Flow-induced vibrations, ASME Symposium,* New Orleans, Louisiana, M. P. Paidoussis, M. K. Au-Yang & S. S. Chen (eds). p. 1.

Griffin, O. M. 1985. Vortex shedding from bluff bodies in a shear flow – A review. *ASME Journal Fluids Engineering*, Vol. 107, p. 298.

Griffin, O. M. 1989. Flow similitude and vortex lock-on in bluff body near wakes. *Physics of Fluids A*, Vol. 1, No. 4, p. 697.

Griffin, O. M. & G. H. Koopmann 1977. The vortex-excited lift and reaction forces on resonantly vibrating cylinders. *Journal Sound and Vibration*, Vol. 54, p. 435.

Griffin, O. M. & S. E. Ramberg 1976. Vortex shedding from a cylinder vibrating in line with an incident flow. *Journal Fluid Mechanics*, Vol. 75, Pt. 2, p. 257.

Griffin, O. M. & S. E. Ramberg 1982. Some recent studies of vortex shedding with application to marine tubulars and risers. *ASME Journ. Energy Resources Technology*, Vol. 104, p. 2.

Griffiths, P. T. A. 1969. Large gates and valves: Vibration. Colloq. for hydraulics laboratory staff, Australian Water Resources Council, Cooma, NSW, p. 40.

Grimminger, G. 1945. The effect of rigid guide vanes on the vibration and drag of a towed circular cylinder. David Taylor Model Basin, Washington, DTMB-Re. 504.

Hamilton, W. S. 1982. Some design factors for the Bath County trashracks. *Water Power and Dam Construction*, p. 16.

Hannoyer, M. J. & M. P. Paidoussis 1978. Instabilities of tubular beams simultaneously subjected to internal and external axial flows. *ASME Journal Mechanical Design*, Vol. 100, p. 328.

Hara, F. 1973. A theory on the two-phase-flow-induced vibrations in piping systems. *2nd Intern. conference structural mechanics in reactor technology*, Berlin, Paper No. F5/1.

Hara, F. 1977. Two-phase flow-induced vibrations in a horizontal piping system. *Bulletin Japanese Soc. Mech. Engineers, JSME*, Vol. 20 (142), p. 419.

Hara, F. 1982a. Two-phase cross-flow-induced forces acting on a circular cylinder. In: *Flow-induced vibration of circular cylindrical structures*, S.S. Chen, M.P. Paidoussis & M.K. Au-Yang (eds). ASME Publ. PVP-63, p. 9.

Hara, F. 1982b. Air-bubble effects on vortex-induced vibrations of a circular cylinder. *ASME Symposium flow-induced vibrations*, New Orleans, Louisiana, Vol. 1, p. 103.

Hara, F. & O. Kohgo 1982. Added mass and damping of a vibrating rod in a two-phase air-water mixed fluid. *Conference flow-induced vibration of circular cylindrical structures*, Orlando, Florida, ASME Publ. PVP Vol. 63, p. 1.

Hara, F., T. Shigeta & H. Shibata 1974. Two-phase flow induced random vibrations. In: *Flow-induced structural vibrations*, E. Naudascher (ed.). Springer, p. 691.

Hardwick, J.D. 1969. Periodic vibrations in model sluice gates. Ph.D. Thesis, Imperial College of Science and Technology, London.

Hardwick, J.D. 1978. Hydraulic model studies of the rising sector gate constructed at Imperial College; Thames Barrier Design. *Proceedings Institution of Civil Engineers*, London, p. 143.

Hardwick, J.D. 1985. Progressive gate modelling for studies of flow-induced vibration. *Proceedings Institution of Civil Engineers*, London, Vol. 79, p. 483.

Hardwick, J.D. & L.R. Wootton 1973. The use of model and full-scale investigations on marine structures. *Intern. symposium vibration problems in industry*. UKAEA & NPL, Keswick, England, Paper 127.

Hardwick, J.D., M.J. Kenn & W.T. Mee 1980. Gate vibrations at El-Chocon hydro-power scheme, Argentina. In: *Practical experiences with flow-induced vibrations*, E. Naudascher & D. Rockwell (eds.). Springer.

Haroun, M.A. 1983. Vibration studies and tests of liquid storage tanks. *Earthquake Engineering and Structural Dynamics*, Vol. 11, No. 2.

Hartlen, R.T. & I.G. Currie 1970. Lift oscillator model of vortex-induced vibration. *ASCE Journal Engineering Mechanics Division*, Vol. 96, No. EM5.

Hawthorne, W.R. 1961. The early development of the Dracone Flexible Barge. *Proceedings Institution of Mechanical Engineers*, London, Vol. 175, p. 52.

Hebaus, G.G. 1983. Hydraulics of closed conduit spillways, Part XVIII. Agricultural Research Service, US Dept. Agriculture; in coop. with SAFHL, University of Minnesota.

Hecker, G.E. & J.A. Dale 1966. Vibration of lock valves. *ASCE structural engineering conference*, Miami Beach, Florida. (Also: Research 1956-66. Tennessee Valley Authority, Engineering Lab. Norris, Tennessee, 1967.)

Heller, H.H. & D.B. Bliss 1975. The physical mechanism of flow-induced pressure fluctuations in cavities and concepts for their suppression. *AIAA 2nd aero-acoustics conference*, Hampton, Virginia, Paper 75-491.

Heskestad, G. & D.R. Olberts 1960. Influence of trailing edge geometry on hydraulic-turbine-blade vibration resulting from vortex excitation. *ASME Journal Engineer for Power*, Vol. 82, p. 103.

Hikami, Y. & N. Shiraishi 1987. Rain-wind induced vibrations of cables in cable stayed bridges. *7th Intern. conference wind engineering*, Aachen, Germany, Vol. 4, p. 293.

Hirsch, G. 1980. Critical comparison of active and passive damping systems for the suppression of wind-induced vibrations of slender structures (in German). Beiträge zur Aeroelastik im Bauwesen. Techn. Universität München, Germany, No. 11.

Hirsch, G., H. Ruscheweyh & H. Zutt 1975. Damage on a 140 m high steel stack due to wind-induced transverse vibration (in German). *Der Stahlbau*, Vol. 2, p. 33.

Holley, E. R. 1969. Surging in laboratory pipeline with steady inflow. *ASCE Journal Hydraulics Division*, Vol. 95, No. HY3, p. 961.

Housner, G.W. 1957. Dynamic pressures on accelerated fluid containers. *Bulletin Seismological Society of America*, Vol. 47.

Hsu, C.C. 1975. Some remarks on the progress of cavity flow studies. *ASME Journal Fluids Engineering*, Vol. 97, p. 439.

Hsu, C.C. & C. F. Chen 1962. On the pulsation of finite, ventilated cavities. Technical report 115-4, Hydronautics, Inc., Laurel, Maryland.

Huerre, P. & P.A. Monkewitz 1990. Local and global instabilities in spatially developing flows. *Annual Review of Fluid Mechanics*, Vol. 22, p. 473.

Hunt, J.C.R. 1973. A theory of turbulent flow round two-dimensional bluff bodies. *Journal Fluid Mechanics*, Vol. 61, p. 625.

Hussain, A.K. & K.B.M.Q. Zaman 1978. The free shear layer tone phenomenon and probe interference. *Journal Fluid Mechanics*, Vol. 87, p. 349.

Hydraulic Institute Standards 1975. *Centrifugal, rotary, and reciprocating pumps*. 13th ed., Cleveland, Ohio.

Igarashi, T. 1978. Flow characteristics around a circular cylinder with a slit. *Bulletin Japanese Society Mechanical Engineers*, Vol. 21, p. 656.

Ingard, U. 1953. On the theory and design of acoustic resonators. *Journal Acoustical Society of America*, Vol. 25, No. 6, p. 1037.

Ingard, U. & V.K. Singal 1975. Effect of flow on the acoustic resonances of an open-ended duct. *Journal Acoustical Society of America*, Vol. 58, No. 4, p. 788.

Ippen, A.T. (ed.) 1966. *Estuary and coastline hydrodynamics*. McGraw-Hill.

Ippen, A.T. & Y. Goda 1963. Wave-induced oscillations in harbors: The solution for a rectangular harbor connected to the open sea. Report No. 59, Hydrodynamic Lab., M.I.T., Cambridge, Massachusetts.

Ishii, N. 1992. Flow-induced vibration of long-span gates. *Journal Fluids and Structures*, Vol. 6, p. 539.

Ishii, N., K. Imaichi & A. Hirose 1977. Instability of elastically suspended Tainter-gate system caused by surface waves on the reservoir of a dam. *ASME Journal Fluids Engineering*, Vol. 99, No. 4, p. 699.

Ishii, N., K. Imaichi & A. Hirose 1980. Dynamic instability of Tainter gates. In: *Practical experiences with flow-induced vibrations*, E. Naudascher & D. Rockwell (eds.). Springer.

Ishii, N. & C.W. Knisely 1992. Flow-induced vibration of shell-type long-span gates. *Journal Fluids and Structures,* Vol. 6, p. 681.

Ishii, N. & E. Naudascher 1992a. A design criterion for dynamic stability of Tainter gates. *Journal Fluids and Structures*, Vol. 6, p. 67.

Ishhi, N., C.W. Knisely & A. Nakata 1993. Coupled-mode vibration of gates with simultaneous over- and underflow. *ASME symposium on flow-induced vibration and noise*. Vol. 6, M. P. Paidoussis, C. Dalton & D.S. Weaver (eds), p. 193.

Iversen, H.W. 1975. An analysis of the hydraulic ram. *ASME Journal Fluids Engineering*, Vol. 97, p. 191.

Jaeger, C. 1949. *Technical hydraulics* (in German). Birkhäuser, Basel.

Jaeger, C . 1963. The theory of resonance in hydropower systems. Discussion of incidents and accidents occurring in pressure systems. *ASME Journal Basic Engineering*, Vol. 85, p. 631.

Johns, D.J. & R.J. Allwood 1968. Wind induced ovalling oscillations of circular cylindrical shell structures such as chimneys. *Symposium wind effects on buildings and structures*, Loughborough, England. Paper 28.

Johns, D.J. & C.B. Sharma 1974. On the mechanism of wind-excited ovalling vibrations of thin circular cylindrical shells. In: *Flow-induced structural vibrations*, E. Naudascher (ed.). Springer, p. 650.

Jones, G.W., J.C. Cincotta & R.W. Walker 1969. Aerodynamic forces on a stationary and oscillating circular cylinder at high Reynolds number. NASA TR R-300.

Jongeling, T.H.G. 1988. Flow-induced self-excited in-flow vibrations of gate plates. *Journal Fluids and Structures*, Vol. 2, p. 541.

Kana, D.D. 1982. Status and research needs for prediction of seismic response in liquid containers. *Nuclear Engineering and Design*, Vol. 69.

Kanne, S. 1989. Vibration of a vertical-lift gate of variable bottom geometry (in German). Diploma Thesis, Institut für Hydromechanic, Universität Karlsruhe, Germany.

Kanne, S., E. Naudascher & Y. Wang 1990. Self-excited vertical vibrations of gates with underflow (in German). Sonderforschungsbereich 210, Universität Karlsruhe, Germany, Report SFB 210/E/61.

Kanne, S., E. Naudascher, M. Castro & K. Stephanoff 1991. Self-excited vibration of an inclined gate plate due to gap fluctuations (in German). Sonderforschungsbereich 210, Universität Karlsruhe, Germany. Report SFB 210/E/73.

Karamcheti, K., A.B. Bauer, W.L. Shields, G.R. Stegen & J.P. Woolley 1969. Some features of an edge-tone flow field. Basic aerodynamic noise research, NASA SP-207.

Katsura, S. 1985. Wind-excited ovalling vibration of a thin circular cylindrical shell. *Journal Sound and Vibration*, Vol. 100, p. 527.

Keim, S.R. 1956. Fluid resistance to cylinders in accelerated motion. *ASCE Journal Hydraulics Division*, Vol. 82, No. HY6, paper 1113.

Keller, J.J., W. Egli & J. Exley 1985. Force- and loss-free transitions between flow states. *Journal Applied Mathematics and Physics (ZAMP)*, Vol. 36, p. 854.

Kemp, A.R. & R.A. Pullen 1961. Report on the vibrating nappe phenomenon. Dept. Civil Engineering, University Witwatersrand, Johannesburg, South Africa.

Kenn, M.J., A.C. Cassell & P. Grootenhuis 1980. Vibrations of the Borlarque Dam. In: *Practical experiences with flow-induced vibrations*, E. Naudascher & D. Rockwell (eds.). Springer, p. 508.

Kimball, A.L. 1929. Vibration damping, including the case of solid damping. *Transactions ASME*, Vol. 51, p. APM-227.

King, R. 1974. Hydroelastic model tests of marine piles. British Hydromechanic Research Association (BHRA), Report RR 1254.

King, R. 1977. A review of vortex shedding research and its application. *Ocean Engineering*, Vol. 4, p. 141.

King, R. 1977a. Vortex-excited oscillations of yawed circular cylinders. *ASME Journal Fluids Engineering*, Vol. 99, p. 495.

King, R. & D.J. Johns 1976. Wake interaction experiments with two flexible circular cylinders in flowing water. *Journal Sound and Vibrations*, Vol. 45, No. 2, p. 259.

King, R. & R. Jones 1980. Flow-induced vibrations of an anchor agitator. In: *Practical experiences with flow-induced vibrations*, E. Naudascher & D. Rockwell (eds.). Springer.

King, R., M.J. Prosser & D.J. Johns 1973. On vortex excitation of model piles in water. *Journal Sound and Vibration*, Vol. 29, No. 2. p. 169.

Kinsler, L.E. & A.R. Frey 1962. *Fundamentals of acoustics*. John Wiley and Sons, 2nd ed.

Kinsler, L.E., A.R Frey, A.B. Coppens & J.V. Sanders 1982. *Fundamentals of acoustics*. Wiley & Sons, 3rd ed.

Klein, R.E. & H. Salhi 1978. Optimal vibration control in structures. *3rd US conference wind engineering research*, University of Florida, Gainesville, Florida, p. 413.

Klein, R.E. & H. Salhi 1979. The time-optimal control of wind-induced structural vibration using active appendages. *IUTAM symposium*, Waterloo, Canada.

Knauss, J. 1987. *Swirling flow problems at intakes*. IAHR Hydraulic Structures Design Manual, Vol. 1. Balkema.

Knisely, C.W. 1985a. Strouhal number of rectangular cylinders at incidence. Sonderforschungsbereich 210, Universität Karlsruhe, Germany, Report SFB 210/E/13.

Knisely, C.W. 1985b. Flow visualization and kinematics of tandem cylinder interaction. Sonderforschungsbereich 210, Universität Karlsruhe, Germany, Report SFB 210/E/15.

Knisely, C.W. 1989. On the acoustics of nappe oscillations. *HYDRO-COMP 89*, Dubrovnik, Yugoslavia.

Knisely, C.W. & M. Kawagoe 1988. Force-displacement measurements on closely spaced tandem cylinders. *Journal Wind Engineering and Industrial Aerodynamics*, Vol. 33, p. 121.

König, A. 1986. Experimental investigation on plunging vibration of a partially obstructed trashrack (in German). Diploma Thesis, Institut für Hydromechanik, Universität Karlsruhe, Germany.

Kolkman, P.A. 1976. Flow-induced gate vibrations. Prevention of self-excitation. Computation of dynamic gate behavior and the use of models. Delft Hydraulics Laboratory, Publication No. 164.

Kolkman, P.A. 1980. Development of vibration-free gate design. Learning from experiment and theory. In: *Practical experiences with flow-induced vibrations*, E. Naudascher & D. Rockwell (eds.). Springer, p. 351.

Kolkman, P.A. 1984a. Vibrations of hydraulic structures. Chapter 1 of: *Developments in hydraulic engineering*, Vol. 2, P. Novak (ed.). Elsevier.

Kolkman, P.A. 1984b. Gate vibrations. Chapter 2 of: *Developments in hydraulic engineering*, Vol. 2, P. Novak (ed.). Elsevier.

Kolkman, P.A. 1988. A simple scheme for calculating the added mass of hydraulic gates. *Journal Fluids and Structures*, Vol. 2, p. 339.

Kolkman, P.A. & A. Vrijer 1977. Gate edge suction as a cause of self-excited vertical vibrations. *17th IAHR Congress*, Baden-Baden, Germany.

Komatsu, S. & H. Kobayashi 1980. Vortex-induced oscillations of bluff cylinders. *Journal Wind Engineering and Industrial Aerodynamics*, Vol. 6, p. 335.

Kornecki, A. 1974. Static and dynamic instability of panels and cylindrical shells in subsonic potential flow. *Journal Sound and Vibration*, Vol. 32, No. 2, p. 251.

Kovasznay, L.S.G. 1949. Hot-wire investigation of the wake behind cylinders at low Reynolds numbers. *Proceedings Royal Society of London*, Series A, Vol. 198, p. 174.

Krol, J. 1976. The automatic hydraulic ram: Its theory and design. ASME paper No. 76-DE-17.

Kubota, T. & H. Aoki 1980. Pressure surge in the draft tube of Francis turbine. In: *Practical experiences with flow-induced vibrations*, E. Naudascher & D. Rockwell (eds.). Springer.

Lamb, H. 1945. *Hydrodynamics*. Dover Publications, 6th ed.

Laneville, A. & L.Z. Yong 1983. Mean flow patterns around two-dimensional rectangular cylinders and their interpretation. *Journal Wind Engineering and Industrial Aerodynamics*, Vol. 14, p. 387.

Lang, J.D. 1976. The experimental results for an airfoil with oscillating spoiler and flap. *Journal of Aircraft*, Vol. 13, No. 9, p. 687.

Langer, W. 1969. Transverse vibrations of high, slender structures with circular cross-section (in German). In: Mitteilungen Institut für Luftfahrt, Dresden, No. 8, p. 184.

Lazan, B.J. 1968. *Damping of materials and members in structural mechanics*. Pergamon Press, New York.

Lee, J.J. 1971. Wave-induced oscillations in harbors of arbitrary geometry. *Journal Fluid Mechanics*, Vol. 45, p. 375.

Lee, J.J. & F. Raichlen 1972. Oscillations in harbors with connected basins. *ASCE Journal Waterways, Harbors, and Coast Engineering Division*, Vol. 91, p. 311.

Leibovich, S. 1984. Vortex stability and breakdown: Survey and extension. *AIAA Journal*, Vol. 22, No. 9, p. 1192.

Leipholz, H.H.E. (ed) 1980. *Structural control*. IUTAM symposium on structural control, University of Waterloo, Ontario. North-Holland Publishing Co., Amsterdam.

Leissa, A.W. 1973. The free vibration of rectangular plates. *Journal Sound and Vibration*, Vol. 31, p. 257.

Lever, J.H. & D.S. Weaver 1982. A theoretical model for fluid-elastic instability in heat-exchanger tube bundles. *ASME Journal Pressure Vessel Technology*, Vol. 104, p. 147.

Lever, J.H. & D.S. Weaver 1986. On the stability behavior of heat exchanger tube bundles. *Journal Sound and Vibration*, Vol. 107, p. 375 (Pt. 1), p. 393 (Pt. 2).

Levin, L. 1957. Hydraulic studies of trashracks. *7th IAHR Congress*, Lisbon, Portugal, Paper C11.

Liiva, J.D. et al. 1969. Two-dimensional tests of airfoils oscillating near stall. *AIAA Journal of Aircraft*, Vol. 6, No. 1, p. 46.

Lindholm, U.S. et al. 1965. Elastic vibration characteristics of cantilever plates in water. *Journal Ship Research*, Vol. 9(1), p. 123.

Locher, F.A. & E. Naudascher 1967. Some characteristics of macro turbulence in flow past a normal wall. *12th IAHR Congress*, Fort Collins, Colorado, Vol. 2, p. 298.

Loiseau, H. & E. Szechenyi 1972. Analyse expérimentale des portances sur un cylindre immobile soumis à un ecoulement perpendiculaire à son axe à des nombres de Reynolds elevès. *La Recherche Aérospatiale*, Sept./Oct., p. 279.

Lucas, M. & D. Rockwell 1987. Effect of nozzle asymmetry on jet-edge oscillations. *Journal Sound and Vibration*, Vol. 116, p. 355.

Lundgren, C.W. 1961. Charts for determining size of surge suppressors for pump-discharge lines. *ASME Journal Engineering Power*, January.

Lundgren, T.S., P.R. Sethna & A.K. Bajaj 1979. Stability boundaries for flow-induced motions of tubes with an inclined terminal nozzle. *Journal Sound and Vibration*, Vol. 64, p. 553.

Lyssenko, P.E. & G.A. Chepaykin 1974. On self-excited oscillations of gate seals. In: *Flow-induced structural vibrations*, E. Naudascher (ed.). Springer, p. 278.

Magnus, K. 1965. *Vibrations*. Blackie & Son.

Mahrenholtz, D. & H. Bardowicks 1980. Wind-induced oscillations of some steel structures. In: *Practical experiences with flow-induced vibrations*, E. Naudascher & D. Rockwell (eds.). Springer, p. 643.

Malamet, S. 1965. Operation of pumped storage schemes. *Intern. symposium waterhammer in pumped storage projects (ASME)*, Chicago, Illinois.

Mankau, H.J. 1980. Unsteady distribution of pressure on vibrating overhanging-roof models (in German). Abhandlungen Aerodyn. Institut, Techn. Hochschule Aachen, Germany, No. 24/25, p. 61/70.

March, P.A. & S. Vigander 1980. Some TVA experiences with flow-induced vibrations. In: *Practical experiences with flow-induced vibrations*, E. Naudascher & D. Rockwell (eds.). Springer, p. 228.

Maresca, C., D. Favier & J. Rebont 1979. Experiments on an airfoil at high angle of incidence in longitudinal oscillations. *Journal Fluid Mechanics*, Vol. 92, p. 671.

Marris, A.W. 1964. A review on vortex streets, periodic wakes, and induced vibration phenomena. *ASME Journal Basic Engineering*, Vol. 86, p. 185.

Martin, H.M. & J.W. Ball 1955. Laboratory and prototype tests for the investigation and correction of excessive downpull forces of large cylinder gates under high heads. *6th IAHR Congress*, The Hague, Netherlands, Paper C6.

Martin, W.W., I.G. Currie & E. Naudascher 1981. Streamwise oscillations of cylinders. *ASCE Journal Engineering Mechanics Division*, Vol. 107, No. EM3, p. 589.

Martin, W.W., E. Naudascher & M. Padmanabhan 1975. Fluid-dynamic excitation involving flow instability. *ASCE Journal Hydraulics Division*, Vol. 101, No. HY6.

Marvaud, M. & N. Ramette 1964. Water level oscillations around bridge piers (in French). *Comptes rendus 8ièmes journées de l'hydraulique*, Tome 1, Paris, p. 170.

Mateescu, D. & M.P. Paidoussis 1987. Unsteady viscous effects on the annular-flow induced instabilities of a rigid cylindrical body in a narrow duct. *Journal Fluids and Structures*, Vol. 1, p. 197.

Matsumoto, M. 1983. Aerodynamic stability and wind-resistant design of the Ohshima Bridge (in Japanese). Research reports of Bridge-Engineering Laboratory, Kyoto University, Kyoto, Japan.

Matsumoto, M., N. Shiraishi & H. Shirato 1987. Bluff body aerodynamics in pulsating flow. *7th Intern. conference wind engineering*, Aachen, Germany, Vol. 2, p. 129.

Matsumoto, M., N. Shiraishi & H. Shirato 1988b. Aerodynamic instabilities of twin circular cylinders. *Journal Wind Engineering and Industrial Aerodynamics*, Vol. 33, p. 131.

Matsumoto, M., N. Shiraishi, M. Kitazawa, C.W. Knisely, H. Shirato, Y. Kim & M. Tsujii 1988a. Aerodynamic behavior of inclined circular cylinders. Cable aerodynamics. *Journal Wind Engineering and Industrial Aerodynamics*, Vol. 33, p. 103.

Maurer, O.F. 1976. Divide to reduce flow-induced pressure oscillations in open cavities. US Patent 3,934,846.

McAlister, K.W. & L.W. Carr 1979. Water-tunnel visualizations of dynamic stall. *ASME symposium on nonsteady fluid dynamics*, Winter Meeting, D.E. Crow & J.A. Miller (eds.), p. 103.

McConnell, K.G. & D.F. Young 1965. Added mass of a sphere in a bounded viscous fluid. *ASCE Journal Engineering Mechanics Division*, Vol. 91, No. EM4.

McCroskey, W.J. 1977. Some current research in unsteady dynamics. The 1976 Freeman Scholar Lecture. *ASME Journal Fluids Engineering*, Vol. 99, p. 8.

McCrosky, W.J. 1982. Unsteady airfoils. *Annual Review of Fluid Mechanics*, Vol. 14, p. 285.

McCroskey, W.J. & S.L. Pucci 1981. Viscous-inviscid interaction on oscillating airfoil. *19th AIAA aerospace meeting*, St. Louise, Missouri, Paper No. 81-0051.

McLachlan, N.W. 1983. The accession to intertia of flexible discs vibrating in a fluid. *Proceedings Physical Society of London*, Vol. 44, p. 546.

McNown, J.S. & G.H. Keulegan 1959. Vortex formation and resistance in periodic motion. *ASCE Journal Engineering Mechanics Division*, Vol. 85, No. EM1.

Mehta, U.B. 1977. Dynamic stall of an oscillating airfoil. *AGARD conference on unsteady aerodynamics* CP-227, Paper No. 23.

Meier-Windhorst, A. 1939. Flutter of cylinders in steady liquid flow (in German). Mittei-lungen Hydraulik Institut, Techn. Hochschule München, No. 9.

Meirovitch, L. 1967. *Analytical methods in vibrations.* MacMillan.

Meirovitch, L. 1975. *Elements of vibration analysis.* McGraw-Hill.

Melbourne, W.H. & J.C.K. Cheung 1988. Reducing the wind loading on large cantilevered roofs. *Journal Wind Engineering and Industrial Aerodynamics*, Vol. 28, p. 401.

Merkli, P. 1978. Acoustic resonance frequencies for a T-tube. *Journal Applied Mathematics and Physics (ZAMP)*, Vol. 29, p. 486.

Merkli, P. & M.P. Escudier 1979. Observation of flow in a ring inlet chamber. *ASME Journal Fluids Engineering*, Vol. 101, p. 135.

Meyer, E. & E.G. Neumann 1972. *Physical and applied acoustics.* Academic Press, p. 191.

Mikojcsak, A.A., R.A. Arnoldi, L.E. Snyder & H. Stargardter 1975. Advances in fan and compressor blade flutter analysis and predictions. *Journal Aircraft*, Vol. 12, p. 32.

Miksad, R.W. 1972. Experiments on the nonlinear stages of free shear layer transition. *Journal Fluid Mechanics*, Vol. 56, p. 695.

Miles, J.W. 1971. Resident response of harbors: An equivalent-circuit analysis. *Journal Fluid Mechanics,* Vol. 46, Pt. 2, p. 241.

Miles, J.W. & Y.K. Lee 1975. Helmholtz resonance of harbors. *Journal Fluid Mechanics*, Vol. 67, Pt. 3, p. 445.

Miller, D.R. 1960. Critical flow velocities for collapse of reactor parallel-plate assemblies. *ASME Journal Engineering for Power*, Vol. 82, p. 83.

Minorski, N. 1962. *Non-linear oscillations.* Van Nostrand.

Misra, A.K., M.P. Paidoussis & K.S. Van 1988. On the dynamics of curved pipes transporting fluid. Part 1: Inextensible theory. *Journal Fluids and Structures*, Vol. 2, p. 221.

Modi, V.J. & F. Welt 1984. Nutation dampers and suppression of wind induced instabilities. In: *Flow-induced vibrations,* ASME Symposium, New Orleans, Louisiana, M.P. Paidoussis, M.K. Au-Yang & S.S. Chen (eds.), Vol. 1, p. 173.

Modi, V.J. & F. Welt 1988. Damping of wind induced oscillations through liquid sloshing. *Journal Wind Engineering and Industrial Aerodynamics*, Vol. 30, p. 85.

Modi, V.J., F. Welt & M.B. Irani 1988. On the suppression of vibrations using nutation dampers. *Journal Wind Engineering and Industrial Aerodynamics*, Vol. 33, p. 547.

Moiseev, N.N. & A.A. Petrov 1966. The calculation of free oscillations of a liquid in a motionless container. In: *Advances in Applied Mechanics*, G. Kuerti (ed.). Academic Press.

Monkewitz, P.A. 1988. The absolute and convective nature of instability in two-dimensional wakes at low Reynolds number. *Physics of Fluids*, Vol. 31, No. 5, p. 999.

Monkewitz, P.A. & L.N. Nguyen 1987. Absolute instability in the near-wake of two-dimensional bluff bodies. *Journal Fluids and Structures*, Vol. 1, No. 2, p. 165.

Morch, A.K. 1964. A theory for the mode of operation of the Hartmann air jet generator. *Journal Fluid Mechanics*, Vol. 20, Pt. 1, p. 141.

Morel, T. 1979. Experimental study of jet-driven Helmholtz oscillator. *ASME Journal Fluids Engineering*, Vol. 101, No. 3, p. 383.

Morkovin, M.V. 1964. Flow around a circular cylinder. A kaleidoscope of challenging fluid phenomena. *ASME Symposium on Fully Separated Flows*, A.G. Hansen (ed.). New York, p. 102.

Mulcahy, T.M. 1983. Leakage flow induced vibrations of reactor components. *The Shock and Vibration Digest*, Vol. 15, p. 11.

Mulcahy, T.M. 1988. One-dimensional leakage-flow vibration instabilities. *Journal Fluids and Structures*, Vol. 2, p. 383.

Müller, O. 1933. Vibrations of gates with underflow (in German). Mitteilungen Preussische Versuchsanstalt Wasserbau und Schiffbau, Berlin, No. 13.

Müller, O. 1937. The vibrating gate with overflow as self-excited, coupled system, using analogies to the reed pipe (in German). Mitteilungen Preussische Versuchsanstalt Wasserbau und Schiffbau, No. 33.

Munk, M.M. 1924. The aerodynamic forces on airship hulls. NACA Report 184.

Nakagawa, K., T. Fujino, Y. Arita & K. Shima 1963. An experimental study of aerodynamic devices for reducing wind-induced oscillatory tendencies of stacks. *Symposium wind effects on buildings and structures*, Nat. Phys. Lab., Teddington, UK, p. 774.

Nakagawa, K., T. Fujino, Y. Arita, Y. Ogata & K. Masaki 1959. An experimental investigation of aerodynamic instability of circular cylinders at supercritical Reynolds numbers. *9th Japan. congress appl. mech.*, Tokyo, p. 235.

Nakajima, H. & H. Kobayashi 1983. Prevention of vortex-induced vibration of retangular cylinders by means of side plates (in Japanese). *38th Annual congress Japan. society of civil engineers*, p. 563.

Nakamura, Y. 1978. An analysis of binary flutter of bridge deck sections. *Journal Sound and Vibration*, Vol. 57, p. 471.

Nakamura, Y. 1979. On the aerodynamic mechanism of torsional flutter of bluff structures. *Journal Sound and Vibration*, Vol. 67, p. 163.

Nakamura, Y. & K. Hirata 1991. Pressure fluctuations on oscillating rectangular cylinders with the long side normal to the flow. *Journal Fluids and Structures*, Vol. 5, p. 165.

Nakamura, Y. & T. Mizota 1975. Torsional flutter of rectangular prisms. *ASME Journal Engineering Mechanics Division*, Vol. 101, No. EM2, p. 125.

Nakamura, Y. & M. Nakashima 1986. Vortex excitation of prisms with elongated rectangular, H and T cross-sections. *Journal Fluid Mechanics*, Vol. 163, p. 149.

Nakamura, Y. & Y. Tomonari 1977. Galloping of rectangular prisms in a smooth and in a turbulent flow. *Journal Sound and Vibration*, Vol. 52, No. 2, p. 233.

Nakamura, Y. & T. Yoshimura 1982. Flutter and vortex excitation of rectangular prisms in pure torsion in smooth and turbulent flows. *Journal Sound and Vibration*, Vol. 84, p. 305.

Nakamura, Y., Y. Ohya & H. Tsuruta 1991. Experiments on vortex shedding from flat plates with square leading and trailing edges. *Journal Fluid Mechanics*, Vol. 222, p. 437.

Narayanan, R. & A.J. Reynolds 1968. Pressure fluctuations in reattaching flow. *ASCE Journal Hydraulics Division*, Vol. 94, No. HY6, p. 1383.

NASA 1972. Panel flutter, NASA space vehicle design criteria. NASA SP-8004.

Naudascher, E. 1959. On the excitation of gates with simultaneous over- and underflow (in German). Doctoral dissertation, Universität Karlsruhe, Germany. (English version: Vibration of gates during overflow and underflow. *ASCE Journal Hydraulics Division*, Vol. 87, 1961, No. HY5, p. 63.)

Naudascher, E. 1964. Hydrodynamic and hydroelastic loading of high-head gates (in German). *Der Stahlbau*, No. 7.

Naudascher, E. 1967. From flow instability to flow-induced excitation. *ASCE Journal Hydraulics Division*, Vol. 93, No. HY4, p. 15.

Naudascher, E. 1987a. *Hydraulics of open channels and open-channel structures* (in German). Springer (updated edition: 1992).

Naudascher, E. 1987b. Flow-induced streamwise vibrations of structures. *Journal Fluids and Structures*, Vol. 1, p. 265.

Naudascher, E . 1991. *Hydrodynamic forces*. IAHR Hydraulic Structures Design Manual, Vol. 3. Balkema.

Naudascher, E. & C. Farell 1965. Form-induced hydraulic forces on three-leaf intake gates. Discussion of paper by R.E. Elder & T.M. Garrison. *ASCE Journal Hydraulics Division*, Vol. 91, No. HY2, p. 320.

Naudascher, E. & C. Farrell 1970. Unified analysis of grid turbulence. *ASCE Journal Engineering Mechanics Division*, Vol. 96, No. EM2, p. 121.

Naudascher, E. & S. Kanne 1990. Pressure pulsations in the culverts of the lock Henrichenburg (in German). Institut für Hydromechanik, Universität Karlsruhe, Germany, Report 674.

Naudascher, E. & F.A. Locher 1974. Flow-induced forces on protruding walls. *ASCE Journal Hydraulics Division*, Vol. 100, No. HY2, p. 295.

Naudascher, E. & D. Rockwell (eds.) 1980. *Practical experiences with flow-induced vibrations*. Springer.

Naudascher, E. & D. Rockwell 1980a. Oscillator-model approach to the identification and assessment of flow-induced vibrations in a system. *Journal Hydraulic Research*, Vol. 18, No. 1, p. 59.

Naudascher, E. & Y. Wang 1993. Flow-induced vibrations of prismatic bodies and grids of prisms. *Journal Fluids and Structures*, Vol. 7, p. 341.

Naudascher, E., A. Richter & D.T. Nguyen 1988. Vibration problem at the cylinder gates of the Jaguas Hydroelectric Project, Colombia. Institut für Hydromechanik, Universität Karlsruhe, Germany, Report 660.

Naudascher, E., J.R. Weske & B. Fey 1981. Exploratory study on damping galloping vibrations. *Journal Wind Engineering and Industrial Aerodynamics*, Vol. 8, p. 211.

Naumann, A. & H. Quadflieg 1974. Vortex generation and its simulation in wind tunnels. In: *Flow-induced structural vibrations*, E. Naudascher (ed.). Springer.

Nayfeh, A.M. & D.T. Mook 1979. *Nonlinear oscillations*. Wiley.

Neilson, F.M. & E.B. Pickett 1980. Corps of Engineers experiences with flow-induced vibrations. In: *Practical experiences with flow-induced vibrations*, E. Naudascher & D. Rockwell (eds.). Springer, p. 399.

Newman, B.G. 1961. The deflection of plane jets by adjacent boundaries. Coanda effect. In: *Boundary layer and flow control, its principles and application*, G.V. Lachman (ed.). Pergamon Press.

Nguyen, D.T. 1982. Added mass behavior and its characteristics at sluice gates. *Intern. Conference Flow Induced Vibrations in Fluid Engineering*, Reading, UK., Paper A-2.

Nguyen, D.T. 1984. Flow-induced vibration of gates with underflow (in German). *Fortschrittsberichte der VDI Zeitschriften*, Reihe 4, No. 66.

Nguyen, D.T. 1990. Gate vibrations due to unstable flow separation. *ASCE Journal Hydraulic Engineering*, Vol. 116, No. 3, p. 342.

Nguyen, D.T. & E. Naudascher 1983. Approach-flow effects on downpull of gates. *ASCE Journal Hydraulics Division*, Vol. 109. No. HY11.

Nguyen, D.T. & E. Naudascher 1986a. Vortex-excited vibrations of underflow gates. *Journal Hydraulic Research*, Vol. 24, No. 2, p. 133.

Nguyen, D.T. & E. Naudascher 1986b. Self-excited vibrations of vertical-lift gates. *Journal Hydraulic Research*, Vol. 24, No. 5, p. 391.

Nguyen, D.T. & E. Naudascher 1991. Vibration of beams and trashracks in parallel and inclined flows. *ASCE Journal on Hydraulic Engineering*, Vol. 117, No. 8, p. 1056.

Nguyen, D.T., Q.H. Lin & E. Naudascher 1987. Flexural streamwise vibration of gate plates under vortex action. *Intern. conference flow-induced vibrations*, Bowness-on-Windermere, England, p. 171. (Also: Naudascher, E. 1987b.)

Newman, J.M. 1978. *Marine hydrodynamics*. MIT Press, Cambridge, Massachusetts.

Nikitina, F.A. & A.I. Kulik 1980. The interaction of separated flows with periodic external perturbations. *Fluid Mechanics – Soviet Research*, Vol. 9, No. 4.

Nishi, M., S. Matsunaga, T. Kubota & Y. Senoo 1984. Surging characteristics of conical and elbow-type draft tubes. *12th IAHR hydraulic machinery symposium*, Stirling, UK, p. 272.

Nishioka, M. & H. Sato 1983. Mechanism of determination of the shedding frequency of vortices behind a cylinder at low Reynolds numbers. *Journal Fluid Mechanics*, Vol. 89, Pt. 1, p. 49.

Novak, J. 1972. Strouhal number and flat plate oscillation in an air stream. *Acta Technica CSAV*, p. 372.

Novak, M. 1969. Aeroelastic galloping of prismatic bodies. *ASCE Journal Engineering Mechanics Division*, Vol. 95, No. EM1, p. 115.

Novak, M. 1972. Galloping oscillations of prismatic structures. *ASCE Journal Engineering Mechanics Division*, Vol. 98, No. EM1, p. 27.

Novak, M. & A.G. Davenport 1970. Aeroelastic instability of prisms in turbulent flow. *ASCE Journal Engineering Mechanics Division*, Vol. 96, No. EM1, p. 17.

Novak, M. & A.G. Davenport 1978. Vibration of towers due to galloping of iced cables. *ASCE Journal of Engineering Mechanics Division*, Vol. 104, No. EM2, p. 457.

Novak, M. & H. Tanaka 1974. Effect of turbulence on galloping instability. *ASCE Journal Engineering Mechanics Division*, Vol. 100, No. EM1, p. 27.

Novak, M. & H. Tanaka 1975. Pressure correlations on a vibrating cylinder. *4th Intern. conference wind effects on buildings and structures*, Heathrow, UK.

Nyborg, W.L., M.D. Burkhard & H.K. Schilling 1952. Acoustical characteristics of jet-edge-resonator systems. *Journal Acoustical Society of America*, Vol. 24, No. 3.

Obasaju, E.D. 1983. An investigation of the effects of incidence on the flow around a square-section cylinder. *Aeronautics Quarterly*, Vol. 34, p. 243.

Obasaju, E.D., R. Ermshaus & E. Naudascher 1990. Vortex-induced streamwise oscillations of a square-section cylinder in a uniform stream. *Journal Fluid Mechanics*, Vol. 213, p. 171.

Oden, J.T. et al. 1974. Finite element methods in flow problems. *Intern. Symposium*, Swansea, Wales.

Oertel, H. Jr. 1990. Wakes behind blunt bodies. *Annual Review of Fluid Mechanics*, Vol. 22, p. 539.

Ogihara, K. & S. Ueda 1980. Flap gate oscillation. In: *Practical experiences with flow-induced vibrations*, E. Naudascher & D. Rockwell (eds.). Springer, p. 466.

Oka, S., Z.G. Kostic & S. Sikmanovic 1972. Investigation of the heat transfer processes in tube banks in cross flow. *Intern. seminar on recent developments in heat-exchangers*, Trogir, Yugoslavia. (Cited in: Zdravkovich, M.M. 1977.)

Okajima, A. 1977. The acrodydnamic characteristics of stationary tandem cylinders at high Re (in Japanese). Bull. Research Inst. Applied Mechanics, Kyu-shu University, Vol. 46, p. 111. (Also: Zdravkovich, M.M. 1984b.)

Otsuki, Y., K. Washizu, H. Tomizawa & A. Ohya 1974. A note on the aeroelastic instability of a prismatic bar with square section. *Journal Sound and Vibration*, Vol. 34, No. 2, p. 233.

Owen, P.R. 1965. Buffeting excitation of boiler tube vibration. *Journal Mechanical Engineering Science*, Vol. 7, p. 431.

Padmanabhan, M. 1987. Pump sumps. In: *Swirling flow problems at intakes*, J. Knauss (ed.). Balkema.

Paidoussis, M. P. 1966. Dynamics of flexible circular cylinder in axial flow; Part 1: Theory, and Part 2: Experiments. *Journal Fluid Mechanics*, Vol. 26, p. 717 and 737.

Paidoussis, M. P. 1968. Stability of towed, totally submerged flexible cylinders. *Journal Fluid Mechanics*, Vol. 34, p. 273.

Paidoussis, M. P. 1970. Dynamics of tubular cantilevers conveying fluid. *Journal Mechanical Engineering Science*, Vol. 12, p. 85.

Paidoussis, M. P. 1973. Dynamics of cylindrical structures subjected to axial flow. *Journal Sound and Vibration*, Vol. 29, No. 3, p. 365.

Paidoussis, M. P. 1975. Flutter of conservative systems of pipes conveying fluid. *Journal Mechanical Engineering Sciences*, Vol. 17, p. 19.

Paidoussis, M. P. 1979. The dynamics of clusters of flexible cylinders in axial flow. Theory and experiments. *Journal Sound and Vibration*, Vol. 65, p. 391.

Paidoussis, M. P. 1980. Flow-induced vibrations in nuclear reactors and heat exchangers. In: *Practical experiences with flow-induced vibrations*, E. Naudascher & D. Rockwell (eds.). Springer.

Paidoussis, M. P. 1981. Fluid elastic vibration of cylinder arrays in axial and cross flow: State of the art. *Journal Sound and Vibration*, Vol. 76, p. 329.

Paidoussis, M. P. 1983. A review of flow-induced vibrations in reactors and reactor components. *Power Industry Research*, Vol. 74, No. 1, p. 31.

Paidoussis, M. P. 1987. Flow-induced instabilities of cylindrical structures. *Applied Mechanics Review*, Vol. 40, p. 163.

Paidoussis, M. P. & P. Besançon 1981. Dynamics of arrays of cylinders with internal and external axial flow. *Journal Sound and Vibration*, Vol. 76, p. 361.

Paidoussis, M. P. & L.R. Curling 1985. An analytical model for vibration of clusters of flexible cylinders in turbulent axial flow. *Journal Sound and Vibration*, Vol. 98, p. 493.

Paidoussis, M. P. & J.P. Denise 1972. Flutter of thin cylindrical shells conveying fluid. *Journal Sound and Vibration*, Vol. 20, p. 9.

Paidoussis, M. P. & J.O. Gagnon 1984. Experiments on vibration of clusters of cylinders in axial flow: Modal and spectral characteristics. *Journal Sound and Vibration*, Vol. 94, p. 341.

Paidoussis, M. P. & C. Helleur 1979. On ovalling oscillation of cylindrical shells in cross-flow. *Journal Sound and Vibration*, Vol. 63, p. 527.

Paidoussis, M. P. & M.T. Issid 1974. Dynamic stability of pipes conveying fluid. *Journal Sound and Vibration*, Vol. 33, No. 3, p. 267.

Paidoussis, M. P. & T.P. Luu 1985. Dynamics of a pipe aspirating fluid such as might be used in ocean mining. *ASME Journal Energy Resources Technology*, Vol. 107, p. 250.

Paidoussis, M. P. & F.C. Moon 1988a. Nonlinear and chaotic fluidelastic vibrations of a flexible pipe conveying fluid. *Journal Fluids and Structures*, Vol. 2, p. 567.

Paidoussis, M. P. & S.J. Price 1988. The mechanism underlying flow-induced instability of cylinder arrays in cross-flow. *Journal Fluid Mechanics*, Vol. 187, p. 45.

Paidoussis, M. P. & D.T.M. Wong 1982. Flutter of thin cylindrical shells in cross flow. *Journal Fluid Mechanics*, Vol. 115, p. 411.

Paidoussis, M. P. & B.K. Yu 1976. Elastohydrodynamics of towed slender bodies: The effect of nose and tail shapes on stability. *Journal Hydronautics*, Vol. 10, No. 4, p. 127.

Paidoussis, M. P., S.P. Chan & A.K. Misra 1984. Dynamics and stability of coaxial shells containing flowing fluid. *Journal Sound and Vibration*, Vol. 97, p. 201.

Paidoussis, M. P., N.T. Issid & M. Tsui 1980. Parametric resonance oscillations of flexible slender cylinders in harmonically perturbed axial flow. Part 1: Theory; Part 2: Experiments. *ASME Journal Applied Mechanics*, Vol. 47, p. 709 and 715.

Paidoussis, M.P., T.P. Luu & B.E. Laithier 1986. Dynamics of finite-length tubular beams conveying fluid. *Journal Sound and Vibration*, Vol. 106, No. 2, p. 311.

Paidoussis, M.P., D. Mateescu & W.G. Sim 1990. Dynamics and stablity of a flexible cylinder in a narrow coaxial cylindrical duct subjected to annular flow. *ASME Journal Applied Mechanics*, Vol. 57, p. 232.

Paidoussis, M.P., A.K. Misra & S.P. Chan 1985. Dynamics and stability of coaxial cylindrical shells conveying viscous fluid. *ASME Journal Applied Mechanics*, Vol. 52, p. 389.

Paidoussis, M.P., S.J. Price & S.Y. Ang 1988. Ovalling oscillations of cylindrical shells in cross-flow: A review and some new results. *Journal Fluids and Structures*, Vol. 2, p. 95.

Paidoussis, M.P., S. Suss & M. Pustejovsky 1977. Free vibration of clusters of cylinders in liquid-filled channels. *Journal Sound and Vibration*, Vol. 55, p. 443.

Palde, U.J. 1972. Influence of draft-tube shape on surging characteristics of reaction turbines. US Bureau of Reclamation, Rep. REC-ERC-72-24.

Parker, R. 1966. Resonance effects in wake shedding from parallel plates: Some experimental observations. *Journal Sound and Vibration*, Vol. 4, p. 62.

Parker, R. 1967. Resonance effects in wake shedding from parallel plates: Calculation of resonance frequencies. *Journal Sound and Vibration*, Vol. 5, p. 330.

Parker, R. 1978. Acoustic resonances in passages containing banks of heat exchanger tubes. *Journal Sound and Vibration*, Vol. 57, p. 245.

Parker, R. & S.A.T. Stoneman 1989. The excitation and consequences of acoustic resonance in enclosed fluid flow around solid bodies. *Proceedings Institution of Mechanical Engineers*, London, Vol. 203, p. 9.

Parker, R. & M.C. Welsh 1981. The effect of sound on flow over bluff bodies. University of Wales, Swansea, Mechanical Engineering Rep. MR/87/81. (Also: *Intern. Journal Heat and Fluid Flow*, 1983, p. 113.)

Parkin, M.W. 1980. Flow-induced vibration problems in gas cooled reactors. In: *Practical experiences with flow-induced vibrations*, E. Naudascher & D. Rockwell (eds.). Springer.

Parkinson, G.V. 1963. Aeroelastic galloping in one degree of freedom. *Symposium wind effects on buildings and structures*, Nat. Phys. Lab., Teddington, UK, Paper 23.

Parkinson, G.V. 1974. Mathematical models of flow-induced vibrations of bluff bodies. In: *Flow-induced structural vibrations*, E. Naudascher (ed.). Springer, p. 81.

Parkinson, G.V. 1980. Nonlinear oscillator modelling of flow-induced vibrations. In: *Practical experiences with flow-induced vibrations*, E. Naudascher & D. Rockwell (eds.). Springer, p. 786.

Parkinson, G.V. 1985. Hydroelastic phenomena of bodies of bluff section in steady flow. *Intern. symposium separated flow around marine structures*. Norwegian Inst. Technology, Trondheim, Norway.

Parkinson, G.V. 1989. Phenomena and modelling of flow-induced vibrations of bluff bodies. *Progress in Aerospace Sciences*, Vol. 26, p. 169.

Parkinson, G.V. & V.J. Modi 1967. Recent research on wind effects on bluff two-dimensional bodies. *Symposium wind effects on buildings and structures*, Nat. Research Council, Ottawa, Canada, Vol. 1, Paper 18, p. 485.

Parkinson, G.V. & J.D. Smith 1964. The square prism as an aeroelastic nonlinear oscillator. *Quarterly Journal Mechanics and Applied Mathematics*, XVII, 2.

Parkinson, G.V. & M.A. Wawzonek 1981. Some considerations of combined effects of galloping and vortex resonance. *Journal Wind Engineering and Industrial Aerodynamics*, Vol. 8, p. 135.

Partenscky, H.W. & I.S. Khloeung 1967. Etude des vibrations des lames déversantes. Rapport soumis au Conseil Nat. de Recherches, Ottawa, Canada.

Patton, K.T. 1965. Tables of hydrodynamic mass factors for translational motion. *Winter Ann. Meeting ASME*, Chicago, Illinois, Paper 65-WA/UNT-2.

Perumal, P.V.K. 1976. Thin airfoil in eddy array and part-stalled oscillating cascade. Doctoral dissertation, Stevens Inst. of Techn., Hoboken, N.J.

Petrikat, K. 1955. Vibrations in hydraulic engineering (in German). *Der Stahlbau*, No. 9 and 12, p. 198 and 272.

Petrikat, K. 1964. Excitation and suppression of vibrations in hydraulic engineering (in German). *Die Wasserwirtschaft*, Vol. 54, No. 8, p. 213.

Petrikat, K. 1980. Seal vibration. In: *Practical experiences with flow-induced vibrations*, E. Naudascher & D. Rockwell (eds.). Springer, p. 476.

Pettigrew, M.J. 1981. Flow-induced vibration phenomena in nuclear power station components. *Power Industry Research*, Vol. 1, p. 97.

Pettigrew, M.J. & D.J. Gorman 1981. Vibration of heat exchanger tube bundles in liquid and two-phase cross-flow. In; *Flow-induced vibration design guidelines (ASME)*, P.Y. Chen (ed.), p. 89.

Pettigrew et al. 1986. Damping of multispan heat exchanger tubes; Part 1: In gases; Part 2: In liquids. *ASME Symposium on flow-induced vibrations*, PVP104, S.S. Chen, J.C. Simonis & Y.S. Shin (eds.), p. 81 and 89.

Pettigrew, M.J., J.H. Tromp, C.E. Taylor & B.S. Kim 1988. Vibration of tube bundles in two-phase cross-flow. In: *Flow-induced vibration and noise (ASME)*, M.P. Paidoussis, S.S. Chen & M.D. Bernstein (eds.), Vol. 2, p. 79 (Pt. 1), Vol. 3, p. 159 (Pt. 2).

Pines, S. 1958. An elementary explanation of the flutter mechanism. *Nat. specialists meeting dynamics and aeroelasticity*, Inst. Aeronautical Sciences, Ft. Worth, Texas, p. 52.

Plesset, M.S. & A. Prosperetti 1977. Bubble dynamics and cavitation. In: *Annual Review of Fluid Mechanics*, M. Van Dyke (ed.). Annual Reviews Inc., Palo Alto, California.

Popović, P. 1970. Vibrations of the Basta Dam outlets. *IAHR symposium hydraulic machinery and equipment*, Stockholm.

Powell, A. 1961. On the edge tone. *Journal Acoustical Society of America*, Vol. 33, No. 4, p. 395.

Powell, A. & H. Unfried 1964. An experimental study of low-speed edgetones. Dept. Engineering, Univ. of California, Los Angeles, Rep. No. 64-49.

Prandtl, L. 1904. On liquid flows with small friction (in German). *3rd Intern. congress of mathematicians*, Heidelberg, Germany.

Price, P. 1956. Suppression of the fluid-induced vibration of circular cylinders. *ASCE Journal Engineering Mechanics Division*, Vol. 82, No. EM3, p. 1030.

Price, S.J. & M.P. Paidoussis 1983. Fluidelastic instability of an infinite double row of circular cylinders subject to a uniform crossflow. *ASME Journal Vibration, Acoustics, Stress, and Reliability in Design*, Vol. 105, p. 59.

Price, S.J. & M.P. Paidoussis 1984. A theoretical investigation of the fluid elastic stability of a single flexible cylinder surrounded by rigid cylinders. In: *Flow-induced vibrations, ASME Symposium*, New Orleans, Louisiana, M.P. Paidoussis, M.K. Au-Yang & S.S. Chen (eds.), Vol. 2.

Price, S.J. & M.P. Paidoussis 1984a. An improved mathematical model for the stability of cylinder rows subject to cross-flow. *Journal Sound and Vibration*, Vol. 97, p. 615.

Price, S.J. & P. Piperni 1988. An investigation of the effect of mechanical damping to alleviate wake-induced flutter of overhead power conductors. *Journal Fluids and Structures*, Vol. 2, p. 53.

Prosser, M.J. 1977. The hydraulic design of pump sumps and intakes. Brit. Hydromech. Res. Association (BHRA), Cranfield, UK.

Quadflieg, H. 1975. Aerodynamic aspects of the mitigation of vortex-induced loads on cylindrical structures (in German). Doctoral dissertation, Techn. Hochschule Aachen, Germany.

Quadflieg, H. 1977. Vortex-induced loads on pairs of cylinders in incompressible flow of large Reynolds numbers (in German). *Zeitschrift Forschung im Ingenieurwesen*, Vol. 43, No. 1, p. 9.

Quadflieg, H. & H. Mankau 1978. Variable damping in cases of vibration of plates and prismatic bodies in stagnant and streaming air (in German). Mitteilung Curt-Risch-Institut, Techn. Universität Hannover, Germany, CRI-KI/78, p. 164.

Quinn, M.C. & M.S. Howe 1984. The influence of mean flow on the acoustic properties of a tube bank. *Proceedings Royal Society of London*, Series A, Vol. 396, No. 1911, p. 383.

Raichlen, F. 1966. Harbor Resonance. In: *Estuary and coastline hydrodynamics*, A.T. Ippen (ed.). McGraw-Hill.

Ramberg, S.E. 1983. The effects of yaw and finite length upon the vortex wakes of stationary vibrating circular cylinders. *Journal Fluid Mechanics*, Vol. 128, p. 81.

Rao, B.C.S. 1989a. Interference effects on the vibration response of a pair of cylinders. Sonderforschungsbereich 210, Universität Karlsruhe, Germany, Report SFB 210/E/57.

Rao, B.C.S. 1989b. A review of trashrack failures and related investigations. *Water Power and Dam Construction*, p. 28.

Rao, B.C.S., D.T. Nguyen & E. Naudascher 1987. Vibration of trashracks in flow with different incidence angles. *Intern. conference flow-induced vibrations*, Bowness-on-Windmere, UK, p. 329.

Rayleigh, Lord 1945. *Theory of sound*, Dover, New York.

Reed, F.E. 1980. Propeller singing. In: *Practical experiences with flow-induced vibrations*, E. Naudascher & D. Rockwell (eds.). Springer.

Relf, E.F. & L.G.F. Simmons 1924. The frequency generated by the motion of circular cylinders through a fluid. Aeronautical Research Council, UK, ARC R&M 917.

Remenieras, G. 1952. Dispositif simple pour réduire le célérité des ondes elastiques dans les conduits en charge. *La Houille Blanche*, No. Spécial A, p. 172.

Richardson, A.S., J.R. Martucelli & W.S. Price 1965. Research study on galloping of electric power transmission lines. *Nat. Phys. Lab. Symposium*, H.M.S.O.

Richter, A. & E. Naudascher 1976. Fluctuating forces on a rigid circular cylinder in confined flow. *Journal Fluid Mechanics*, Vol. 78, p. 561.

Roberts, B.W. 1966. Low frequency, aeroelastic vibrations in a cascade of circular cylinders. Institution of Mechanical Engineers, Mechanical Engineering Science Monograph No. 4.

Rockwell, D. 1972. External excitation of planar jets. *ASME Journal Applied Mechanics*, Paper No. 72-WA/APM-21.

Rockwell, D. 1977a. Prediction of oscillation frequencies for unstable flow past cavities. *ASME Journal Fluids Engineering*, Vol. 99.

Rockwell, D. 1977b. Organized fluctuations due to flow past a square crossed-section cylinder. *ASME Journal Fluids Engineering*, Vol. 99, p. 511.

Rockwell, D. & C.W. Knisely 1979. The organized nature of flow impingement upon a corner. *Journal Fluid Mechanics*, Vol. 93, p. 413.

Rockwell, D. & E. Naudascher 1978. Review – Self-sustaining oscillations of flow past cavities. *ASME Journal Fluids Engineering*, Vol. 100, p. 152.

Rockwell, D. & E. Naudascher 1979. Self-sustained oscillations of impinging shear layers. *Annual Review of Fluid Mechanics*, Vol. 11, p. 67.

Rockwell, D. & E. Naudascher 1990. A guide to flow-induced oscillations in engineering systems. A lecture-note series. Department of Mechanical Engineering, Lehigh University, Bethlehem, Pennsylvania.

Ronneberger, D. 1967/1968. Experiments on the acoustic reflection factor for air flow in a pipe with abrupt changes in cross-section (in German). *Acustica*, Vol. 19, p. 222.

Rosenberg, G.S. & C.K. Youngdahl 1962. A simplified dynamic model for the vibration frequencies and coolant flow velocities for reactor parallel plate fuel assemblies. *Nuclear Science and Engineering*, Vol. 13, p. 91.

Roshko, A. 1953. On the development of turbulent wakes from vortex streets. NACA TN 2913 (also Techn. Rep. No. 1191, 1954).

Roshko, A. 1954. On the drag and shedding frequency of two-dimensional bluff bodies. NACA TN 3169. (See also Roshko, A. 1955: On the wake and drag of bluff bodies. *Journal of Aeronautical Sciences*, Vol. 22.)

Roshko, A. 1961. Experiments on the flow past a circular cylinder at very high Reynolds numbers. *Journal Fluid Mechanics*, Vol. 10, p. 345.

Roshko, A. & K. Koenig 1978. Interaction effects on the drag of bluff bodies in tandem. In: *Aerodynamic drag mechanisms*, Sovran, Morel & Mason (eds.). Plenum Press.

Rossiter, J.E. 1964. Wind tunnel experiments on the flow over rectangular cavities at subsonic and transonic speeds. RAE Techn. Rep. 64037 and Reports and Memoranda No. 3438.

Rouse, H. 1938. *Fluid mechanics for hydraulic engineers*. Dover Publ.

Runyan, H.L. 1951. Single-degree-of-freedom flutter calculations for a wing in subsonic potential flow and comparison with an experiment. NACA TN 2396.

Ruscheweyh, H. 1978. Measures to control dangerous stack vibrations (in German). Mitteilung Curt-Risch-Institut, Techn. Universität Hannover, Germany, p. 519.

Ruscheweyh, H. 1981. Straked in-line steel stacks with low mass-damping parameter. *Journal Wind Engineering and Industrial Aerodynamics*, Vol. 8, p. 203.

Ruscheweyh, H. 1982. *Dynamic wind action on structures*, Vol. 2: Practical applications (in German). Bauverlag.

Ruscheweyh, H. 1983. Aeroelastic interference effects between slender structures. *Journal Wind Engineering and Industrial Aerodynamics*, Vol. 14, p. 129.

Ruscheweyh, H. 1985. Vortex-excited vibrations of yawed cantilevered circular cylinders with different Scruton numbers. *6th Colloq. industrial aerodynamics*, Fachhochschule Aachen, Germany, Pt. 2, p. 157.

Ruscheweyh, H. 1988. Practical experiences with wind-induced vibrations. *Journal Wind Engineering and Industrial Aerodynamics*, Vol. 33, p. 493.

Sakomoto, H., H. Haniu & Y. Obata 1987. Fluctuating forces acting on two square prisms in tandem arrangement. *Journal Wind Engineering and Industrial Aerodynamics*, Vol. 26, p. 85.

Sallet, D.W. 1970. A method of stabilizing cylinders in fluid flow. *Journal Hydronautics*, Vol. 4, No. 1, p. 40.

Sallet, D.W. 1980. Transverse motion of a buoy by surface wave. In: *Practical experiences with flow-induced vibrations*, E. Naudascher & D. Rockwell (eds). Springer, p. 587.

Sandifer, J.B. & R.T. Bailey 1984. Turbulent buffeting of tube arrays in liquid cross-flow. *ASME symposium on flow-induced vibrations*, Vol. 2, p. 211.

Sarohia, V. 1977. Experimental investigation of oscillations in flows over shallow cavities. *AIAA Journal*, Vol. 15, p. 984.

Sarohia, V. & P.F. Massier 1976. Control of cavity noise. *3rd AIAA Aerocoustics Conference*, Palo Alto, California, Paper 75-528.

Sarpkaya, T. 1971. On stationary and travelling vortex breakdowns. *Journal Fluid Mechanics*, Vol. 45, p. 545.

Sarpkaya, T. 1976. Vortex shedding and resistance in harmonic flow about smooth and rough circular cylinders at high Reynolds numbers. Naval Postgraduate School, Monterey, California, Report No. NPS-59SL76021.

Sarpkaya, T. 1978. Fluid forces on oscillating cylinders. *ASCE Journal Waterway, Port, Coastal and Ocean Division*, No. WW4, p. 275.

Sarpkaya, T. 1979. Vortex-induced oscillations. *ASME Journal Applied Mechanics*, Vol. 46, No. 2, p. 241.

Sarpkaya, T. 1982. Flow-induced vibration of roughened cylinders. *Intern. conference flow-induced vibration in fluid engineering (BHRA)*, Reading, England.

Sarpkaya, T. & C.J. Garrison 1963. Vortex formation and resistance in unsteady flow. *ASME Journal Applied Mechanics*, p. 16.

Sarpkaya, T. & M. Isaacson 1981. *Mechanics of wave forces on offshore structures*. Van Nostrand Reinhold.

Savkar, S.D. 1977. A survey of flow-induced vibrations of cylindrical arrays in cross flow. *ASME Journal Fluids Engineering*, Vol. 99, No. 3.

Scanlan, R.H. & J.J. Tomko 1971. Airfoil and bridge deck flutter derivatives. *ASCE Journal Engineering Mechanics Division*, Vol. 97, No. EM6, p. 1717.

Scavuzzo, R.J. 1965. Hydraulic instability of flat plate parallel plate assemblies. *Nuclear Science and Engineering*, Vol. 21, p. 463.

Schewe, G. 1989. Nonlinear flow-induced resonances of a H-shaped section. *Journal Fluids and Structures*, Vol. 3, p. 327.

Scruton, C. 1956. Note on the aerodynamic stability of truncated circular cylinders and tapered stacks. Nat. Phys. Lab., Teddington, UK, Aero Report 305.

Scruton, C. 1963. On the wind-excited oscillations of stacks, towers, and masts. *1st Conference wind effects on buildings and structures*, Teddington, UK, Paper 16.

Scruton, C. 1969. Some consideration of wind effects on large structures. In: *Structures technology for large radio and radar telescope systems*, J.W. Mar & H. Liebowitz (eds.). MIT Press.

Scruton, C. & D.E.J. Walshe 1957. A means of avoiding wind-excited oscillations of structures with circular or nearly circular cross-section. Nat. Phys. Lab., UK. Aero Rep. No. 335. (Also: Scruton, C. 1963, Nat. Phys. Lab., Teddington, UK, Aero Note 1012.)

Sears, W.R. 1941. Some aspects of nonstationary airfoil theory and its practical application. *Journal Aeronautical Sciences*, Vol. 3, p. 104.

Sedov, L.I. 1965. *Two-dimensional problems in hydrodynamics and aeronautics*. John Wiley, Interscience, p. 24.

Seifert, R. 1940. On the vibration of gates (in German). *Zeitschrift Verein Deutscher Ingenieure*, Vol. 84, p. 105.

Seifert, R. 1941. Nappe oscillations (in German). *Zeitschrift Verein Deutscher Ingenieure*, Vol. 85, p. 626.

Sensburg, O., J. Becker & H. Hönlinger 1979. Active control of flutter and vibration of aircraft. *IUTAM symposium on structural control*, Waterloo, Canada.

Sethna, P.R. & X.M. Gu 1985. On global motions of articulated tubes carrying fluid. *Intern. Journal Non-Linear Mechanics*, Vol. 20, p. 453.

Seybert, T.A., W.S. Gearhart & H.T. Falvey 1978. Studies of a method to prevent draft-tube surge in pump-turbines. *ASCE/IAHR/ASME joint symposium fluid machinery*, Fort Collins, Colorado, Vol. 1, p. 151.

Shaw, T.L. 1971. Wake dynamics of two-dimensional structures in confined flows. *14th IAHR Congress*, Paris, Paper B6.

Shiraishi, N. & M. Matsumoto 1983. On classification of vortex-induced oscillation and its application for bridge structures. *Journal Wind Engineering and Industrial Aerodynamics*, Vol. 14, p. 419.

Shiraishi, N., M. Matsumoto & H. Shirato 1985. On aerodynamic instabilities of tandem structures. *6th Colloquium industrial aerodynamics*, Fachhochschule Aachen, Germany, Pt. 2, p. 179.

Shiraishi, N., M. Matsumoto, H. Shirato & H. Ishizaki 1988. On aerodynamic stability effects for bluff rectangular cylinders by their corner-cut. *Journal Wind Engineering and Industrial Aerodynamics*, Vol. 28, p. 271.

Shirakashi, M., Y. Ishida & S. Wakiya 1985. Higher-velocity resonance of circular cylinder in crossflow. *ASME Journal Fluids Engineering*, Vol. 107, p. 392.

Shirato, H. 1988. Study on the aerodynamic behavior of multiple structures. Dissertation, University of Kyoto, Kyoto, Japan.

Silberman, E. & C.S. Song 1959. Instability of ventilated cavities. Univ. of Minnesota, St. Anthony Falls Hydraulic Laboratory, Techn. Paper No. 29, Series B.

Simiu, E. & R.H. Scanlan 1978. *Wind effects on structures: An introduction to wind engineering*. John Wiley & Sons.

Simmons, W.P. 1965. Experiences with flow-induced vibrations. *ASCE Journal Hydraulics Division*, Vol. 91, No. HY4, p. 185.

Simpson, A. 1965. Aerodynamic instability of long-span transmission lines. *Proceedings Institution of Electrical Engineers*, Vol. 112, No. 2, p. 315.

Simpson, A. 1971. On the flutter of a smooth circular cylinder in a wake. *Aeronautical Quarterly*, Vol. 22, p. 25.

Simpson, A. 1972. Determination of the natural frequencies of multi-conductor overhead transmission lines. *Journal Sound and Vibration*, Vol. 20, p. 417.

Simpson, A. 1979. Fluid-dynamic stability aspects of cables. In: *Mechanics of wave-induced forces on cylinders*, T.L. Shaw (ed.). Pitman, p. 90.

Singh, K.P. & A.I. Soler 1984. *Mechanical design of heat exchangers and pressure vessel components*. Arcutus Publishers, New Jersey.

Sisto, F. 1977. A review of the fluid mechanics of aeroelasticity in turbomachines. *ASME Journal Fluids Engineering*, Vol. 99, p. 40.

Skarecky, R. 1975. Yaw effects on galloping instability. *ASCE Journal Engineering Mechanics Division*, Vol. 101, No. EM6, p. 739.

Slater, J.E. 1969. Aeroelastic instability and aerodynamics of structural angle sections. Dissertation, Univ. British Columbia, Vancouver, Canada.

Smith, C.R. & S.J. Kline 1974. An experimental investigation of the transitory stall regime in two-dimensional diffusors. *ASME Journal Fluids Engineering*, Vol. 96, p. 11.

Smith, J.D. 1962. An experimental study of the aeroelastic instability of rectangular cylinders. M.A.Sc. Thesis, Univ. British Columbia, Vancouver, Canada.

Snowdon, J.C. 1968. *Vibration and shock in damped mechanical systems*. John Wiley & Sons.

So, R.M.C. & S.D. Savkar 1981. Buffeting forces on rigid circular cylinders in cross-flows. *Journal Fluid Mechanics*, Vol. 105, p. 397.

Soliman, R. 1986. Vortex shedding from two-dimensional bodies with non-circular cross sections. DFVLR Aerodyn. Versuchsanstalt Göttingen, Germany, Report 1B 222-86A16.

Song, C.S. 1962. Pulsation of two-dimensional cavities. *4th Symposium naval hydrodynamics*, Washington, p. 1033.

Spivack, H.M. 1946. Vortex frequency and flow pattern in the wake of two parallel cylinders at varied spacing normal to an air stream. *Journal Aeronautical Science*, Vol. 13, p. 289.

Stansby, P.K. 1976. The locking-on of vortex shedding due to cross-shear vibration of circular cylinders in uniform and shear flows. *Journal Fluid Mechanics*, Vol. 74, Pt. 4, p. 641.

Staubli, T. 1983. Investigation of oscillating forces on a vibrating cylinder in cross-flow (in German). Doctoral dissertation ETH Zürich. (Also: Staubli, T. 1981. Calculation of the vibration of an elastically mounted cylinder using experimental data from forced oscillation. *ASME Symposium fluid/structure interactions in turbomachinery*, Washington D.C.)

Stenning, A.H. 1980. Rotating stall and surge. *ASME Journal Fluids Engineering*, Vol. 102, p. 14.

Stokes, A.N. & M.C. Welsh 1986. Flow-resonant sound interaction in a duct containing a plate, II: Square leading edge. *Journal Sound and Vibration*, Vol. 104, p. 55.

Streeter, V.L. & E.B. Wylie 1967. *Hydraulic transient*. McGraw-Hill. (Updated edition: *Fluid transients*. McGraw-Hill, 1978.)

Sturm, R.G. 1936. Vibration of cables and dampers – I and II. *Electrical Engineering*, Vol. 55, p. 455 and 673.

Surry, J. & D. Surry 1967. The effect of inclination on the Strouhal number and other wake properties of circular cylinders at subcritical Reynolds numbers. Institute of Aerospace Studies, Univ. of Toronto, Toronto, Canada, Techn. Note 116.

Tanner, T. 1972. A method for reducing the base drag of wings with blunt trailing edges. *Aeronautical Quarterly*, Vol. 23, p. 15.

Taylor, C.E., M.J. Pettigrew, F. Axisa & B. Villard 1986. Experimental determinations of single and two-phase cross-flow-induced forces on tube rows. *ASME Symposium on flow-induced vibrations*, PVP 104, S.S. Chen, J.C. Simonis & Y.S. Shin (eds.), p. 31.

Tennekes, H. & J.L. Lumley 1972. *A first course in turbulence*. MIT Press.

Thang, D. Nguyen: See Nguyen, D.T.

Thompson, P.A. 1964. Jet-driven resonance tube. *AIAA Journal*, Vol. 2, No. 7, p. 1230.

Toebes, G.H. 1965. Flow-induced structural vibrations. *ASCE Journal Engineering Mechanics Division*, Vol. 91, No. EM6, p. 39.

Toebes, G.H. & P.S. Eagleson 1961. Hydroelastic vibrations of flat plates related to trailing edge geometry. *ASME Journal Basic Engineering*, Series D, p. 671.

Torum, A. & N.M. Anand 1985. Free span vibrations of submarine pipelines in steady flows. Effect of free-stream turbulence on mean drag coefficients. *Journal Energy Resources Technology*, Vol. 107, p. 415.

Treiber, B. 1974. Theoretical study of nappe oscillation. In: *Flow-induced structural vibrations*, E. Naudascher (ed.). Springer.

Tyler, J.M. & T.G. Sofrin 1962. Axial flow compressor noise studies. *Society of Automotive Engineers Transactions*, Vol. 70, p. 309.

Uematsu, Y. & K. Uchiyama 1985. An experimental investigation of wind-induced ovalling oscillation of thin, circular cylindrical shells. *Journal Wind Engineering and Industrial Aerodynamics*, Vol. 18, p. 229.

Ulith, P. 1968. A contribution to influencing the part-load behavior of Francis turbines by aeration and σ-value. *IAHR Hydraulic machinery symposium*, Lausanne, Switzerland, Paper B1.

US Corps of Engineers 1956. Vibration and pressure-cell tests, flood-control intake gates Fort Randall Dam. US Waterways Experiment Station, Vicksburg, Mississippi, TR No. 2-435.

Valembois, J. 1957. Study of the effect of resonant structures on wave propagation. Translation No. 57-6. US Waterways Experiment Station, Vicksburg, Mississippi.

Van Dyke, M. 1981. *An album of fluid motion*. Parabolic Press.
Verdon, J.M. & J.E. McCure 1975. Unsteady supersonic structures. *AIAA Journal*, Vol. 13, No. 2, p. 193.
Vickery, B.J. 1966. Fluctuating lift and drag on a long cylinder of square cross-section in a smooth and in a turbulent stream. *Journal Fluid Mechanics*, Vol. 25, Pt. 3, p. 481.
Vickery, B.J. & A.W. Clark 1972. Lift or across-wind response of tapered stacks. *ASCE Journal Structural Division*, Vol. 98, No. ST 1, p. 1.
Vickey, B.J. & A.G. Davenport 1967. A comparison of theoretical and experimental determination of the response of elastic structures to turbulent flow. *Symposium wind effects on buildings and structures*, Ottawa, Canada.
Vickery, B.J. & R.D. Watkins 1964. Flow-induced vibrations of cylindrical structures. *1st Australasian conference hydraulics and fluid mechanics*, McMillan.
Vogel, H. 1920. The Reed pipe as a coupled system (in German). *Annalen der Physik*, Vol. 62, No. 11.
Vugts, J.H. 1968. The hydrodynamic coefficients for swaying, heaving, and rolling cylinders in a free surface. *Intern. Ship Building Progress*, Vol. 15, p. 251.
Walker, D. & R. King 1988. Vortex excited vibrations of tapered and stepped cylinders. *7th Intern. conference offshore mechanics and arctic engineering (ASME)*, J.S. Chung (ed.), Vol. II, p. 229.
Walker, E.M. & G.F.S. Reising 1968. Flow-induced vibrations in cross-flow heat exchangers. *Chemical Process and Engineering*, Vol. 49, p. 95.
Waller, E.J. 1959. Pressure surge control in pipeline systems. Oklahoma State University, Stillwater, Oklahoma, Publ. 102. (Also: Publ. 101, 1958, and 107, 1959.)
Walshe, D.E. & C.F. Cowdrey 1972. A brief study of the effect of shrouds on buffet amplitudes of chimney stacks. Nat. Phys. Lab. Maritime Sciences, Techn. Memo 2-72.
Wambsganns, M.W. 1967. Second-order effects as related to critical coolant flow velocities and reactor parallel plate fuel assemblies. *Nuclear Engineering and Design*, Vol. 5, p. 268.
Wambsganns, M.W. & J.A. Jendrzejczyk 1979. The effect of trailing end geometry on the vibration of a circular cantilevered rod in nominally axial flow. *Journal Sound and Vibration*, Vol. 65, No. 2.
Wang, Y. 1992a. Vibration of grid bars transverse to and in-line with the flow (in German). Sonderforschungsbereich 210, Universität Karlsruhe, Germany, Report SFB 210/E/74.
Wang, Y. 1992b. Vibrations of beams and trashracks in parallel and inclined flows. Discussion of paper by D.T. Nguyen and E. Naudascher. *Journal of Hydraulic Engineering*, Vol. 118, No. 10, p. 1454.
Wang, Y. & E. Naudascher 1992. Scale effects in tests on flow-induced vibrations, Parts 1 and 2. *Intern. symposium hydraulic research in nature and laboratory*, Wuhan, PR China.
Wardlaw, R.L. 1980. Unresolved problems. Bridge decks, beams, and cables. In: *Practical experiences with flow-induced vibrations*, E. Naudascher & D. Rockwell (eds.). Springer.
Wardlaw, R.L. & K.R. Cooper 1978. Dynamic vibration absorbers for suppressing wind-induced motion of structures. *3rd Colloquium on industrial aerodynamics*, Fachhochschule Aachen, Germany, Vol. 2, p. 205.
Wardlaw, R.L., K.R. Cooper & R.H. Scanlan 1973. Observations on the problem of subspan oscillation of bundled power conductors. *Intern. symposium vibration problems in industry*, Keswick, UK. Atomic Energy Authority, Paper No. 323.
Washizu, K., A. Ohya, Y. Otsuki & K. Fujii 1978. Aeroelastic instability of rectangular cylinders in a heaving mode. *Journal Sound and Vibration*, Vol. 59, No. 2, p. 195.
Weaver, D.S. 1974. On flow-induced vibrations in hydraulic structures and their alleviation.

2nd Symposium applications of solid mechanics, McMaster University, Hamilton, Ontario, Canada.

Weaver, D.S. 1974a. On the non-conservative nature of gyroscopic conservative systems. *Journal Sound and Vibration*, Vol. 36, p. 435.

Weaver, D.S. 1980. Flow-induced vibrations in valves operating at small openings. In: *Practical experiences with flow-induced vibrations*, E. Naudascher & D. Rockwell (eds.). Springer, p. 305.

Weaver, D.S. & J.A. Fitzpatrick 1988. A review of cross-flow induced vibrations in heat-exchanger tube arrays. *Journal Fluids and Structures*, Vol. 2, p. 73.

Weaver, D.S. & H.G.D. Goyder 1990. An experimental study of fluidelastic instability in a three-span tube array. *Journal Fluids and Structures*, Vol. 4, p. 429.

Weaver, D.S. & B. Myklatun 1973. On the stability of thin pipes with an internal flow. *Journal Sound and Vibration*, Vol. 31, p. 399.

Weaver, D.S. & M.P. Paidoussis 1977. On collapse and flutter phenomena in thin tubes conveying fluid. *Journal Sound and Vibration*, Vol. 50, p. 117.

Weaver, D.S. & J. Parrondo 1990. Fluidelastic instability in multispan heat exchanger tube arrays. In: *Flow-induced vibrations (ASME, PVP)* S.S. Chen, K. Fujita & M.K. Au-Yang (eds.), Vol. 189, p. 79.

Weaver, D.S. & T.E. Unny 1972. The influence of a free surface on the hydroelastic stability of a flat panel. *ASME Journal Applied Mechanics*, Vol. 39, p. 53.

Weaver, D.S. & T.E. Unny 1973. On the dynamic stability of fluid-conveying pipes. *ASME Journal Applied Mechanics*, Vol. 40, p. 48.

Weaver, D.S. & H.C. Yeung 1984. The effect of tube mass on the flow response of various tube arrays in water. *Journal Sound and Vibration*, Vol. 93, p. 409.

Weaver, D.S. & S. Ziada 1980. A theoretical model for self-excited vibrations in hydraulic gates, valves, and seals. *ASME Journal Pressure Vessel Technology*, Vol. 102, p. 146.

Weaver, D.S., F.A. Adubi & N. Kouwen 1978. Flow-induced vibrations of a hydraulic valve and their elimination. *ASME Journal Fluids Engineering*, Vol. 100, p. 239.

Weaver, W. 1961. Wind-induced vibrations in antenna members. *ASCE Journal Engineering Mechanics Division*, Vol. 87, No. EM1, p. 141.

Wedding, J.B., J.M. Robertson, J.A. Peterka & R.E. Akins 1978. Spectral and probability-density nature of square-prism separation-attachment wall pressures. *ASME Journal Fluids Engineering*, Paper 78-WA/FE 3.

Wehausen, J.V. 1971. The motion of floating bodies. *Annual Review of Fluid Mechanics*, p. 237.

Welsh, R.I. Jr. 1966. The effectiveness of a splitter plate in reducing transverse oscillations of finite circular cylinder in turbulent flow. US Navy Underwater Sound Laboratory, New London, Connecticut, Report 759.

Wendel, K. 1950. Hydrodynamic masses and hydrodynamic mass moments of inertia (in German). *Jahrbuch der Schiffbautechnischen Gesellschaft*, Vol. 44.

Whitney, A.K., J.S. Chung & B.K. Yu 1981. Vibrations of long marine pipes due to vortex shedding. *ASCE Journal Engineering Resources Technology*, Vol. 103, p. 103.

Wiesner, K.B. 1979. Tuned mass dampers to reduce building wind motions. *ASCE Convention and Exposition*, Boston, Massachusetts, Preprint 3510.

Wille, R. 1974. Generation of oscillatory flows. In: *Flow-induced structural vibrations*, E. Naudascher (ed.). Springer.

Willmarth, W.W., R.F. Gasparovic, J.M. Maszatics, J.L. McNaughton & D.J. Thomas 1978. Management of turbulent shear layers in separated flow. *AIAA Journal Aircraft*, Vol. 15, p. 385.

Wong, H.Y. 1977. An aerodynamic means of suppressing vortex-induced oscillations. *Proceedings Institution of Civil Engineers*, London, Vol. 63, p. 693.

Wong, H.Y. 1985. Wake flow stabilization by the action of base bleed. *ASME Journal Fluids Engineering*, September issue.

Wong, H.Y. & R.N. Cox 1978. The suppression of vortex-induced oscillations of circular cylinders by aerodynamic devices. *3rd Colloquium industrial aerodynamics*, Fachhochschule Aachen, Germany, Vol. 2.

Wood, C.J. 1964. The effect of base bleed on a periodic wake. *Journal Royal Aeronautical Society*, Vol. 68, p. 477.

Wood, C.J. 1967. Visualization of an incompressible wake with base bleed. *Journal Fluid Mechanics*, Vol. 29, p. 259.

Woodgate, L. & J.F.M. Mabey 1959. Further experiments on the use of helical strakes for avoiding wind-excited oscillations of structures of circular or nearly circular section. Nat. Phys. Lab., Teddington, UK, Aero Rep. No. 381.

Woods, L.C. 1966. On the instability of ventilated cavities. *Journal Fluid Mechanics*, Vol. 26.

Wootton, L.R. 1969. The oscillations of large circular stacks in wind. *Proceedings Institution of Civil Engineers*, London, Vol. 43, p. 573.

Wootton, L.R. & C. Scruton 1971. Aerodynamic stability. *CIRIA Seminar modern design of wind-sensitive structures*, Construct. Ind. Res. Inf. Assoc., London.

Wootton, L.R., M.H. Warner & D.H. Cooper 1974. Some aspects of the oscillations of full-scale piles. In: *Flow-induced structural vibrations*, E. Naudascher (ed.). Springer.

Worraker, W.J. 1980. Grazing flow effects on the impedance of cavity liners. Institute Sound and Vibration Research, University Southampton, UK, Memo No. 600.

Wylie, E.B. 1965. Resonance in pressurized piping systems. *ASME Journal Basic Engineering*, Vol. 87, No. 4, p. 960.

Wylie, E.B. & V.L. Streeter 1993. *Fluid transients in systems*. Prentice Hall.

Yashima, S. & H. Tanaka 1978. Torsional flutter in stalled cascade. *Journal of Engineering for Power*, Vol. 100, No. 2, p. 317.

Yeh, G.C.K. 1966. Sloshing of a liquid in connected cylindrical tanks owing to U-tube free oscillations. *Journal Acoustic Society of America*, Vol. 40, p. 807.

Yeung, H.C. & D.S. Weaver 1983. The effect of approach flow direction on the flow-induced vibrations of a triangular tube array. *ASME Journal Vibration, Acoustics, Stress, and Reliability in Design*, Vol. 105, p. 76.

Young, J.O. & J.W. Holl 1966. Effects of cavitation on periodic wakes behind symmetric wedges. *ASME Journal Basic Engineering*, Vol. 88, p. 163.

Zdravkovich, M.M. 1974. Flow-induced vibrations of two cylinders in tandem, and their suppression. In: *Flow-induced structural vibrations*, E. Naudascher (ed.). Springer.

Zdravkovich, M.M. 1977. Review of flow interference between two circular cylinders in various arrangements. *ASME Journal Fluids Engineering*, Vol. 99, p. 618.

Zdravkovich, M.M. 1981. Review and classification of various aero- and hydrodynamic means for suppressing vortex shedding. *Journal Wind Engineering and Industrial Aerodynamics*, Vol. 7, p. 145.

Zdravkovich, M.M. 1983. Interference between two circular cylinders forming a cross. *Journal Fluid Mechanics*, Vol. 128, p. 231.

Zdravkovich, M.M. 1984a. Classification of flow-induced oscillations of two parallel circular cylinders in various arrangements. In: *Flow-induced vibrations, ASME Symposium*, New Orleans, Louisiana, M.P. Paidoussis, M.K. Au-Yang & S.S. Chen (eds.), Vol. 2.

Zdravkovich, M.M. 1984b. Reduction of effectiveness of means for suppressing wind-induced oscillation. *Engineering Structures*, Vol. 6, p. 344.

Zdravkovich, M.M. 1985. Flow-induced oscillations of two interfering circular cylinders. *Journal Sound and Vibration*, Vol. 101, No. 4, p. 511.

Zdravkovich, M.M. 1987. The effects of interference between circular cylinders in cross flow. *Journal Fluids and Structures*, Vol. 1, p. 239.

Zdravkovich, M.M. 1988. Review of interference-induced oscillations in flow past two parallel circular cylinders in various arrangements. *Journal Wind Engineering and Industrial Aerodynamics*, Vol. 28, p. 183.

Zdravkovich, M.M. 1990. On origins of hysteretic responses of circular cylinder induced by vortex shedding. *Zeitschrift Flugwissenschaften und Weltraumforschung*, Vol. 14, p. 47.

Zdravkovich, M.M. & J.E. Namork 1979. Structures of interstitial flow between closely spaced tubes in staggered array. In: *Flow-induced vibrations (ASME)*, New York, S.S. Chen & M.D. Bernstein (eds.), p. 41.

Zdravkovich, M.M. & J.E. Namork 1980. Excitation, amplification, and suppression of flow-induced vibrations in heat exchangers. In: *Practical experiences with flow-induced vibration*, E. Naudascher & D. Rockwell (eds.). Springer, p. 109.

Zdravkovich, M.M. & J.A. Nuttall 1974. On the elimination of aerodynamic noise in a staggered tube bank. *Journal Sound and Vibrations*, Vol. 34, p. 173.

Zdravkovich, M.M. & J.R. Volk 1972. Effect of shroud geometry on the pressure distribution around a circular cylinder. *Journal Sound and Vibrations*, Vol. 20, p. 751.

Ziada, S. & D. Rockwell 1982. Oscillations of an unstable mixing layer impinging upon an edge. *Journal Fluid Mechanics*, Vol. 124, p. 307.

Name index

Subject index

Acoustic resonance: see standing wave in closed-conduit systems
Added
  coefficients 56-59, 179-181, 186, 189-190, 198, 242-245, 251-252
  damping or stiffness: see fluid damping or fluid stiffness
Added-mass
  coefficient (see also added coefficients) 31-48, 53, 357-358
  effect on: fluid oscillators 69, 73, 77, 82-84, 87-88, 270, 292; gates 38-39, 47
  flow 219, 237, 361
Admittance
  mechanical 12-13, 17-18, 101-103, 106
  fluid-dynamic 97-98, 101, 106
Aeration 143, 145, 152, 162, 167-168, 174, 175, 237, 293, 300, 305, 361, 363-364
Air-chamber system 70, 74
Air entrainment: see aeration
Airfoil: see plate
Airship 224
Air-water mixture (see also two-phase flow) 48, 71, 134
Amplification
  factor 10-11, 13, 20-21, 85, 87, 113, 249-250, 271, 274-275, 278-281, 298, 306
  fluid-dynamic 114-116, 121, 142
  fluid-elastic 116-120, 123-127, 139, 142, 282-283, 345, 347-348, 354
  fluid-resonant 116-120, 139-142, 282-283, 285, 348, 351, 354
  mechanisms 4, 111-120
Analysis: see vibration analysis
Argand diagram 182-185, 222-223

Aspect ratio
  effect on: galloping 204, 205; vortex shedding 128, 318
Assessment
  of possible excitation concerning: fluid oscillators 2-6, 67, 179, 273, 292-293; structures (see also indicator, stability criteria) 2-6, 91, 110, 122, 177-179, 228, 331, 333, 339-342, 345-347, 364-365
Attenuation
  basic concepts 104-105, 156-157, 260-261
  related to body oscillators with: extraneously induced excitation 104-106, 109, 164; instability-induced excitation 134, 150, 156-176, 263, 265-267, 292-297, 315-318, 327, 338-342, 345-347, 350, 352, 356-357, 364; movement-induced excitation 224, 231, 235-237, 245-246, 255, 260-268, 317-318, 327, 338-342, 345, 349-350, 356-357, 359-360
  related to fluid oscillators 292-307
Autooscillation (see also movement-induced excitation) 287-292
Auxiliary valving 293

Baffle block or plate 105, 255, 293-294, 339
Bar: see prism
Barge 221, 224
Base bleed 167-168, 171, 173
Basin; see harbor, open basin
Beam (see also prism) 26-30, 35
Bellow 248

405

Errata list

Page IX, 5th to the last line, replace coridal with cordial

Page 38, Fig 3.9b: axis of abscissa: replace s/d with s/D

Page 43, Fig. 3.15 a & b: replace fd^2/γ with fd^2/ν

Bottom of page 52 (last line) and first paragraph of page 53, to be replaced by the following text: Instability-induced excitation, IIE, of an elastically-mounted cylinder yields the type of response indicated in Figures 3.23(a) and (b). The vortex shedding frequency is f_o, and in the locked-in range, the cylinder vibrates at its natural frequency f_n. Within this locked-in range, the amplitude of vibration of the cylinder reaches its maximum. In contrast, for the case of extraneously-induced excitation, EIE, the excitation frequency is determined by the frequency content of the inflow, which may involve a single frequency component at one extreme, or a wide range of frequency components, associated with broadband turbulence, at the other extreme. The vibrational response of the cylinder will be enhanced when an EIE frequency component is at its natural frequency f_n.

Page 94, Figure 5.3 and subsequent locations in text: Replace cospectrum with cross-spectrum and co-spectral to cross-spectral

Page 122, Fig. 6.12, axis of ordinate: Change to $Sh = f_o d/V_\infty$ (add ∞)

Page 210, figure caption of Fig. 7.28: replace f_n with f_N

Page 220, figure caption of Fig. 7.38: replace Aquirre with Aguirre

Page 236, bottom below Fig. 7.49: add the following footnote:

Concerning instability of Tainter gates *without* eccentric trunnion, see: Anami, K.& Ishii, N. 2003. Model tests for non-eccentricity dynamic instability closely related to Folsom Dam Tainter-gate failure. *2003 ASME PVP Conf. Cleveland, USA*; and Anami, K., Ishii, N.& Knisely, C. W. 2004. Validation of theoretical analysis of Tainter gate instability using full-scale field tests. *Proceedings of the Eighth International Conference on Flow-Induced Vibration FIV 2004*, Paris, France, 6-9 July (edited by E. deLangre and F.Axisa).

Page 325, Figs. 9.17, a & b: omit hyphen in Threshold-amplitude

Page 326, 2nd line in 3rd paragraph: replace (Wang, 1922a) with (Wang, 1992a)

Page 342, 10th line: add 6.31, in front of 6.34, 6.68, 6.69

Page 350, Equation (9.11): replace $(V_r)_{cr}$ with $(V_r)_{max}$

Page 355, Fig. 9.44 a, axis of ordinate should be: $100\,(2\pi f x_o)/V$ (add: /)

Page 394, 8th line: replace Vickey with Vickery

414